Th**o** week
week loan

Please return on or before the last
date stamped below.
Charges are made for late return.

D0275227

The Theory of Linear Systems

ELECTRICAL SCIENCE
A Series of Monographs and Texts

Editors: Henry G. Booker and Nicholas DeClaris
A complete list of titles in this series appears at the end of this volume

THE THEORY OF LINEAR SYSTEMS

J. E. RUBIO

Department of Electrical Engineering
Carnegie-Mellon University
Pittsburgh, Pennsylvania

ACADEMIC PRESS   New York and London   1971

ACADEMIC PRESS, INC.
111 Fifth Avenue, New York, New York 10003

United Kingdom Edition published by
ACADEMIC PRESS, INC. (LONDON) LTD.
Berkeley Square House, London W1X 6BA

LIBRARY OF CONGRESS CATALOG CARD NUMBER: 70-127699

PRINTED IN THE UNITED STATES OF AMERICA

Contents

1. The Search for an Internal Structure

2. An Introduction to the Theory of Linear Spaces

3. Differential Systems, I

4. Differential Systems, II

5. Controllability, Observability

6. Synthesis

7. Difference Systems

8. Stability

9. Infinite Dimensions

Appendix. An Introduction to Computational Methods

Preface

This book is concerned with the state-space analysis of linear systems, and is intended for use in a two-semester course, the first semester dealing with background and basic material and the second with somewhat more advanced topics. Due to the wide variety of preparation, needs, and prospects of the students wishing to pursue studies in this subject matter, it is certainly wise to offer first an introductory course serving the needs of advanced seniors, of beginning graduate students in the areas of systems and control, and perhaps of more advanced graduate students in areas such as bioengineering and economics. Students majoring in systems and control who intend to do research would then move on into a course dealing with advanced topics.

In spite of this double function which the book is expected to fulfill, it has been my intention to have the basic material serve as preparation and motivation for the more advanced subjects, so that the reader is led naturally and smoothly from the more elementary concepts into deeper matters. Two main aspects of the book have been dictated by this philosophy. First, since the treatment of infinite-dimensional systems does require the full-dress apparatus of linear spaces and linear transformations, these concepts have been used throughout, even when, as is the case while dealing with some aspects of finite-dimensional systems, a simpler approach, based on matrix algebra, would have perhaps sufficed. Secondly, the simpler finite-dimensional systems are used as models for a class of abstract objects, the dynamical systems; in turn, systems described by partial differential equations are treated as special cases of these abstract entities. References to this modeling process (essentially a process of induction) are to be found therefore in the chapters dealing with the simpler finite-dimensional systems.

It has been my objective to put the advanced topics at the reach of graduate students with a good knowledge of advanced calculus. The level attained in these matters is, therefore, limited; many interesting subjects are simply touched upon. Some of the sections treating advanced matters have as a purpose the motivation of the students to acquire the mathematics required to study the corresponding subjects in depth.

Chapter 1 is a bridge between the usual elementary undergraduate courses in control and systems and the subject matter of this book. The language, style, and level of treatment follow, therefore, the manner which is customary in such courses. Several important concepts are introduced here by means of simple examples.

Chapter 2 presents a concise treatment of vector spaces, transformations, matrices, norms, inner products, etc. It has been my experience that most students taking introductory courses in linear systems lack a solid basis on even such simple matters as the definition of a function, or of a set. These concepts, and many more advanced ones, are introduced here carefully.

Chapters 3, 4, and 7 are a basic treatment of finite-dimensional linear systems, and should be included in a first course, with the exception of Sections 3.7 and 7.4, which deal mainly with the inductive process used to define dynamical systems. In Chapter 3 the existence and uniqueness theorem for the solutions of a homogeneous linear differential system is given in detail. It is hoped that such treatment will illustrate for the student several concepts pertaining to his studies of calculus.

Chapters 5 and 6 are quite related, and form a core of engineering applications of the material in the book. Many sections treat or introduce the reader to advanced topics; a first course should include only Sections 5.1–5.8 and 6.1–6.4.

Chapter 8 deals mainly with two topics, Lyapunov methods for the study of the stability of linear systems, and the concept which we have chosen to call external stability. A first course should include at least Sections 8.1 and 8.2.

In Chapter 9 we introduce first the abstract concept of a dynamical system, and then go on to derive properties of these systems, applying the results obtained to some simple systems described by partial differential equations. This chapter should be considered only as an introduction to the subject; it is hoped that students will be motivated by the material presented into improving their mathematical background so as to be able to go deeper into these matters.

Acknowledgments

I thank Professors E. M. Williams and A. G. Jordan, past and present Heads respectively of the Department of Electrical Engineering of Carnegie–Mellon University for their encouragement and help in this undertaking. Thanks also to my colleague, Dr. C. P. Neuman, for many fruitful discussions, and to Mrs. Betty Williams, Mrs. Darlene Westlake, Miss Claudia Schumann, and Miss Debbie Paul who typed the manuscript. Thanks to the staff of Academic Press, who through their editorial advice and professional skills contributed greatly to give this book its final form. And last, thanks to the reviewers, Professor C. A. Desoer and Professor Fawzi Emad, whose many comments were very helpful.

I dedicate this book to two of my professors, Professor Rubén Toro of the School of Engineering of the Catholic University of Chile, and Professor Sid Deutsch of the Electrical Engineering Department of the Polytechnic Institute of Brooklyn, who made this book possible.

1

The Search for an Internal Structure

1.1. Introduction

This century has seen some remarkable developments in scientific thought. For instance, the conceptual evolution which led to the birth of the new mechanics of the quanta stands as one of the truly great accomplishments of the human spirit.

This century has also witnessed interesting, although more modest, developments in engineering science. Long gone are the days in which the engineer designed blindly; the need for understanding has become apparent to everybody engaged in engineering endeavors. Our field, systems engineering, is a good example of the outcome of this evolution of engineering practices from blind groping to scientifically based design. Perhaps the turning point in the development of systems concepts was reached in the middle fifties, when Fourier and Laplace transforms [1][†] started to be known and used by engineers involved in the design of systems. Many interesting and useful things were achieved with this method [2].

As time went by, however, not many new developments appeared, and the suspicion started to grow in some people that something was being missed. Just what this missing element was can perhaps be seen best by means of an example.

Consider a simple system, known as an inertial plant, that has as

---

† Numbers in square brackets indicate references in the corresponding section at the end of the chapter.

transfer function $Y(s)/U(s) = 1/s^2$. Of course, the output y and the input u are related by the differential equation $\ddot{y} = u$. The reader should remember from his studies of ordinary differential equations that given an input u for, say, $t \geqslant t_0$, the solution of this equation is completely determined at any time $t_1 \geqslant t_0$ when the values of y and $\dot{y}$, position and velocity say, are specified at time t_0. These two numbers, the position and velocity at time t_0, determine, together with the values of the input for $t \geqslant t_0$, the future behavior of the system; they will be referred to as its *state* at time t_0. Note that t_0 is an arbitrary real number, so that actually at each instant of time a state is defined for the inertial plant. It should not be surprising that a knowledge of the state of a system provides us with better means of interaction with it, so that more efficient controls can be devised when the state is available.

Indeed, suppose an inertial plant of transfer function $1/s^2$ has at an initial time t_0 the position and velocity corresponding to point 1 on the position-velocity plane of Figure 1.1; and assume further that the

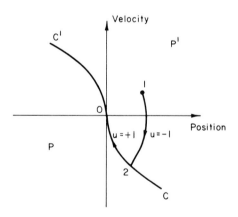

FIGURE 1.1. The position-velocity plane for a system with transfer function $1/s^2$, showing an optimal trajectory 1–2–0; the curves C' and C divide the plane into regions P, where the optimal control takes a value equal to $+1$, and P', where it takes a value of -1.

admissible inputs to this plant are characterized by the fact that their absolute values are to be no higher than unity. It is desired to find an input, or control, such that the position and velocity of the plant under the influence of this control become jointly zero in the minimum possible time; that is, a control is sought to drive the state of the system from the point 1 to the origin of the position-velocity plane in the least possible time. We will not derive the form of this control, but rather show the way in which the system is driven by it to the origin; in this manner we will demonstrate how this control can be built only if the position and velocity of the system resulting from its application are known at all times; we can say, therefore, that the control which drives the system to the origin in the least possible time is a function of the position and velocity, that is, of the state of the system. It follows, of course, that in order to implement this control, the state variables, position and velocity, have to be available; it would not be sufficient for this purpose to have knowledge of the output of the system (that is, of the position y) alone.

The behavior of the inertial plant under the optimal control is based on a simple fact: There is a trajectory, labeled C in Figure 1.1, which corresponds to $u = +1$, and which passes through the origin 0; in other words, if the system is given an initial condition on this curve, and the control is made equal to $+1$ thereafter, then the state of the system will follow the curve C to the origin. The given initial condition in our problem, point 1, is not on C, so we have to reach C, at point 2, by means of the trajectory 1–2, corresponding to $u = -1$. It so happens that the trajectory 1–2–0 obtained in this way is optimal, in the sense that, under the conditions imposed on u, there is no trajectory between point 1 and the origin for which the time to reach this point is less than for the trajectory 1–2–0. Indeed, the system follows this trajectory by accelerating as hard as permitted in the appropriate direction.

A similar analysis can be made with initial points other than the point 1; the conclusion is reached that the value of the optimal control is to be $+1$ on the curve C and the region P, and -1 on the curve C' and the region P'. Clearly the optimal control is a function of the state of the system, and this state has to be available for the implementation of the optimal control.

Observe that to the traditional concepts of input and output of a system we have added another, the state. In the example just presented, the state is a tuple (that is, a set of two numbers) composed of the output and its first derivative; this is, of course, a simple case, and the relationship between state and output (or rather between output and state) is generally not as simple, as we will show presently.

Let us go back now to our historical discussion. Transform methods are certainly based on input–output descriptions of systems; what engineers engaged in systems design realized in the late fifties was exactly what we have just learned from our example above: that in order to interact with a system it may be necessary to have information available concerning its *internal* behavior rather than just the output.

This information at any given time is what we have chosen to call the state of the system.

Once this was recognized, the theory of linear systems started to develop in a scientific manner. To be fair, the concept of state was implicit in the treatment given to a traditional engineering discipline, that of linear network theory. As we will see in detail in the next section, the set of mesh currents of a network, together with the input sources which may be present, determine any output whatsoever that we might choose in the network; the set of mesh currents at a given time is the *state* of the network at that time. Note that a state description of a known network is possible in this way because complete information is available about its *internal structure*, that is, the way in which its (known) branches are interconnected. Note as well that in the case in which the internal structure and/or the branches of a system are not known, we still may want to have at our disposal a state description of the system; that is, we may have to develop such a state description from input–output information, such as a set of transfer functions or impulse responses which have been experimentally determined. This problem is crucial in the theory of linear systems which we are presenting in this book, and as such will be amply treated in the following chapters. It could be described as the problem of finding an internal structure for a system whose input–output behavior is known.

A word concerning the mathematics of our state descriptions. The underlying mathematical structures will no longer be the theory of complex variables and transform theory, but the theory of vector spaces and the time-domain treatment of systems of differential and difference equations. It is interesting to consider that these techniques of time-domain analysis have been known to mathematicians for the best part of a hundred years [3]. Our objective is to adapt this body of knowledge to the treatment of engineering situations such as the problem of providing a state description of a system known only through input–output measurements. The theory of linear systems resulting from this development, which has taken place during the last ten years or so, is the subject matter of this book.

A word of caution seems in order now. The transform theory of linear systems had most of its success when applied to time-invariant systems. The method to be studied in this book can be applied to time-dependent systems with considerable success. However, the analysis and synthesis of this type of system is indeed a very difficult matter—witness the large amount of work devoted to seemingly simple equations such as the Mathieu equation [4]—and we cannot expect to answer all the questions raised by the study of such systems. At least, though, a framework will be provided on which can rest the deeper study which may be necessary in some particular case.

Another advantage of the present formulation is what we would like to call its "continuity." If a linear time-invariant system is modified so that it becomes slightly nonlinear, methods based on transforms cannot be used to study the new system; state space methods can provide useful information about its behavior. We will not, however, treat these extensions of the method in this text.

In the following two sections we present examples of the formulation of state models. The first system is a simple electrical network, and we will use it to define some of our terms as well as to investigate properties of the state representation. Next, we provide an example of the process of assigning a state representation to a system known only through its transfer function; this will serve to illustrate several characteristics of the state description as well as to give an introduction to this type of process, to be studied in detail in Chapter 6 of this book.

1.2. The State of a System: An Example

Consider the network in Figure 1.2. It is a simple undergraduate exercise [5] to write the mesh equations for this network

$$R_1 i_1 + L_1 \frac{d}{dt}(i_1 - i_2) = v$$

$$L_1 \frac{d}{dt}(i_2 - i_1) + L_2 \frac{di_2}{dt} + R_2 i_2 = 0$$

(1.1)

This is a set of two differential equations. Provided that the input function v possesses some properties, such as being piecewise continuous,

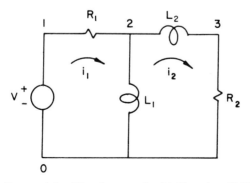

FIGURE 1.2. Electric network with Equations (1.1).

which make it Riemann integrable, the following fact is well known [5]: For a given input, defined for all values of $t \geq t_0$, t_0 being an arbitrary instant of time, the mesh currents $i_1(t)$ and $i_2(t)$ can be evaluated for any value of $t \geq t_0$, provided their values are known at $t = t_0$, that is, if the "initial conditions" of the problem are known.

Once these currents are determined, any output of the circuit can be evaluated. For a particular application, the output might be the voltage from node 3 to node 0, or the one from node 2 to node 0, or an arbitrary linear combination of the currents i_1 and i_2, such as $7i_1 - 98.4i_2$, or even perhaps a combination involving i_1, i_2, and the input v, such as $i_1 + i_2 - 3v$. This is clearly a choice dictated by the particular application; what is important is that any output can be determined once the currents are known. The set of the two numbers $i_1(t_1)$ and $i_2(t_1)$ computed at an arbitrary instant of time $t_1 \geq t_0$ will be called the *state* of the system in Figure 1.2 at time t_1; i_1 and i_2 are the *state variables*.

Let us make a more systematic list of the properties of this set of the two numbers i_1 and i_2, the state of the system in Figure 1.2:

i. Provided that the input function v satisfies some kind of integrability condition, such as being piecewise continuous, the state at any time $t_1 \geq t_0$ is well defined and determined by (a) the values of the state variables at $t = t_0$; (b) the values which the input function v takes on in the interval $t_0 \leq t \leq t_1$. Note that this does not preclude finite jumps in the state variables, even if in this case, and for the particular set of inputs which have been selected, the state variables are indeed continuous. Of course, infinite jumps are not allowed, since in this case the state variables would not be well defined at those values of time at which the jumps occur.

ii. The manner in which the initial conditions $i_1(t_0)$ and $i_2(t_0)$ have been established is of no consequence; only the values themselves have influence in the future values of i_1 and i_2.

iii. The output of the system is a function of the state and of the input; that is, the output can be determined uniquely once the state and the input are known.

iv. The choice made above for the description of the circuit in Figure 1.2 is not unique; a set of node voltages could have been chosen instead, or perhaps a mixed set composed of some node voltages and some mesh currents.

The system of differential equations (1.1) will be transformed slightly so as to put it in a standard form which will be used extensively in the remainder of the book. This consists in isolating each derivative di_1/dt and di_2/dt in the left-hand side of an equation, as follows:

$$\frac{di_1}{dt} = -\left(\frac{1}{L_1} + \frac{1}{L_2}\right) R_1 i_1 - \frac{R_2}{L_2} i_2 + \left(\frac{1}{L_1} + \frac{1}{L_2}\right) v$$

$$\frac{di_2}{dt} = -\frac{R_1}{L_2} i_1 - \frac{R_2}{L_2} i_2 + \frac{1}{L_2} v$$

(1.2a)

Note that the right-hand sides of these equations contain only the state variables themselves, together with the input function v and the parameters of the circuit.

Suppose now that the output y of the network in Figure 1.2 is the voltage between nodes 3 and 0. Clearly,

$$y = R_2 i_2$$

(1.2b)

The set of equations (1.2) will be called the *state equations* of the network in Figure 1.2.

At last, it is interesting to consider the fact that the resistor R_1, say, in the circuit of Figure 1.2 could be time-varying, that is, its value could change with time; for instance, the resistance could be given at time t by $R_1 = C \cos \omega t$. We conclude that the coefficients of the state variables and of the input function in the right-hand side of Equations (1.2) can indeed be functions of time.

1.3. Transfer Functions and State

We consider in this section systems with one input and one output, so that the input–output information is contained in one transfer function, $G(s) = Y(s)/U(s)$; here y is the output and u is the input. The aim, of course, is to derive a set of state equations such as (1.2) for a system described by a given transfer function $G(s)$.

The simplest case, which will be dealt with first, is when $G(s)$ has no finite zeros. An example will introduce the procedure to follow in this case. Suppose

$$G(s) = \frac{1}{s^2 + 3s + 2}$$

(1.3)

Clearly, y and u are related by the differential equation

$$\ddot{y} + 3\dot{y} + 2y = u$$

Let us define two new variables, x_1 and x_2

$$x_1 = y, \qquad x_2 = \dot{y}$$

and let us show that these are legitimate state variables. Our criterion of legitimacy is, of course, the extent to which the would-be state variables share the relevant properties of the mesh currents which were studied in the previous section. This is the first example of the use of a technique which will be extensively employed in the rest of the book, by means of which we will take simpler systems with desirable properties as models of general classes of systems.

Let us write the equations

$$\dot{x}_1 = \dot{y} = x_2$$
$$\dot{x}_2 = \ddot{y} = -(3\dot{y} + 2y) + u = -2x_1 - 3x_2 + u \qquad (1.4)$$
$$y = x_1$$

which are of the form (1.2). Assume that, at $t = t_0$, $x_1(t_0) = x_1{}^0$, $x_2(t_0) = x_2{}^0$. For simplicity in solving (1.4) by means of Laplace transforms, we will choose $t_0 = 0$.

Transforming the first two equations, we obtain

$$sX_1 - x_1{}^0 = X_2$$
$$sX_2 - x_2{}^0 = -2X_1 - 3X_2 + U \qquad (1.5)$$

The expression for $x_1(t)$, $t \geqslant 0$, will be derived in some detail. From (1.5) it follows that

$$X_1(s) = \frac{x_1{}^0(s + 3) + x_2{}^0}{(s + 1)(s + 2)} + \frac{U(s)}{(s + 1)(s + 2)} \qquad (1.6)$$

By means of a partial fraction expansion of (1.6), it can easily be shown that

$$x_1(t) = (2e^{-t} - e^{-2t}) x_1{}^0 + (e^{-t} - e^{-2t}) x_2{}^0$$
$$+ \int_0^t [e^{-(t-\tau)} - e^{-2(t-\tau)}] u(\tau) \, d\tau \qquad (1.7)$$

where the integral in the right-hand side of (1.7) is the convolution of u with the inverse transform of $1/(s + 1)(s + 2)$, as follows from (1.6). In a similar way, $x_2(t)$ can be shown to be

$$x_2(t) = -(2e^{-t} + 2e^{-2t}) x_1{}^0 + (-e^{-t} + 2e^{-2t}) x_2{}^0$$
$$+ \int_0^t [e^{-(t-\tau)} + 2e^{-2(t-\tau)}] u(\tau) \, d\tau \qquad (1.8)$$

Can we say, from (1.7) and (1.8), that x_1 and x_2 qualify as state variables? Before answering this question, it must be recognized that $x_1(t_1)$ and $x_2(t_1)$ depend on the values of u for $0 \leqslant t \leqslant t_1$; therefore, no meaningful statements about the behavior of the state variables can be made before defining clearly the class of functions u which we are willing or forced to admit as inputs. For instance, if in a particular problem one has to consider input functions which are not Riemann integrable [6] (all integrals in this book will be considered to be Riemann, unless identified otherwise), then certainly x_1 and x_2 do not qualify as state variables, since in such a case they would not be well defined for any $t \geqslant t_0$. This matter will be discussed in detail in Chapter 3; it suffices for the purposes of the present section to define the admissible inputs as those described either by piecewise continuous functions† or by impulse functions or by linear combinations of both types. The derivatives of the impulse function are definitely not admissible. Clearly, then, $x_1(t)$ and $x_2(t)$ are well defined for every $t \geqslant t_0$ provided that u is admissible; at those instants of time at which an impulse is present in the input, some state variables will experience a finite jump. It should be clear now why the derivatives of the impulse function are not admissible, since in such a case x_1 or x_2 or both would become undefined when these functions occur; this is a property which we most definitely do not want the state variables to have.

Since the variables x_1 and x_2 are well defined for each $t \geqslant 0$ for any admissible u and a set of initial values $x_1{}^0$ and $x_2{}^0$ and since the output y is uniquely defined by x_1 and x_2, we conclude that x_1 and x_2 are indeed state variables and (1.4) is a set of state equations for the system described by the transfer function (1.3).

If the transfer function under consideration has finite zeros, the technique used in the previous example cannot be used. Consider the transfer function

$$\frac{Y(s)}{U(s)} = G(s) = \frac{s^2 + s + 2}{s^3 + 5s^2 + s + 7} \tag{1.9}$$

equivalent to the third-order differential equation

$$\dddot{y} + 5\ddot{y} + \dot{y} + 7y = \ddot{u} + \dot{u} + 2u \tag{1.10}$$

---

† By a piecewise continuous function we understand a function of a real variable which is continuous at all but a finite number of points in every finite interval contained in its domain of definition, and such that at each point of discontinuity the jump in value of the function is finite.

Put, as before, $y = x_1$, $\dot{y} = x_2$, $\ddot{y} = x_3$. Then, (1.10) is transformed into

$$\dot{x}_1 = \dot{y} = x_2$$

$$\dot{x}_2 = \ddot{y} = x_3$$

$$\dot{x}_3 = \dddot{y} = -(5\ddot{y} + \dot{y} + 7y) + \ddot{u} + \dot{u} + 2u \qquad (1.11)$$

$$= -7x_1 - x_2 - 5x_3 + \ddot{u} + \dot{u} + 2u$$

$$y = x_1$$

Suppose now that we consider as admissible, input functions which are piecewise continuous (these input functions occur quite naturally in many areas of system theory). Then, $\dot{u}$ would exhibit impulse functions, and $\ddot{u}$ derivatives of impulse functions, the so-called doublet functions. The solution of (1.11) is given by expressions similar to (1.7) and (1.8); at those instants of time when u has a discontinuity, $\ddot{u}$ exhibits a doublet, and the variables x_1, x_2, and x_3 become undefined for those values of the time. We conclude that the previous construction does not give rise to a state description of the system given by the transfer function (1.9); (1.11) is not a set of state equations. The following construction, however, does provide a set of state variables:

$$\begin{aligned}\dot{x}_1 &= x_2 \\ \dot{x}_2 &= x_3 \\ \dot{x}_3 &= -(7x_1 + x_2 + 5x_3) + u \\ y &= 2x_1 + x_2 + x_3\end{aligned} \qquad (1.12)$$

Note that now the derivatives of u do not appear in the right-hand side of the differential equations. Clearly, for any set of initial conditions at $t = t_0$ and any admissible function u (among which functions that exhibit impulses can be included), the variables x_1, x_2, and x_3 are well defined for any $t \geqslant t_0$. The output y is uniquely determined by x_1, x_2, and x_3; we conclude that these are a set of state variables for the system described by (1.9), provided of course that the system of equations (1.12) and the transfer function (1.9) are equivalent, in the sense that if the transfer function $Y(s)/U(s)$ is computed from (1.12), it is equal to the transfer function (1.9). This can be seen by inspection from the flow graph [7] of (1.12) in Figure 1.3.

The procedure used above is quite general in the sense that it can be applied to the derivation of state equations for any transfer function with n poles and $(n - 1)$ zeros. Indeed, if the transfer function is given by

$$\frac{Y(s)}{U(s)} = \frac{a_n s^{n-1} + \cdots + a_2 s + a_1}{s^n + b_n s^{n-1} + \cdots + b_2 s + b_1} \qquad (1.13)$$

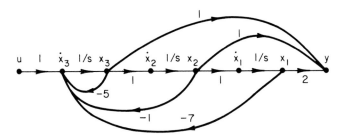

FIGURE 1.3.   Flow graph corresponding to the system of differential equations (1.12).

then put

$$\dot{x}_1 = x_2$$

$$\dot{x}_2 = x_3$$

$$\vdots$$

$$\dot{x}_{n-1} = x_n \qquad\qquad (1.14)$$

$$\dot{x}_n = -(b_1 x_1 + b_2 x_2 + \cdots + b_n x_n) + u$$

$$y = a_1 x_1 + a_2 x_2 + \cdots + a_n x_n$$

It can be proved, just as was done in the case of the transfer function (1.9), that the variables x_1, x_2, ..., x_n are well defined for any $t \geqslant t_0$ by a set of initial conditions at $t = t_0$ and an admissible function u which is piecewise continuous and might also exhibit impulses. It is obvious that the output is uniquely determined by these variables. Therefore, (1.14) is a set of state equations and x_1, x_2, ..., x_n is a set of state variables for the system described by (1.13). The equivalence of this transfer function and the set of equations (1.14) is easily proved by transforming (1.14) or by inspection of the flow graph of Figure 1.4. It happens, though, that the set of equations (1.14). in spite of indeed

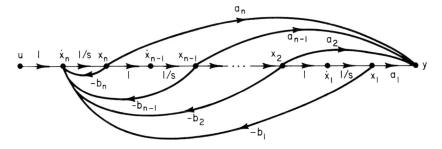

FIGURE 1.4.   Flow graph corresponding to the system of differential equations (1.14).

being a state description of the system described by the transfer function (1.13), is not a very good one in some cases. We will show this deficiency by means of an example, postponing the general treatment of the subject until Chapter 6.

Consider a system with transfer function

$$G(s) = \frac{s + \beta}{s^2 - \alpha^2}, \qquad \alpha \neq 0 \tag{1.15}$$

The set of state equations for this system derived by the method shown in (1.14) is

$$\dot{x}_1 = x_2$$
$$\dot{x}_2 = \alpha^2 x_1 + u \tag{1.16}$$
$$y = \beta x_1 + x_2$$

Let us now solve the first engineering problem of the book. Suppose that a record exists of the output y over the interval $[0, t_1]$, this output being due to initial conditions x_1^0, x_2^0 in the state variables specified at $t = 0$. The input is known to be identically equal to zero over $[0, t_1]$. It is desired to ascertain the initial condition of the state variables by means of the record of the output.

After performing some elementary algebra, we find that the output y is defined by

$$y(t) = x_1^0(\beta \cosh \alpha t + \alpha \sinh \alpha t) + x_2^0(\cosh \alpha t + (\beta/\alpha) \sinh \beta t)$$

$$\equiv F(t, x_1^0, x_2^0)$$

for t in $[0, t_1]$, $\alpha \neq 0$. We can now solve the problem quite simply: Since we want to determine two numbers x_1^0 and x_2^0 from the record of y over $[0, t_1]$ we choose two instants of time in this interval, t_a and t_b, $t_a \neq t_b$, and set up two linear algebraic equations in x_1^0 and x_2^0,

$$y(t_a) = F(t_a, x_1^0, x_2^0), \qquad y(t_b) = F(t_b, x_1^0, x_2^0)$$

The values $y(t_a)$ and $y(t_b)$ are of course read from the record of the output. The problem proposed above has a solution if and only if this system of equations has itself a solution; note that if this system has no solution, a new system constructed by taking more samples of the output, and thus consisting of more equations than unknowns, has no solution either.

We are led, therefore, to investigate the determinant of this system of equations. This can be computed to be

$$(\alpha - \beta^2/\alpha) \sinh \alpha(t_a - t_b)$$

Since α is not zero, and $t_a \neq t_b$, $\sinh \alpha(t_a - t_b) \neq 0$; we conclude that the problem has a solution if and only if $\alpha^2 \neq \beta^2$, that is, if β is different from both α and $-\alpha$. Note that $s^2 - \alpha^2 = (s + \alpha)(s - \alpha)$, so that we can say that the problem has a solution if and only if the zero of the transfer function (1.15) does not coincide with any of its poles.

This property happens to be a general characteristic of the state equations (1.14); whenever there is coincidence in the original transfer function of some poles with some zeros, the initial conditions cannot be ascertained from a knowledge of the output over an interval; we say in such a case that the system (1.14) is not *observable*. The study of the concept of observability, as well as of the closely related idea of controllability, will be done in Chapter 5. It can be concluded now, though, that the procedure we have been using for the setting up of state equations is not a good one if poles happen to coincide with zeros in the original transfer function.

There are, however, other ways of setting up state equations for systems such as the one described by Equation (1.13). Consider (1.15), and the system of equations

$$\dot{x}_1 = x_2 + u$$

$$\dot{x}_2 = \alpha^2 x_1 + \beta u \qquad (1.17)$$

$$y = x_1$$

The transfer function $Y(s)/U(s)$ corresponding to this system equals (1.15). The determinant of the system of two equations in $x_1{}^0$ and $x_2{}^0$ referred to above is now

$$(1/\alpha) \sinh \alpha(t_b - t_a)$$

which is never zero. We conclude that if the engineering situation is such that it is necessary to determine initial conditions from records of the output, the set of state equations (1.17) is preferable to the set (1.16) for the system described by the transfer function (1.15).

On the other hand, the set (1.17) has some disadvantages with respect to the set (1.16). Suppose that our aim now is *to control* the system (1.17), so that we require, reasonably enough, that this system has the property of *controllability*, that is, that the state variables can be made to take any value at a later time by the application of an admissible control given that they were equal to zero at $t = 0$.

If we write down the equations for the transforms of the state variables with their initial conditions (at $t = 0$) equal to zero, we see that

$$X_1(s) = \frac{s + \beta}{s^2 - \alpha^2} U(s)$$

$$X_2(s) = \frac{\alpha^2 + \beta s}{s^2 - \alpha^2} U(s)$$

(1.18)

so that if β equals either α or $-\alpha$ the transforms of the state variables, and then the state variables themselves, are related by a linear relationship, regardless of the values of the control u; for instance, if $\beta = \alpha$,

$$\alpha X_1(s) = X_2(s), \qquad \alpha \neq 0$$

or

$$\alpha x_1(t) = x_2(t), \qquad \alpha \neq 0, \quad t \geqslant 0 \tag{1.19}$$

for every admissible control u, so that states which do not satisfy the relationship (1.19) between the state variables, such as the state $x_1 = 0$, $x_2 = 1$, cannot be reached from the origin by the use of a control when $\beta = \alpha$ (or $-\alpha$). It is left to the reader to show that this is not the case for the system (1.16); it follows that if the aim is the control of the system, this set of state equations should be used for the transfer function (1.15).

There is, however, a set of state equations for the transfer function (1.15) when β equals either α or $-\alpha$ which has both properties, observability and controllability. Let $\beta = \alpha$. Then the system

$$\dot{x}_1 = \alpha x_1 + u, \qquad y = x_1 \tag{1.20}$$

can easily be proved to have both properties as well as having the transfer function $Y(s)/U(s) = 1/(s - \alpha)$.

An interesting fact emerges from the lack of uniqueness of the state descriptions of a system known solely by its transfer function; this is that no given description can be taken to represent the true internal structure of the system itself. As we produce state descriptions of a system, we therefore generate what in a sense are false internal structures of it. The purpose of constructing these may escape the reader. As shown above, the knowledge of the state description of a system is of great value in the design of controls which are to interact with the system in a somewhat sophisticated manner. This is true regardless of the origin of the state description; any of our false structures would serve the purpose of providing a set of state variables which would help in the construction of controls, even though, as shown in the examples above, some structures may be more appropriate than others for some particular tasks. A particular structure can be generated by simulation of the state equations determined from the transfer function, as shown in Figure 1.5 for the system described by the transfer function (1.9).

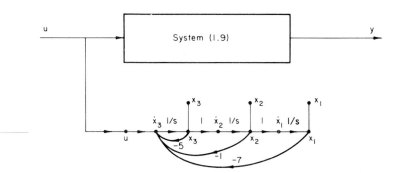

FIGURE 1.5.   Generation by means of analog simulation of a set of state variables for the system described by the transfer function (1.9).

Means other than straight simulation exist to generate the state variables in this situation; the corresponding theory, which deals with the so-called *observer*, will be treated at the end of Chapter 6.

Exercises[†]

1. Obtain a set of state equations for the transfer functions

$$G(s) = \frac{s + 1}{s^2 + as + 1}$$

so that this set is observable for all values of a.

2. Obtain a set of state equations for the transfer function in Problem 1, so that this set is controllable for all values of a.

3. Is it possible to find a set of state equations for the transfer function in Problem 1 so that this set is controllable and observable regardless of the value of a? If this is possible, find this set.

4. Suppose that the output y of the system described by the transfer function (1.15) is not available directly, so that it cannot be sampled and thus used to determine

---

[†] The exercises marked with an asterisk * illustrate important aspects of the concepts introduced in the text. The student *should* try and solve these exercises before proceeding any further.

the initial conditions of the state as explained in the text. However, the *moments* of the output are available,

$$m_k = \int_0^{t_1} t^k y(t) \, dt, \qquad k = 0, 1, \dots$$

Devise a method for the determination of the initial conditions of the state by the use of some of these moments. Find conditions on α and β such that this is possible.

*5. Suppose a system is controllable. Would this imply that the state $x_i = 0, i = 1, \dots, n$, can be transferred to any state at a later time by the use of a control with a magnitude not higher than unity?

6. Characterize explicitly the set of all points which can be reached from the point $x_1 = 0, x_2 = 0$ at $t = 0$ by means of free trajectories (that is, trajectories without inputs) of the system (1.16) at $t \geqslant 3$. Take $\alpha = 1, \beta = 2$.

1.4. Some Conclusions and a Program

In the previous pages we have introduced a new major concept, that of the state of a linear system. The reader should be aware that this has been an important step forward; by associating this idea to the rather shopworn concepts related to input–output descriptions, we are in a position to develop in the following chapters a theory of linear systems which is useful (only by the use of methods springing from this theory can we control well systems of complex characteristics) as well as intellectually satisfying in its rich set of axioms and theorems. System theory has come of age by the introduction of the state-space approach, and the resulting body of knowledge has already some of the classic feel encountered in some theories of the physical world such as quantum mechanics.

It is to be noted that this is not a mathematical theory. In fact, the input–output descriptions, which are often the only observables in a system, and thus of great engineering importance, have not disappeared from the picture but have indeed been integrated with the state concept, forming what is clearly one of the most important areas of our study, which we will treat in Chapter 6 under the name of synthesis. The concepts of controllability and observability, which have a pure, undiluted engineering motivation, take over in Chapter 5 the description of linear systems done from a purely mathematical standpoint in Chapters 3 and 4; it is truly amazing how, when we take an engineering viewpoint, the systems acquire rich and interesting structures and properties. Ours is, fundamentally, an engineering theory.

We have not as yet defined axiomatically the concept of state, and actually will not do so until Chapter 9. The method to be followed is in essence the same used in Sections 1.2 and 1.3; there we studied a system, actually a simple electrical network, and decided that the particular method used for this study, the setting up of a set of mesh equations, and the particular physical entities used to characterize its state, the mesh currents, had several useful qualities; therefore, when developing descriptions of the systems which appear in Section 1.3, we required that these shared those useful qualities encountered in Section 1.2. This is, of course, a process of induction.

A similar process will be used in the rest of the book. We will study simple differential and difference systems in the first eight chapters, and find that these have properties (which are essentially the same as the properties of the network in Section 1.2) which make their study easy and useful. By simple systems we mean those which are described by a finite number of differential equations such as (1.2), or by a finite number of difference equations, and which, as we shall show later, have a state composed of a finite number of state variables. It so happens that the interest of the system engineer is not circumscribed to such simple systems; there are linear systems, of great practical and theoretical interest, which cannot be described by a finite number of state variables; an example is shown below. Not all such systems have an interesting behavior which can be studied with ease, while we are only interested in those which do have such behavior; we need, therefore, to construct a set of guidelines in which to embody those properties of the simple systems studied in Chapters 1 to 8 which we want to be maintained in more complex systems. This is done in Chapter 9 by the definition of a new entity, the dynamical system, whose behavior summarizes those properties by means of a set of axioms. After the setting of these axioms, it is not difficult to study the not-so-simple systems themselves; this is also done in Chapter 9.

We present now an example of a system with a state which does *not* consist of a finite set of numbers. Consider a simple partial differential equation, the heat or diffusion equation:

$$\frac{\partial x}{\partial t}(z, t) = \frac{\partial^2 x}{\partial z^2}(z, t) \tag{1.21}$$

where t is as usual time and z is a spatial coordinate. The quantity x could be visualized as the temperature along a one-dimensional bar; as such, it depends on the position z of the point in the bar being considered and on the time t.

Of course, Eq. (1.21) is not complete; one must specify the ranges of variation of the coordinates z and t, as well as initial and boundary conditions. Also, a method has to be defined to interact with the system; this can take the form on an extra term $g(z, t)$ added to the right-hand side (what is the physical significance of such a term?), or it can be embodied in the boundary conditions; if the bar has a finite length, the temperature at any point of it can be manipulated by varying the temperature at one or both of its ends in an appropriate manner.

Suppose that the spatial coordinate z ranges from $-\infty$ to ∞; there are no boundary conditions for finite values of z, and the control, if any, is effected by a term $g(z, t)$ in the right side of (1.21). What is the state of this system? Assume that the temperature along the bar is specified at an arbitrary time t_0; that is, assume that the values $x(z, t_0)$, $-\infty < z < \infty$, are given. We shall show in Chapter 9 that the values of the temperature along the rod for $t > t_0$ (that is, the whole set of values $x(z, t)$, $t > t_0$, $-\infty < z < \infty$) are determined by $x(z, t_0)$, $-\infty < z < \infty$, and of course by the control $g(z, t)$. It follows that the state of this system at a time t_1 is simply the whole of the values of the function $x(z, t_1)$, $-\infty < z < \infty$; as time changes, these values change, as shown in Figure 1.6.

This situation should be compared with the one met in connection with Eq. (1.14). In that case, the dynamic behavior of the system was shown in the change with time of the n state variables x_1, x_2, ..., x_n. In the present situation, the dynamical behavior of the bar is shown

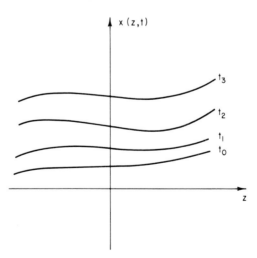

FIGURE 1.6.   As time proceeds, the state of the system (1.21) changes; the values of $x(z, t)$ are modified as the time undergoes a change, $t_0 \rightarrow t_1 \rightarrow t_2 \rightarrow t_3$.

in the change with time of an infinity of state variables, the whole of the values of the temperature along the bar. The state space is not, in this case, finite-dimensional, that is, the state cannot be specified by a finite set of numbers.

Similar considerations apply when the spatial coordinate ranges over a finite interval, $0 \leqslant z \leqslant L$, say; the control action may be specified by the values of x at $z = 0$ and $z = L$; that is, $x(0, t) = F_1(t)$ and $x(L, t) = F_2(t)$ can be considered as the controlling inputs. Other combinations are, of course, possible; for instance, the control can be specified, as in the previous case, by a term $g(z, t)$ in the right-hand side of (1.21). In all of these cases, the state at t_0 is the function $x(z, t_0)$, this time defined only for $0 \leqslant z \leqslant L$; the state space is, again, not finite dimensional.

SUPPLEMENTARY NOTES AND REFERENCES

§1.1. The problem of determining the transfer function of a system from a set of external measurements is treated in detail in Truxal's book [2]. Other methods are described in

E. M. Mishkin and L. Braun, "Adaptive Control Systems." McGraw-Hill, New York, 1961.

A modern treatment of the application of Laplace transforms to engineering analysis is given in

A. H. Zemanian, "Distribution Theory and Transform Analysis." McGraw-Hill, New York, 1965.

This excellent text also gives a graduate-level approach to the study of impulse functions and related matters.

A pioneering work in the development of the methods of study of differential equations, and which is perhaps of more importance to the modern theory than the classical treatise of Poincaré, is

M. A. Liapunov, Problème général de la stabilité du mouvement. *Ann. Fac. Sci. Univ. Toulouse* **9** (1907), 203–474. This is a translation of the original paper published in 1893 in *Comm. Soc. Math. Kharkow*.

A very good reference on optimal control is

E. B. Lee and L. Markus, "Foundations of Optimal Control Theory." Wiley, New York, 1967.

The reader is cautioned, however, that a considerable background on linear and nonlinear systems is needed to comprehend the material presented in this book.

It is not our purpose to show how to write state equations of physical systems. This is a rather undergraduate-level subject, which is treated, for instance, in

H. R. Martens and D. R. Allen, "Introduction to Systems Theory." Merrill, Columbus, Ohio, 1969.

§1.2. The student interested in the methods and concepts of modern network theory should read

H. J. Carlin and A. B. Giordano, "Network Theory: An Introduction to Reciprocal and Nonreciprocal Circuits." Prentice-Hall, Englewood Cliffs, New Jersey, 1964.

M. R. Wholers, "Lumped and Distributed Passive Networks." Academic Press, New York, 1969.

V. Doležal, "Dynamics of Linear Systems." Academia, Prague, 1967.

The book by Doležal deals not only with network theory, but is also a deep treatise on some aspects of the theory of linear systems. Of special interest to the reader is that a very good treatment can be found here of the theory of distributions and its relationships with linear systems theory; we shall not even touch upon this subject.

§1.3. When a system is described by a transfer function with the same number of zeros as poles, the determination of a set of state equations for the system is a lengthy process which will not be treated in this book, since the models of physical systems seldom exhibit this property.

The subject is treated at length in a number of texts, such as

P. M. DeRusso, R. J. Roy, and C. M. Close, "State Variables for Engineers." Wiley, New York, 1965.

K. Ogata, "State Space Analysis of Control Systems." Prentice-Hall, Englewood Cliffs, New Jersey, 1967.

§1.4. Another important example of a system with an infinite-dimensional state space is shown in Figure 1.7. The following equation relates the input u to the output y:

$$\dot{y}(t) + y(t - T) = u(t) \tag{1.22}$$

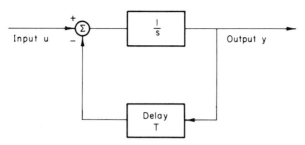

FIGURE 1.7. A linear system in which the feedback signal is delayed by T seconds before being fed back to the plant with transfer function $1/s$.

Our interest is to find a mathematical entity which shares the properties the state was found to have in those systems of Sections 1.2 and 1.3. We must, therefore, find the equivalent of the initial conditions of the state equations which appear in those sections. A consideration of the structure of the system in Figure 1.7 will convince the reader that if the contents of the delay line, that is, the values of the output for $t_0 - T \leqslant t \leqslant t_0$, are known at a time t_0, then this, together with a knowledge of the input for $t \geqslant t_0$, determine the output for all time after t_0. Therefore the state of this system at a time t_0 is composed simply of the values of the output from $t_0 - T$ to t_0. This is not a finite set of numbers, and we say that the state space of this system, which is the set of all possible outputs defined from $T - t_0$ to t_0, for all t_0, is not finite dimensional.

Unfortunately, we will not be able to present in this book a study of this class of systems, mainly because of lack of space. The reader is referred to the following texts:

R. Bellman and K. L. Cooke, "Differential-Difference Equations." Academic Press, New York, 1963.

M. N. Oğuztöreli, "Time-Lag Control Systems." Academic Press, New York, 1966,

if he desires to pursue further the concepts introduced above.

In connection with our inductive approach to the development of a theory of linear systems, we can say that the texts

L. A. Zadeh and C. A. Desoer, "Linear Systems Theory." McGraw-Hill, New York, 1963.

L. K. Timothy and B. E. Bona, "State Space Analysis: An Introduction." McGraw-Hill, New York, 1968.

take an axiomatic approach to the development of the theory of linear systems, since they start by defining in general the concept of state

and studying its properties, and then applying these ideas to particular cases such as simple differential or difference systems.
    See also,

R. W. Brockett, "Finite-Dimensional Linear Systems." Wiley, New York, 1970.

REFERENCES

1. A. Papoulis, "The Fourier Integral and Its Applications." McGraw-Hill, New York, 1962.
2. J. G. Truxal, "Automatic Feedback Control Systems Synthesis." McGraw-Hill, New York, 1955.
3. H. Poicaré, "Les Méthodes Nouvelles de la Mécanique Celeste." Gauthier-Villars, Paris, 1873.
4. W. J. Cunningham, "Introduction to Nonlinear Analysis." McGraw-Hill, New York, 1958.
5. M. E. Van Valkenburg, "Network Analysis." Prentice-Hall, Englewood Cliffs, New Jersey, 1964.
6. A. E. Taylor, "Advanced Calculus." Ginn, Boston, Massachussets, 1955.
7. S. J. Mason and H. J. Zimmermann, "Electronic Circuits, Signals and Systems." Wiley, New York, 1960.

2

An Introduction to the Theory of Linear Spaces

2.1. Introduction

We must pause now in our construction of a theory of linear systems to build a basis of fundamental mathematical concepts and techniques on which to erect the theory. This chapter is concerned mainly with the concepts rather than with the techniques, with the theory of linear spaces rather than with matrix algebra and numerical methods. Matrix algebra, the technical aspect of the theory of linear spaces, would have been almost sufficient as a mathematical basis for the first eight chapters; it is possible to treat matrix algebra without going through the precise, axiomatic, demanding theory of linear spaces. It has been our intention, however, to be consistent with our original plan of considering these first chapters not only as the treatment of linear systems with finite-dimensional state spaces, but also as an introduction to the study of more complex systems in Chapter 9. It so happens that the underlying mathematical structure of systems with infinite dimensional spaces does not admit simple representation for operators such as matrices, so that the full-dress apparatus of the theory of linear spaces becomes in this case a necessity.

2.2. Sets and Operations on Sets

We start this chapter on the fundamentals of the theory of linear spaces by introducing some basic mathematical concepts which are to be used extensively throughout the rest of the book.

The first, and most basic, of such notions is the concept of a set [1]. By definition, a set is simply a collection of objects. For example, the collection of vertices of a triangle, the points of the (x, y) plane such that $x^2 + y^2 \leqslant 1$ or the collection of solutions of the differential equation $\dot{x} + 3x + 7 = 0$, for $0 \leqslant x(t_0) \leqslant 1$. The objects which form a set will be called the *elements* of the set; if A is a set and x is an element of it, we write $x \in A$. If we want to indicate that an object x is not an element of a set A we write $x \notin A$.

Sometimes we will use the following type of notation to define a set

$$S = \{x : x \geqslant 0\}$$

By this we mean that the set S has as elements all the nonnegative real numbers. On other occasions we will use a simple enumeration of the elements of a set to define it, displaying them between braces. The set

$$A = \{2, 4, 6, 8, 10\}$$

is the set of all even positive integers which are less than twelve.

The set of real numbers, to be used extensively in this book, will be denoted by R. Likewise, the set of all positive real numbers will be denoted by R_+ and the set of all negative real numbers by R_-.

For mathematical convenience, we will admit a set which contains no elements, to be called the *empty set*, and to be denoted by $\varnothing$.

Definition. A set A is a *subset* of a set B if each element of A is also contained in B; that is, if $x \in A$ implies that $x \in B$. We write $A \subseteq B$. If there is at least one element of B which is not contained in A, we say that A is a *proper subset* of B, and write $A \subset B$.

EXAMPLE

The set $A = \{x : x \geqslant 1\}$ is a proper subset of the set $B = \{x : x \geqslant 0\}$.

EXAMPLE

Consider the set A of all real-valued functions of a real variable t defined for $0 \leqslant t \leqslant 1$. The set B of all continuous functions defined there is a subset of A.

An important class of sets of real numbers are the *intervals*. An *open interval* is described by the inequality $a < x < b$, with $a < b$, and is denoted by (a, b). A *closed* interval is described by $a \leqslant x \leqslant b$, $a \leqslant b$ and denoted by $[a, b]$. Clearly, $(a, b) \subset [a, b]$.

If the following relations hold for two sets A and B, we say that the two sets are *equal*:

$$A \subseteq B, \qquad B \subseteq A \qquad\qquad (2.1)$$

That is, two sets are equal if they consist of the same elements. The reader should have noted that the properties of the relation defined above bear some analogy with the relation $\leqslant$ between real numbers. If x and y are real numbers, either $x \leqslant y$ or $y \leqslant x$; if A and B are sets, however, it is perfectly possible that neither A is a subset of B nor B one of A.

The union of two sets is a basic operation.

Definition. The union of two sets A and B is a set, denoted by $A \cup B$, defined by the relation

$$A \cup B = \{x : x \in A \quad \text{or} \quad x \in B\} \qquad\qquad (2.2)$$

EXAMPLE

If $A = \{2, 4, 6, 8, 10\}$ and $B = \{1, 7\}$, then

$$A \cup B = \{2, 4, 6, 8, 10, 1, 7\}$$

Note that the order in which the elements of a set are enumerated is of no consequence, unless we choose to make it important, in which case we talk about *ordered* sets, to be defined later in this section.

Another important operation on sets is the *intersection*.

Definition. The *intersection* of two sets A and B is a set, denoted by $A \cap B$, defined by the relation

$$A \cap B = \{x : x \in A \quad \text{and} \quad x \in B\} \qquad\qquad (2.3)$$

EXAMPLE

If $A = \{x : x \geqslant 1\}$, $B = \{x : x \leqslant 2\}$, then

$$A \cap B = \{x : 1 \leqslant x \leqslant 2\}$$

Definition. Two sets A and B are *disjoint* if $A \cap B = \varnothing$, that is, if they have no elements in common.

EXAMPLE

If $A = \{x : x > 0\}$, $B = \{x : x < -7\}$, then $A \cap B = \varnothing$; A and B are disjoint.

The operations of union and intersection can easily be defined for several sets. If A_1 , ..., A_i , ..., A_n are sets, then

$$\bigcup_{i=1}^{n} A_i = \{x : x \in A_i \text{ for } some\ i, \quad 1 \leqslant i \leqslant n\}$$

$$\bigcap_{i=1}^{n} A_i = \{x : x \in A_i \text{ for } all\ i, \quad 1 \leqslant i \leqslant n\}$$

We can also talk about the union and intersection of an infinite number of sets. For instance, consider the sets

$$A_n = [0, 1/n]$$

defined for $n = 1, 2, \ldots$. Then,

$$\bigcup_{i=1}^{\infty} A_n = [0, 1]$$

$$\bigcap_{i=1}^{\infty} A_n = \{0\}$$

Definition.   The *difference* of a set B and a set $A \subseteq B$ is a set, denoted by $B - A$, and defined by the relation

$$B - A = \{x : x \in B \quad \text{and} \quad x \notin A\} \tag{2.4}$$

EXAMPLE

If $A = \{x : x \geqslant 1\}$, $B = \{x : -7 \leqslant x\}$, then $B - A = \{x : -7 \leqslant x < 1\}$

Sometimes all sets occurring in a particular problem are subsets of a given set, say S, called the *space*. This situation occurs, for instance, in problems dealing with sets of real numbers; in this case, the space is the set of all real numbers.

Definition.   The difference $S - A$, $A \subseteq S$, is called the *complement* of A, and is denoted by A'.

EXAMPLE

If $A = \{x : x \geqslant 2\}$, and the space S is the set of all real numbers, $A' = \{x : x < 2\}$.

Definition. A set is called *finite* if it consists of a finite number of elements. Otherwise it is called *infinite*. A set is said to be *denumerable* if its elements can be put into one-one correspondence with the integers.

EXAMPLE

The set of all vertices of a triangle is a finite set. The set $\{x : x \geqslant 0\}$ is infinite. The set $\{\frac{1}{2}, \frac{1}{3}, \frac{1}{4}, ..., \frac{1}{n}, ...\}$ is denumerable.

It is useful to consider the *Cartesian product*, or simply the *product*, of two sets A and B. With this in mind, we introduce the concept of an *ordered set*. An *ordering* of a finite set S having n elements is simply a rule which indicates which is to be taken as the first element of S, which the second, etc. An *ordered set* is a set which has an ordering. We may indicate the ordering of a set by a subindex, such as in $A = (x_1, x_2, ..., x_n);$[†] here the first element has subindex 1, etc. If $A = (x_1, x_2, ..., x_n)$ and $B = (y_1, y_2, ..., y_n)$, then we say that $A = B$ if and only if $x_i = y_i$, $i = 1, ..., n$.

Definition. The *Cartesian product* of two sets A and B, denoted by $A \times B$, is the set of *ordered pairs* defined by

$$A \times B = ((x, y) : x \in A, \quad y \in B) \qquad (2.5)$$

EXAMPLE

If $A = \{x : 0 \leqslant x \leqslant 1\}$, $B = \{y : 0 \leqslant y \leqslant 1\}$, then

$$A \times B = ((x, y) : 0 \leqslant x \leqslant 1, \quad 0 \leqslant y \leqslant 1)$$

The Cartesian product of two intervals in the real axis is a rectangle in the real plane.

If $A = B = R$,

$$A \times B = R \times R = ((x, y) : x \in R, \quad y \in R)$$

is the set of *tuples* (x, y) of real numbers. This set will also be denoted by R^2.

In an obvious way, the notion of Cartesian product can be extended to a finite number of sets. Then

$$A_1 \times A_2 \times A_3 \times \cdots \times A_n = ((x_1, x_2, x_3, ..., x_n) : x_i \in A_i)$$

† The elements of ordered sets are usually enclosed in parentheses.

Exercises

1. a. Evaluate $A \cup A$ and $A \cap A$.
    b. Compare $A \cup (B \cup Z)$ with $(A \cup B) \cup Z$.
    c. Prove or give a counterexample

$$A \cup (B \cup Z) = (A \cap B) \cup (A \cap Z)$$

2. a. Evaluate $A \cup A'$ and $A \cap A'$
    b. What is $(A')'$?
    c. Find the complement of $(A' \cup B')' \cap (A \cup B')$. [Hint: Find first $(A \cup B)'$ and $(A \cap B)'$.]

3. Prove the general laws

$$A \cup \left(\bigcap_{i=1}^{n} B_i \right) = \bigcap_{i=1}^{n} (A \cup B_i)$$

$$A \cap \left(\bigcup_{i=1}^{n} B_i \right) = \bigcup_{i=1}^{n} (A \cap B_i)$$

Are these laws valid when the left-hand sides of these expressions show the intersection and union of an infinite, countable number of sets B_i?

4. Let N_1 and N_2 be two sets of integers, not necessarily finite, and let $\{B_i\}$ be a sequence of sets. Prove that if $N_1 \subset N_2$,

$$\bigcap_{i \in N_1} B_i \supset \bigcap_{i \in N_2} B_i, \qquad \bigcup_{i \in N_1} B_i \subset \bigcup_{i \in N_2} B_i$$

5. Prove that $A \subseteq B'$ if and only if A and B are disjoint.

6. Prove that a set A is empty if and only if for a given set B

$$B = (A \cap B') \cup (A' \cap B)$$

2.3. Functions

The concept of function is used sloppily in the engineering literature. This is perhaps excusable in undergraduate textbooks, where the aim is mainly to appeal to the intuition of the young student; in the present book, however, the object is the development of a mathematical theory of linear systems; to achieve this aim, clear definitions, clear notations, and, above all, clear thinking, are of importance.

Definition. Let X and Y be two nonempty sets, not necessarily different, and suppose a correspondence f is given which associates with each element x in X a single element $f(x)$ in Y. Then this correspondence or rule f is called a *function from X into Y*. The set X is called the *domain* of the function, and the set $\{y : y = f(x)$ for some x in $X\}$ is called the *range* of the function f; it is denoted by the expression $f(X)$. Clearly, $f(X) \subseteq Y$. It is said that the element x is *mapped* or *transformed* into $f(x)$; the words *mapping* and *transformation* are generally used synonymously with function.

EXAMPLE

Suppose X is the set R of all real numbers, and that $Y = X$. We associate with each $x \in R$ the number $f(x) = x^2$. Note that the function f is the rule which says, to each $x \in X$ associate its square; $f(x)$ is the *value* of the function at a particular $x \in X$. The domain of f is the set of all real numbers, its range $f(X)$ is the set of all nonnegative real numbers, $f(X) = \{y : y \geqslant 0\}$.

A function from X into Y is sometimes denoted by

$$f: X \to Y$$

Occasionally we will use the notation $f(\cdot)$ for $f: X \to Y$. The dot indicates that f is indeed a function with values $f(x)$, $x \in X$. This notation is especially useful when the domain of the function is a product space; for instance, assume $f: R^2 \to R$. We could denote this function as $f(\cdot)$, with the dot symbolizing the set of two numbers $(x, y) \in R^2$. However, we will generally denote this function as $f(\cdot, \cdot)$, with the first dot symbolizing the number x and the second the number y. Of course, R^2 is a product space, $R^2 = R \times R$, and we can extend this convention to more general product spaces. If $f: X \to Y$, and

$$X = X_1 \times X_2 \times \cdots \times X_p$$

we write for f

$$f(\underbrace{\cdot, \cdot, \cdot, ..., \cdot}_{p \text{ times}})$$

The objective of this notation is not only to indicate that f is a function of p variables. It is often necessary to consider *another* function, generated from $f(\cdot, ..., \cdot)$ by fixing the value of one of the variables. Assume that the value of the first variable is fixed at $\bar{x}_1$. The new function maps $X_2 \times \cdots \times X_p$ into Y; we shall denote this function by

$$f(\bar{x}_1, \cdot, ..., \cdot)$$

Of course, this function is defined by

$$f(\bar{x}_1, \cdot, ..., \cdot)(x_2, ..., x_p) = f(\bar{x}_1, x_2, ..., x_p)$$

Let us go back to our definition of a function. The last example puts well-known facts in a rather different light. However, the importance of the definition can be appreciated even more when some unusual examples of functions are introduced. Suppose X is the set of all continuous real-valued functions defined on the interval $[0, 1]$. A rule which assigns to each of the members of this space a member of the space Y of all real numbers is defined by

$$f(x) = \int_0^1 x(t)\,dt \tag{2.6}$$

Note that we are associating a real number $f(x)$ with each function x defined on $[0, 1]$. We can define another function on X by associating with each $x \in X$ another real-valued continuous function $f(x)$ defined on $[0, 1]$ by the expression

$$y(t) = f(x)(t) = \int_0^1 \exp(s^2 + t^2)\,x(s)\,ds \tag{2.7}$$

Note that the value of the function f, which maps X into itself, is $f(x)$; this value is itself a function defined on $[0, 1]$, which we have chosen to call y; this function is defined at each point $t \in [0, 1]$ by the equality given above.

Definition. A function f defined from X into Y is said to be defined from X *onto* Y if $f(X) = Y$, that is, if the range of f is all of Y. Sometimes it is said simply that f is *onto*.

EXAMPLE

If X and Y are again all of the real numbers, the function f defined by $f(x) = x^3$ is onto; clearly, by choosing $x = y^{1/3}$ for any number y a number $x \in X$ can be chosen such that $y = x^3$.

The function defined by (2.6) is onto, since certainly continuous functions can be defined on $[0, 1]$ which have any arbitrary area. The function defined by (2.7) is not onto, since clearly there are continuous functions defined on $[0, 1]$ which are not of the form (2.7), that is, which cannot be obtained by transforming some function x in the manner defined by (2.7).

Definition. A function $f: X \to Y$ is said to be *one-one* if $f(x_1) = f(x_2)$ implies that $x_1 = x_2$, that is, if each element of $f(X)$ is the map of only one element of X.

EXAMPLE

Let X and Y be again the set R. The function defined by $f(x) = x^3$ is one-one. The function defined by $y = \sin x$ is not, since if $y = \sin x_1$, y is also equal to $\sin(x_1 + 2\pi)$; the two elements x_1 and $x_1 + 2\pi$ are mapped into y.

Neither of the functions defined by (2.6) and (2.7) is one-one. Many continuous functions defined on $[0, 1]$ can be found so that their area equals a given real number. It can be proved that there are many functions which are mapped into a given y in the range of the function defined by (2.7).

Well known to the reader is the device by which the values of a function f with domain and range in the set of real numbers are plotted on a plane with coordinates x and $f(x)$. We will generalize this concept as follows.

Definition. The *graph* of a function f from X into Y is an ordered set G_f defined by

$$G_f = \{(x, y) : x \in X, \quad y = f(x)\}$$

EXAMPLE

The graph of the function defined by $f(x) = x^3$ is shown in Figure 2.1. Note that this graph is a set of points in the (x, y) plane.

FIGURE 2.1. Graph of the function f defined by $f(x) = x^3$. This function maps the set of real numbers onto itself.

Definition. Let $f: X \to Y$. If there is a function *from Y into X*, to be called the *inverse function* of f and to be denoted by f^{-1}, such that $f^{-1}(f(x)) = x$ for all $x \in X$, then f is said to be *invertible*.

If we attempt to construct the inverse function of the function whose graph appears in Figure 2.1, we simply follow the rule: With each $y \in Y$ we associate *the* value x such that $f(x) = y$. Note that this is possible only if (a) the range of f is all of Y, that is, if f is onto [Otherwise, there would be some values of $y \in Y$ to which no values of x could be associated such that $y = f(x)$.] (b) to each y corresponds only one x such that $y = f(x)$, that is, if f is one-one. The function defined by $y = \sin x$ is neither onto nor one-one; it is not invertible. The following theorem, whose formal proof is left as an exercise, makes clear the value of the previous arguments.

THEOREM 2.1.   A function f from X into Y is invertible if and only if it is one-one and onto.

Exercises

*1.   If f is a function from X into Y, and $A \subseteq X$, the set $f(A) = \{y : y = f(x), x \in A\}$ is called the *image* of the set A under the transformation f. Show that f is one-one if and only if for any pair of subsets A_1, A_2 of X,

$$f(A_1 \cap A_2) = f(A_1) \cap f(A_2) \qquad (*)$$

Give an example of a function which maps the interval $[0, 1]$ into itself which is not one-one and for which the expression $(*)$ above is not true.

*2.   Show that $f : X \rightarrow Y$ is one-one if and only if for any pair of disjoints subsets A_1 and A_2 in X,

$$f(A_1) \cap f(A_2) = \varnothing$$

3.   Let f be a function which maps the real line onto itself. Show that if this function is *increasing*, that is, if $t_1 > t_2$ implies that $f(t_1) > f(t_2)$, then it is invertible.

4.   Consider the expression

$$\ddot{y} + \dot{y} + y = u, \qquad y(0) = \dot{y}(0) = 0 \qquad (*)$$

Here y is a twice-differentiable function of the time which satisfies the initial conditions given, defined for $t \geqslant 0$. The expression $(*)$ can be interpreted as defining a function whose domain is the set of all such functions y. What is its range? Is this function onto and one-one? If so, find its inverse.

2.4. Linear Spaces

A set, as defined in the previous section, is an entity without internal mathematical structure, and, as such, of limited usefulness and interest. Nothing very much can be done, at least from our standpoint, with sets such as that of all left-handed policemen in Altoona, Pa. The sets in which we are interested, such as point sets on a line or plane, sets of functions defined on intervals such as $[0, 1]$, sets of ordered numbers, etc, can, however, be given structures which render them at the same time interesting and useful.

It should be recognized that the properties of the set of real numbers stem from two different kinds of structures possesed by it. The first is an algebraic structure; we add two real numbers and obtain another real number, or we multiply a real number by another, and again obtain a real number. We will define below an entity, to be called a linear (or vector) space, which is a set which has been given an algebraic structure closely modeled on the algebraic structure of real numbers. In many cases the operations of addition and the one which we will call scalar multiplication suggest themselves as natural extensions of the corresponding operations with real numbers; for instance, if we want to add two ordered sets of real or complex numbers, say (a_1, a_2) and (b_1, b_2), it is natural to define their sum as the ordered set $(a_1 + b_1, a_2 + b_2)$.

The second structure which the real numbers possess is a *metric* one; we will refer to it in a later section. We define now the concept of linear space. First, however, we must introduce the concept of a *field*.

A *scalar*, to appear below in the definition of a vector space, is for us a number, either real or complex. When we talk about the set of real numbers, we will say "the field R;" likewise, "the field C" will refer to the set of complex numbers. In a situation in which scalars may be real or complex, we will talk about "the field F." The name "field" is given to these sets because they are examples of a general algebraic structure known as a field [2].

Definition. *A linear space X over a field F* is a set, whose elements are to be called *vectors*, in which are defined:

    i. An operation called *vector addition*, which associates with each pair of vectors x and y in X a unique vector in X, denoted by $x + y$, in such a way that

    a. $x + y = y + x$.
    b. $x + (y + z) = (x + y) + z$.
    c. There is a unique vector $0 \in X$, called the zero vector, such that $x + 0 = x$ for all $x \in X$.
    d. For each vector x in X there is a unique vector in X, to be called $-x$, such that $x + (-x) = 0$.

    ii. An operation, called *multiplication*, which associates with each scalar c in F and each x in X a unique vector in X, denoted by cx, in such a way that

    a. $1x = x$ for all x in X.

b. $(c_1 c_2) x = c_1(c_2 x)$, c_1, $c_2 \in F$.

c. $c(x + y) = cx + cy$.

d. $(c_1 + c_2) x = c_1 x + c_2 x$.

An important linear space is the one whose elements are ordered sets of n real numbers, also called n-tuples or n-vectors. This space will be denoted by R^n and its elements by

$$x = \begin{bmatrix} x_1 \\ x_2 \\ \vdots \\ x_n \end{bmatrix} \tag{2.8}$$

We define the addition operation in R^n as

$$x + y = \begin{bmatrix} x_1 + y_1 \\ x_2 + y_2 \\ \vdots \\ x_n + y_n \end{bmatrix} \tag{2.9}$$

and multiplication by a scalar $c \in R$ as

$$cx = \begin{bmatrix} cx_1 \\ cx_2 \\ \vdots \\ cx_n \end{bmatrix} \tag{2.10}$$

It is simple to prove that these operations in R^n indeed satisfy the conditions given above. R^n is therefore a linear space. A similar definition can be made of the linear space of n-tuples whose entries are complex numbers; this is a vector space over C, to be called C^n. These spaces will be encountered in this book generally in connection with the time-domain analysis of systems of linear differential and difference equations.

Another vector space of interest is that in which the elements are continuous, real-valued functions of a real variable t defined on an interval such as $[0, 1]$. Let x and y be in this space. We define the element $x + y$ by the equality

$$(x + y)(t) = x(t) + y(t), \qquad t \in [0, 1]$$

Obviously, $x + y$ belongs to the same space as x and y since the sum of two continuous functions is itself continuous. In a similar way, we define the element cx, $c \in R$, by

$$(cx)(t) = cx(t), \qquad t \in [0, 1]$$

The reader can easily show that this set, under the vector space operations just defined, is indeed a vector space.

Exercises

1. Show that, in any vector space, $cx = 0$ implies either $c = 0$ or $x = 0$.

*2. It happens sometimes that a subset of a linear space is itself a linear space with respect to the operations already defined. Which of the following subsets of R^n are linear spaces?

    i.   The set of all $x \in R^n$ such that $x_1 x_n = 0$.
    ii. The set of all $x \in R^n$ such that $x_1 + x_2 = 0$.
    iii. The set of all $x \in R^n$ such that all their entries are rational.
    iv. The set of all $x \in R^n$ such that $x_4 \geqslant 0$.

*3. Consider now the space of all real-valued continuous functions defined on $[0, 1]$. Which of the following sets are linear spaces with respect to the operations defined in the space itself?

    i.   All functions such that $f(.5) = 2$.
    ii. All functions such that $f(.5) = 0$.
    iii. All differentiable functions.
    iv. All functions such that $f(.5) = -f(.75) - 3$.

4. Consider the space of all polynomials defined over the whole of the real line. Which of the following sets of polynomials is a linear space?

    i.   All polynomials of degree 5.
    ii. All polynomials which are positive for positive values of the variable.
    iii. All polynomials which tend to zero as the variable tends to zero.

*5. It is necessary to give the structure of a linear space to the set composed of all n-tuples whose entries are functions of a real variable t, say, defined over the interval $[0, 1]$, with range in the real numbers; that is, the set of n-vector-valued functions mapping $[0, 1]$ into R^n. Define properly vector space operations, and show that they satisfy the corresponding axioms.

2.5. Basis and Dimension†

If $x^1, ..., x^m$ are vectors in a linear space X over a field F, and if $c_1, ..., c_m$ are scalars in F, it is possible to define by a simple process of induction vectors of the form $c_1 x^1 + \cdots + c_m x^m$.

Definition. We say that a vector x in X is a *linear combination* of the vectors x^i, $i = 1, ..., m$ if there are scalars $c_1, ..., c_m$ such that

$$x = c_1 x^1 + \cdots + c_m x^m \qquad (2.11)$$

† Note that symbols such as x^1, x^2, refer to vectors; the superscript should *not* be confused with an exponent; x^2 *does not* mean x-squared, but the second vector in an ordering of vectors.

EXAMPLE

Let $X = R^3$. The vector

$$x = \begin{bmatrix} 1 \\ 2 \\ 4 \end{bmatrix}$$

is a linear combination of the vectors

$$x^1 = \begin{bmatrix} 0 \\ 1 \\ 2 \end{bmatrix} \quad \text{and} \quad x^2 = \begin{bmatrix} 1 \\ 1 \\ 2 \end{bmatrix}$$

since $x = x^1 + x^2$.

EXAMPLE

Let X be the space of all continuous real-valued functions defined on $[0, 1]$, and let x, x^1, and x^2 be in this space. Suppose $x = x^1 + x^2$. This implies that, *for all* $t \in [0, 1]$, $x(t) = x^1(t) + x^2(t)$. Note that if this equality were not satisfied just for one value of t, the vector x would not be a linear combination of x^1 and x^2; this follows from the definition of addition of vectors in this space. (Of course, it is not possible for the equality $x(t) = x^1(t) + x^2(t)$ to fail to hold for just one value of t, since the functions x, x^1, and x^2 are continuous. Why?)

Definition. A set $\{x^1, x^2, ..., x^m\}$ of vectors in X is *linearly dependent* if there exist scalars c_1, ..., c_m in F, not all of which are zero, such that

$$c_1 x^1 + \cdots + c_m x^m = 0 \tag{2.12}$$

Otherwise, the set is said to be *linearly independent*.

EXAMPLE

The set $\{x, x^1, x^2\}$ in the example above is linearly dependent, since $x - x^1 - x^2 = 0$.

EXAMPLE

The set of vectors $\{e^1, e^2, ..., e^n\}$ in R^n, where all the entries in e^i are 0 except the ith entry, which is 1, is independent. A subset of this set is also independent; but a set formed by an arbitrary vector x in R^n and all the vectors e^1, ..., e^n, is linearly dependent; let x_i be the ith entry in the arbitrary vector; clearly $x - \sum_{i=1}^{n} x_i e^i = 0$.

This last example suggests that perhaps one cannot construct sets of vectors belonging to R^n which have more than n elements and are linearly independent. These ideas and other related concepts are clarified by the following definition and the conclusions to be deduced from it.

Definition. A *basis* for a linear space X is a linearly independent set of vectors in X such that every vector in X can be written as a linear combination of elements in the set.

EXAMPLE

The set of vectors $\{e^1, ..., e^n\}$ defined above is a basis for R^n. It will be called the *standard* basis for this vector space. Note that a linear space may have more than one basis.

The definition of a basis does not imply that a basis of a linear space, if it exists, is a finite set. In this sense, R^n, which has a finite basis, is to be considered as a special case; this prompts us to make the following:

Definition. A vector space X is *finite-dimensional* if it has a finite basis. The *dimension* of the vector space is the number of elements in a basis of X.

The definition of dimension makes sense only if all bases of a finite dimensional vector space have the same number of elements. This is indeed the case; the proof of this fact would take us too far from our path; the reader should consult [2, pp. 42–43], where the proof can be found.

The vector space R^n is indeed finite-dimensional, its dimension being n. The linear space of real-valued, continuous functions defined on $[0, 1]$ is not finite dimensional since no *finite* set of such functions can be found so that every function in the space can be expressed as a linear combination of them.

It is said that a finite-dimensional linear space is *spanned* by its basis; such a space is the set of all possible linear combinations of the vectors in its basis. Indeed, a linear space can be defined by specifying its basis. Consider the functions x^i, $i = 1, ..., n$, defined by

$$x^i(t) = t^i, \qquad i = 1,..., n, \quad t \in [0, 1] \tag{2.13}$$

We can construct a linear space over R by defining addition and multiplication by a scalar just as was done in the case of the space of all continuous, real-valued functions defined on $[0, 1]$ and by defining the set of function x^i, $i = 1, ..., n$, as the basis of this space. Of course, these functions have to be independent, so that no set of scalars $c_1, ..., c_n$ at least one of which is to be not zero, should exist for which

$$c_1 t + c_2 t^2 + \cdots + c_n t^n = 0 \tag{2.14}$$

for all $t \in [0, 1]$. This is indeed the case; take for instance $2n$ values of time, $t_1, t_2, \ldots, t_{2n}$, all of them in $[0, 1]$. If $c_1, \ldots, c_n$ satisfy (2.14) for all $t \in [0, 1]$, they certainly satisfy this relation at $t_1, \ldots, t_{2n}$. However, the only solution of the resulting homogeneous $2n$ equations with n unknowns is $c_1 = c_2 = \cdots = c_n = 0$. It follows that the functions $x^1, \ldots, x^n$ are independent.

Note that the space spanned by these functions is a subset of the space of all continuous, real-valued functions defined on $[0, 1]$. We will say that the space spanned by x^i, $i = 1, \ldots, n$, is a *subspace* of the original, larger space; subspaces will be defined in detail in Section 2.7. Note that the larger space is not finite dimensional, while the subspace has dimension n.

At last, we introduce the important concept of an *ordered basis* in a finite-dimensional linear space; this is simply a basis which is also an ordered set, so that we explicitly state (usually by means of subindexes), which is to be regarded as the first element of the basis, which as the second, etc. The standard basis for R^n is an ordered basis. All bases which we use in this book are ordered; it happens that ordered bases have interesting properties, to be explored below. We shall omit most times the adjective "ordered."

Let X be a finite-dimensional linear space of dimension n, and let $\alpha = \{x^1, \ldots, x^i, \ldots, x^n\}$ be an ordered basis for this space. If a vector x is

$$x = \sum_{i=1}^{n} x_i x^i$$

we say that the scalars x_i, $i = 1, \ldots, n$ are the *coordinates of the vector x* relative to the ordered basis α. We can easily prove that these are unique. Assume that a vector x has two sets of coordinates with respect to a basis $\alpha = \{x^1, \ldots, x^i, \ldots, x^n\}$, that is, it can be written as

$$x = \sum_{i=1}^{n} a_i x^i = \sum_{i=1}^{n} b_i x^i$$

It follows that $\sum_{i=1}^{n} (a_i - b_i) x^i = 0$; since the vectors x^i are independent because they form a basis for the linear space X, it follows that $a_i - b_i = 0$, $i = 1, \ldots, n$, so that indeed the coordinates of a vector with respect to an ordered basis are unique.

In this development, the fact that the basis α is ordered is of crucial importance. In fact, we have managed to establish a one-one correspondence between the vectors in X and the set of n-tuples $(a_1, \ldots, a_n)$, so that many times we will consider the space of n-tuples rather than the vector space X itself. We have actually contrived to abstract the fundamental fabric of finite-dimensional linear spaces, in a way that makes all n-dimensional vector spaces, for any n, essentially identical. Consider a 4-tuple in R^4

$$\begin{bmatrix} 1 \\ 3 \\ 8 \\ -1 \end{bmatrix} \tag{*}$$

The coordinates of this vector with respect to the standard basis are of course 1, 3, 8, and -1. Consider now a vector in the space of polynomials described above, defined by

$$x(t) = 1t + 3t^2 + 8t^3 - 1t^4, \qquad t \in [0, 1]$$

The coordinates of this vector with respect to the basis $\{x^1, x^2, x^3, x^4\}$, $x^i(t) = t^i$, $i = 1, ..., 4$, $t \in [0, 1]$, are indeed 1, 3, 8, and -1; in a sense this vector is identical to the vector $(*)$ above; note that this is, however, a continuous function, a polynomial, defined on $[0, 1]$, while the vector $(*)$ is a 4-tuple. The two spaces, R^4 and the space of polynomials spanned by the basis $\{x^1, x^2, x^3, x^4\}$, have certainly the same algebraic structure; if one does not pay attention to the actual characteristics of the elements composing the two bases, they are in a sense identical.

A word concerning notation. We will denote, as we have been doing so far, different vectors by x, y, z, etc., or by x^1, x^2, etc. The components of a vector x with respect to a given basis will be denoted by x_1, x_2, etc. In this way we avoid the use of special symbols for vectors, such as $\mathbf{x}$, which tend rather to confuse the presentation.

Exercises

*1. Let V be a finite set of vectors in a linear space X. Let W be the intersection of all subspaces of X which contain V. Show that W is a subspace of X spanned by the vectors in V, that is, W is the set of all possible linear combinations of vectors in V.

2. Let W be the set of all vectors in R^5 such that

$$9x_1 - x_2 + \tfrac{4}{3}x_3 - x_4 = 0$$

$$x_1 + \tfrac{2}{3}x_2 - x_5 = 0$$

$$9x_1 - 3x_2 + 6x_3 - 3x_4 - 3x_5 = 0$$

Find a finite set of vectors which spans W. Choose the independent vectors from your set which span W and construct thus a basis for this subspace.

3. Find two bases in R^4 that have no vectors in common so that one of them contains the vector $(2, 0, 0, 0)$ and the other contains the vectors $(2, 3, 1, 0)$ and $(1, 1, 1, 1)$.

4. Try and define the concept of linear independency for an infinite set of vectors. Can you prove some interesting theorems about these sets?

5. Prove or give a counterexample: Let x^1, x^2, and x^3 be independent vectors in a vector space X. Then the vectors $x^1 + x^2$, $x^2 + x^3$, and $x^1 + x^3$ are also independent.

2.6. Linear Transformations and Matrices

We shall now present the first part of our study of linear transformation and matrices. Many important concepts will be treated in Chapter 4 as well as in other parts of the book.

Definition. Let X and Y be linear spaces over F, not necessarily distinct. A *linear transformation* from X into Y is a function $T: X \to Y$ such that

$$T(c_1 x^1 + c_2 x^2) = c_1 T(x^1) + c_2 T(x^2) \tag{2.15}$$

for all x^1 and x_2 in X and all scalars c_1, c_2 in F.

If $X = Y$, then a function $T: X \to X$ which satisfies (2.15) is said to be a *linear operator*. That is, a linear operator is a linear transformation which maps a vector space into itself.

EXAMPLE

Let $X = R^3$, $Y = R^2$, and let T be defined by the expression

$$T(x) = \begin{bmatrix} 2x_1 + 7x_2 \\ x_3 - 8x_2 \end{bmatrix}$$

Clearly, T is a linear transformation from R^3 into R^2.

EXAMPLE

The transformation T maps R^3 into R^2 according to

$$T(x) = \begin{bmatrix} ax_1 + bx_2 + c \\ \alpha x_1 + \beta x_2 \end{bmatrix}$$

This is not a linear transformation unless $c = 0$. If c is zero, this is a linear transformation for all values of a, b, α, and β.

EXAMPLE

Let X be the space of continuous, real-valued functions defined on $[0, 1]$ and let $Y = X$. If $\psi(\lambda, t)$ is a continuous function of λ and t, defined for $\lambda \in [0, 1]$ and $t \in [0, 1]$, then the transformation defined by

$$T(f)(\lambda) = \int_0^1 \psi(\lambda, t) f(t) \, dt, \qquad \lambda \in [0, 1]$$

is a linear operator which maps X into itself.

Let $T: X \to Y$. The range of T, $T(X)$, is certainly a subset of Y; it is not, however, in general all of Y; linear transformations are not necessarily onto. This fact, of great importance, is illustrated in the following example.

EXAMPLE

a.   Let X be R^2, and T be the linear operator mapping R^2 into itself defined by

$$T(x) = \begin{bmatrix} x_1 \\ 0 \end{bmatrix}, \qquad x \in R^2$$

Vectors in R^2 for which the entry x_2 is not zero are not in the range of T; we conclude that this is not an onto function.

b.   Let $X = R^2$ and $Y = R^3$, and T be the linear transformation defined by

$$T(x) = \begin{bmatrix} x_1 \\ 2x_1 + x_2 \\ x_2 \end{bmatrix}, \qquad x \in R^2$$

The 3-vectors in the range of T are characterized by the fact that the sum of twice their first entry minus the second plus the third is zero. Not all vectors in R^3 satisfy this condition; the vector e^1, with entries 1, 0, and 0 in this order, for instance, does not. T is not onto.

We turn now to the description of transformations by means of a class of mathematical objects to be called *matrices*. Let a finite-dimensional linear space X have a basis $\alpha = \{x^1, ..., x^j, ..., x^n\}$. Let T be a linear transformation from X into another finite-dimensional linear space Y, with basis $\beta = \{y^1, ..., y^i, ..., y^m\}$. Since $T(x^j) \in Y, j = 1, ..., n$, we can write

$$T(x^j) = \sum_{i=1}^{m} a_{ij} y^i \qquad (2.16)$$

where a_{ij} is the ith coordinate of $T(x^j)$ relative to the basis β. A vector $x \in X$ given by $x = \sum_{j=1}^{n} c_j x^j$ is transformed by T into

$$T(x) = \sum_{j=1}^{n} c_j T(x^j) = \sum_{j=1}^{n} c_j \sum_{i=1}^{m} a_{ij} y^i$$

$$= \sum_{i=1}^{m} \left(\sum_{j=1}^{n} a_{ij} c_j \right) y^i \qquad (2.17)$$

Two facts are worth noting here. First, the coordinates of $T(x)$ with respect to the basis β are completely specified, for any $x \in X$, by a knowledge of the coordinates of x in X and by the set of scalars $\{a_{ij}, i = 1, ..., m, j = 1, ..., n\}$. In this sense, T is completely specified by this set of $m \cdot n$ scalars. Second, this set is dependent on the bases α and β which have been chosen for X and Y respectively, while the transformation T itself does not depend on the bases chosen. This is an important point, which deserves further consideration. Let an operator T map R^2 into itself. Since vectors in R^2 can be associated in a one-one way with points in a cartesian plane, the operator T is the rule which associates with every point P in the plane another point P^1 also in the plane, in such a way that the linearity condition (2.15) is satisfied. On the other hand, the reader should be aware that many axes of reference can be drawn in the plane, each corresponding to a new basis and a new set of coordinates for a given point in the plane. However, the rule T, which concerns itself with associating points in the plane, does not depend on which axes we happen to project the points. The set of scalars a_{ij} described above will be written in an array as follows:

$$\begin{bmatrix} a_{11} & a_{12} & \cdots & a_{1n} \\ a_{21} & a_{22} & \cdots & a_{2n} \\ \vdots & \vdots & & \vdots \\ a_{m1} & a_{m2} & & a_{mn} \end{bmatrix} \qquad (2.18)$$

This array will be called a *matrix*; as a matter of fact, this is *the matrix of the transformation T relative to the bases α and β*.

The scalars a_{ij} are referred to as the *entries* of the matrix, the horizontal subarrays such as $a_{11}\,a_{12}\,\cdots\,a_{1n}$ or $a_{21}\,a_{22}\,\cdots\,a_{2n}$ are the *rows* of the matrix, the vertical subarrays such as

$$\begin{matrix} a_{11} & & a_{12} \\ a_{21} & & a_{22} \\ \vdots & & \vdots \\ a_{m1} & & a_{m2} \end{matrix}$$

are the *columns*. The matrix (2.18) has m rows and n columns; it is an $m \times n$ matrix. It is said that two matrices are equal if all their entries are equal (matrices are indeed ordered sets).

We will often denote the matrix of a transformation T relative to some bases by the same symbol, T. If necessary we will write $T_{\alpha\beta}$ to indicate explicitly the bases. Sometimes we will write $T = (a_{ij})$ as shorthand for expressions such as (2.18). The number of rows and columns (the *dimensions* of the matrix) are supposedly known when such a notation is used.

The matrix $T_{\alpha\beta}$ of an operator which maps a linear space X into itself can be computed with respect to any pair of bases α and β in X. However, the same basis can be used in this case to serve both roles; the matrix $T_{\alpha\alpha}$ will be referred to as *the matrix of the operator T relative to the basis α*.

Note that the n-tuple (2.8) can be considered as an $n \times 1$ matrix with real entries. It could have been written as

$$\begin{bmatrix} x_{11} \\ x_{21} \\ \vdots \\ x_{n1} \end{bmatrix}$$

It is customary, however, to use the notation in (2.8).

EXAMPLE

Let $X = R^3$, $Y = R^2$, and T be a linear transformation mapping X into Y defined by

$$T(x) = \begin{bmatrix} 3x_1 - x_2 \\ x_3 \end{bmatrix}, \qquad x \in R^3$$

Let us compute the matrix of T relative to the bases α and β,

$$\alpha = \left\{ \begin{bmatrix} 2 \\ 0 \\ 1 \end{bmatrix}, \begin{bmatrix} 0 \\ -1 \\ 0 \end{bmatrix}, \begin{bmatrix} 0 \\ -2 \\ 3 \end{bmatrix} \right\}$$

$$\beta = \left\{ \begin{bmatrix} 1 \\ 0 \end{bmatrix}, \begin{bmatrix} 0 \\ -2 \end{bmatrix} \right\}$$

$$T(x^1) = \begin{bmatrix} 6 \\ 1 \end{bmatrix} = 6y^1 - \tfrac{1}{2}y^2$$

$$T(x^2) = \begin{bmatrix} 1 \\ 0 \end{bmatrix} = y^1 + 0y^2$$

$$T(x^3) = \begin{bmatrix} 2 \\ 3 \end{bmatrix} = 2y^1 - \tfrac{3}{2}y^2$$

The matrix of T relative to α and β is

$$T = \begin{bmatrix} 6 & 1 & 2 \\ -\frac{1}{2} & 0 & -\frac{3}{2} \end{bmatrix}$$

An important operator is the operator I which maps every vector in a vector space X into itself. Let X be n-dimensional, and $\alpha = \{x^1, ..., x^n\}$ be a basis for X. Then

$$I(x^j) = x^j, \qquad j = 1, ..., n,$$

so that the matrix of I with respect to any basis is the $n \times n$ *identity matrix* I_n

$$\begin{bmatrix} 1 & & & \\ & 1 & & \\ & & \cdot & \\ & & & \cdot \\ & & & & 1 \end{bmatrix}$$

where all the entries not shown are zeros.

The zero operator, which maps every vector in X into the vector 0, has as matrix relative to any basis the $n \times n$ zero matrix 0_n, all of whose entries are zero. This matrix will also be denoted sometimes simply by 0.

Let T_1 and T_2 be two transformations which map a finite-dimensional vector space X into a finite-dimensional vector space Y. We define the sum of T_1 and T_2 as a transformation, denoted by $T_1 + T_2$, such that

$$(T_1 + T_2)(x) = T_1(x) + T_2(x) \tag{2.19}$$

Let X and Y have bases α and β as above. If the matrix of the operator T_1 relative to the bases α and β is $T_1 = (a_{ij})$ and the matrix of the operator T_2 is $T_2 = (b_{ij})$, then

$$T_1(x) = \sum_{i=1}^{m} \left(\sum_{j=1}^{n} a_{ij}c_j \right) y^i \tag{2.20}$$

$$T_2(x) = \sum_{i=1}^{m} \left(\sum_{j=1}^{n} b_{ij}c_j \right) y^i \tag{2.21}$$

where c_j, $j = 1, ..., n$, are the coordinates of x relative to the basis α. Clearly,

$$(T_1 + T_2)(x) = \sum_{i=1}^{m} \left(\sum_{j=1}^{n} (a_{ij} + b_{ij}) c_j \right) y^i \tag{2.22}$$

we see that the set of numbers $\{a_{ij} + b_{ij}, \; i = 1, ..., m, \; j = 1, ..., n\}$ completely specify the behavior of the operator $T_1 + T_2$. This leads us to make the following:

Definition. The *sum* of two matrices of the same dimensions, $T_1 = (a_{ij})$ and $T_2 = (b_{ij})$, is a matrix with entries $a_{ij} + b_{ij}$, that is,

$$T_1 + T_2 = (a_{ij} + b_{ij}) \tag{2.23}$$

The matrix of the transformation $T_1 + T_2$ relative to the bases α and β is therefore $T_1 + T_2$.

EXAMPLE

Let $X = Y = R^3$, and T_1 and T_2 linear operators defined by

$$T_1(x) = \begin{bmatrix} 3x_1 \\ x_1 + x_2 \\ x_3 \end{bmatrix}, \qquad T_2(x) = \begin{bmatrix} x_2 \\ x_1 - x_2 \\ 3x_3 \end{bmatrix}, \qquad x \in R^3$$

Let $\alpha = \beta = \{x^1, x^2, x^3\}$, where

$$x^1 = \begin{bmatrix} 1 \\ 0 \\ 0 \end{bmatrix}, \qquad x^2 = \begin{bmatrix} 0 \\ 2 \\ 0 \end{bmatrix}, \qquad x^3 = \begin{bmatrix} 0 \\ 0 \\ 3 \end{bmatrix}$$

The matrices of T_1 and T_2 relative to these bases are

$$T_1 = \begin{bmatrix} 3 & 0 & 0 \\ 1 & 2 & 0 \\ 0 & 0 & 3 \end{bmatrix}, \qquad T_2 = \begin{bmatrix} 0 & 2 & 0 \\ 1 & -2 & 0 \\ 0 & 0 & 9 \end{bmatrix}$$

The matrix of the operator $T_1 + T_2$ relative to the same bases is

$$T_1 + T_2 = \begin{bmatrix} 3 & 2 & 0 \\ 2 & 0 & 0 \\ 0 & 0 & 12 \end{bmatrix}$$

The reader should try to compute the matrices of T_1, T_2, and $T_1 + T_2$ relative to the standard basis in R^3.

In a similar way, multiplication of a matrix by a scalar c in F is defined as follows:

Definition. Let an $m \times n$ matrix $T = (a_{ij})$ be given. The $m \times n$ matrix cT is defined as

$$cT = (ca_{ij}) \tag{2.24}$$

It should be clear that the matrix relative to the bases α and β of a transformation cT defined by

$$(cT)(x) = cT(x) \tag{2.25}$$

is cT, where T is the matrix of T relative to the bases α and β.

The matrix of a linear combination of transformations such as $\sum_{i=1}^{n} a_i T_i$ can be found in a trivial way from the preceding results.

Consider now the following situation: Let X, Y, and Z be finite-dimensional vector spaces of dimensions n, m, and r respectively; let $\alpha = \{x^1, ..., x^n\}$, $\beta = \{y^1, ..., y^m\}$, and $\gamma = \{z^1, ..., z^r\}$ be bases for X, Y, Z in this order. Two transformations are defined, $T_1: X \rightarrow Y$ and $T_2: Y \rightarrow Z$, with matrices T_1 and T_2 relative to the corresponding bases. Consider the *composite* transformation $T_2 T_1$ defined by

$$(T_2 T_1)(x) = T_2(T_1(x)) \qquad \text{for all} \quad x \in X \tag{2.26}$$

This transformation maps X into Z, and is well-defined and linear. If $x = \sum_{j=1}^{n} c_j x^j$,

$$T_1(x) = \sum_{i=1}^{m} \left(\sum_{j=1}^{n} a_{ij} c_j \right) y^i$$

where the scalars a_{ij} are the elements of T_1. Then

$$T_2(T_1(x)) = \sum_{k=1}^{r} \left(\sum_{i=1}^{m} \left(\sum_{j=1}^{n} a_{ij} c_j \right) b_{ki} \right) z^k \tag{2.27}$$

where the scalars b_{ki} are the elements of T_2. Then, rearranging (2.27)

$$T_2(T_1(x)) = \sum_{k=1}^{r} \left(\sum_{j=1}^{n} \left(\sum_{i=1}^{m} b_{ki} a_{ij} \right) c_j \right) z^k \tag{2.28}$$

The behavior of the transformation $T_2 T_1$ is completely defined by the set of scalars $\{\sum_{i=1}^{m} b_{ki} a_{ij}, j = 1, ..., n, k = 1, ..., r\}$; that is, by the $r \times n$ matrix $(\sum_{i=1}^{m} b_{ki} a_{ij})$. We will call this matrix the *product* of T_2 and T_1, and write

$$T_2 T_1 = \left(\sum_{i=1}^{m} b_{ki} a_{ij} \right) \tag{2.29}$$

It is worth noting in (2.29) the way in which $\sum_{i=1}^{m} b_{ki} a_{ij}$ is constructed. There are two matrices (b_{ki}) and (a_{ij}); to obtain the entry in the kth row and jth column of the product we add all terms of the type $b_{ki}a_{ij}$; this addition is performed over the index i, and runs from $i = 1$ to $i = m$. It should be clear that multiplication of two matrices A and B is defined only when the number of columns of A is equal to the number of rows of B. This is of course a reflection of the deeper fact that T_1 describes a mapping *into* the space Y while T_2 describes one *from* the same space Y.

A word of caution regarding subindexes. These play a role similar to the independent variable when one writes an integral such as $\int_0^1 f(x)\, dx$. One could have written as well $\int_0^1 f(t)\, dt$; the symbol by which the independent variable (or the subindexes) is designated is of no importance. The reader should have this in mind when studying the following summary of the rules which govern matrix multiplication.

Given an $n \times m$ matrix $A = (a_{ij})$ and an $m \times p$ matrix $B = (b_{jk})$, the product of A and B, AB, is defined as an $n \times p$ matrix with entries

$$AB = \left(\sum_{j=1}^{m} a_{ij}b_{jk} \right)$$

It may happen that even if the product AB is well defined, because the number of columns of A equals the number of rows of B, the product BA is not: if A is a 3×2 and B a 2×4 matrix, AB exists but BA does not. Even if both exist, however, it generally happens that AB does not equal BA. Matrix multiplication is not a commutative operation. If A and B are such that $AB = BA$, these matrices are said *to commute*. The reader will easily find instances of matrices which do not commute. Two matrices A and B commute if and only if the corresponding operators A and B commute.

EXAMPLE

If

$$A = \begin{bmatrix} 0 & 1 \\ 1 & 0 \end{bmatrix}, \qquad B = \begin{bmatrix} 0 \\ 3 \end{bmatrix}$$

$$AB = \begin{bmatrix} 0 & 1 \\ 1 & 0 \end{bmatrix}\begin{bmatrix} 0 \\ 3 \end{bmatrix} = \begin{bmatrix} 3 \\ 0 \end{bmatrix}$$

This example is intended to show a characteristic of the space R^n (and of course also of C^n); if x is an n-tuple (or n-vector) with real or complex entries such as (2.8), it can be considered as an $n \times 1$ matrix, and the multiplication rule for matrices applies to the multiplication of a vector x by a matrix A. We write this product as Ax. Clearly, since x is an $n \times 1$ matrix, A has to have dimensions $m \times n$, with m

arbitrary. The product has dimensions $m \times 1$; it is, of course, an m-vector. Vectors such as (2.8) are called *column vectors*. Vectors such as

$$x = [x_1 \, x_2 \, \cdots \, x_n] \tag{2.30}$$

which can be considered as $1 \times n$ matrices, are *row vectors*. If x is a column vector such as (2.8), the symbol x' will be used to denote the row vector (2.30) associated with it. This is an example of a more general operation on the columns of a matrix to be defined later on.

Let T map X, an n-dimensional space, into Y, an m-dimensional space. Let α and β be bases for X and Y respectively, and T be the matrix of T relative to the bases α and β. If $x \in X$, its coordinates relative to the basis α can be used to construct an n-vector x; similarly, the coordinates of $y \in Y$ relative to the basis β can be used to construct an m-vector, y. If $y = T(x)$, then it follows from (2.17) that

$$y = Tx$$

Instead of working with general finite-dimensional linear spaces, it is often sufficient to work with the space of n-tuples formed by the coordinates.

Exercises

1. Which of the following functions T from R^2 into itself are linear transformations?

   i. $T(x) = \begin{bmatrix} 1 + x_1 \\ x_2 \end{bmatrix}$

   ii. $T(x) = \begin{bmatrix} x_2 \\ x_1 \end{bmatrix}$

   iii. $T(x) = \begin{bmatrix} 9(x_1 - x_2) \\ 0 \end{bmatrix}$

2. A linear transformation T is defined as follows

$$T\left(\begin{bmatrix} 2 \\ 2 \end{bmatrix}\right) = \begin{bmatrix} 0 \\ -1 \\ -2 \end{bmatrix}, \qquad T\left(\begin{bmatrix} -1 \\ 1 \end{bmatrix}\right) = \begin{bmatrix} -2 \\ -1 \\ 0 \end{bmatrix}$$

   Find a matrix for T. Which are the domain and range of this transformation?

*3. Prove that the set of all linear transformations mapping a linear space X into a linear space Y is itself a linear space, and define the corresponding linear space operations.

4. Show that the set of all $n \times n$ matrices with complex entries is a linear space over C and define the corresponding operations. Find a basis for this space. What is its dimension?

*5. A square matrix is said to be *triangular* if all the entries below the diagonal are zero. Show that if A is triangular there is a positive integer k such that $A^k = 0$.

6. All entries in A other than those in the diagonal are zero. Which matrices commute with A?

7. Let x be a row vector with n entries, and y a column vector with n entries. Compute the matrix product yx. This is said to be the outer product of the vectors.

8. Let $A = (a_{ij})$ be an $n \times n$ matrix such that $|a_{ij}| < M$ for all i, j. Show that the entries of A^k, defined by

$$A^k = \underset{k \text{ times}}{A \cdot A \cdots A}$$

are bounded by $n^{k-1}M^k$.

2.7. Inverse Operators and Inverse Matrices

We will treat in this section the important concept of *inverse operator*. Before doing so, however, we will need some further background on some properties of vector spaces.

Definition. A *subspace* W of a vector space X is a subset of X which is itself a vector space with respect to the operations of addition and scalar multiplication in X.

Several examples of subspaces have already been encountered in the previous pages. The space of continuous functions defined on $[0, 1]$ is a linear space; the space of *polynomials* defined on $[0, 1]$ is a subset of the space of continuous function, while being at the same time a (smaller) linear space in its own right. It is worth remembering that this space was constructed by specifying a basis and then defining the space as the set of all linear combinations of the elements of the basis. This is a standard way of constructing subspaces, as shown by the following theorem.

THEOREM 2.2.   Let X be a linear space. The set of all linear combinations of any set of vectors in X is a subspace of X.

Proof.   Let $\{x^1, ..., x^p\}$ be a set of vectors in X, and let W be the set of all linear combinations of those vectors. W is a subset of X, since X is a linear space and thus contains all linear combinations of vectors in it. The postulates in the definition of a vector space are satisfied by the set W, since (i) the operation of addition of vectors in W satisfies obviously postulates ia and ib; the zero vector is in W because it is a linear combination of $x^1, ..., x^p$, namely,

$$0 = 0 \cdot x^1 + 0 \cdot x^2 + \cdots + 0 \cdot x^p$$

If $x \in W$, $-x$ is in W too, since if x is a linear combination of x^1, ..., x^p, $-x$ is also one, and W is the set of all linear combinations of these vectors. (ii) The postulates related to scalar multiplication are obviously satisfied. The theorem follows. If X is finite-dimensional, and $\{x^1, ..., x^p\}$ is a basis for X, then of course $W = X$.

Let W be a proper subspace of a linear space X. If $\alpha = \{x^1, ..., x^n\}$ is a basis for X, α is not a basis for W, since in such a case W would be equal to the set of all linear combinations of vectors in α, and thus equal to X. We have proved that a basis for W has to have fewer elements than any basis for X; that is,

THEOREM 2.3.   The dimension of a proper subspace W of a finite-dimensional vector space X is less than the dimension of X.

We consider now the concepts of null space, rank, and nullity of a linear transformation, which are of importance in the study of the conditions under which an operator is invertible. We remind the reader that if a linear transformation maps a linear space *into itself*, then it is said to be a linear operator. We prove first:

THEOREM 2.4.   Let T be a linear transformation from X into Y. Then the range of T is a subspace of Y.

Proof.   Let y^1, y^2 be in the range of T. Then, there are x^1, x^2 in X such that $T(x^1) = y^1$, $T(x^2) = y^2$. Since T is linear, for any $c \in F$, $T(cx^1 + x^2) = cy^1 + y^2$; it follows that this vector, $cy^1 + y^2$, is also in the range of T. The range of T, $T(X)$, is therefore a subspace of X: If $y \in T(X)$, $-y + y = 0 \in T(X)$; similarly, $-y = (-1)y + 0$ is in $T(x)$ for any $y \in T(x)$. The other postulates from the definition of a vector space are satisfied trivially.

THEOREM 2.5.   Let T be a linear transformation from X into Y. Then the set of vectors $N_T = \{x : x \in X, Tx = 0\}$ is a subspace of X, called the *null space* of T.

Proof.   Let x^1, $x^2 \in N_T$. If c is any scalar, then $T(cx^1 + x^2) = 0$, so that $(cx^1 + x^2) \in N_T$. It follows, just as in the proof of Theorem 2.4, that N_T is a subspace of X.

EXAMPLE

Let T be a linear transformation of R^3 into itself (that is, a linear operator), whose matrix with respect to the standard basis in R^3 is

$$T = \begin{bmatrix} 1 & 0 & 3 \\ 2 & 1 & 0 \\ -2 & 0 & -6 \end{bmatrix}$$

Let $x \in N_T$. Then $Tx = 0$; the components of x with respect to the standard basis satisfy:

$$x_1 + 3x_3 = 0$$
$$2x_1 + x_2 = 0 \qquad (*)$$
$$-2x_1 - 6x_3 = 0$$

The third equation is redundant; it is simply the first equation multiplied by (-2). Therefore, the solution of $(*)$ is given by:

$$x_1 = -3x_3$$
$$x_2 = 6x_3$$

for any x_3.

$$N_T = \left\{ x : x = \begin{bmatrix} -3x_3 \\ 6x_3 \\ x_3 \end{bmatrix}, \; x_3 \in R \right\}$$

is the null space of T, and is clearly a subspace of R^3.

Note that the null space of the operator T has dimension one. It is not difficult to construct operators mapping R^3 into itself with null spaces of dimension two. An interesting question, which we will answer presently, is whether it is possible to find operators with null spaces of dimension three.

Definition. The *rank* of a linear transformation from a finite-dimensional vector space X into Y is the dimension of the range of T. The *nullity* of T is the dimension of its null space.

For instance, the nullity of the operator T in the example above is equal to 1. In a similar way as the one used in that example, it can be proved that the rank of this operator equals 2. Note that, in this case, rank + nullity $= 3$, the dimension of R^3. The following theorem shows that this is no accident.

THEOREM 2.6. If T is a linear transformation from a finite-dimensional vector space X into a finite-dimensional vector space Y, the rank + nullity of T = dimension of X.

Proof. Let s be the nullity of T and r its rank. Then N_T, the null space of T, has a basis $\{x^1, ..., x^s\}$, where $s \leqslant n$, n being the dimension of X. There are vectors $x^{s+1}, ..., x^n$ in X such that $\{x^1, ..., x^s, x^{s+1}, ..., x^n\}$ is a basis for X. It follows that the vectors

$$\{T(x^1), ..., T(x^s), T(x^{s+1}), ..., T(x^n)\}$$

span the range of T. However, since $T(x^i) = 0$, $i = 1, ..., s$, the vectors $T(x^i)$, $i = s + 1, ..., n$, alone also span the range of T. We prove now that these vectors are independent. Assume otherwise, that is, that there are scalars c_{s+1} , ..., c_n such that

$$\sum_{j=s+1}^{n} c_j T(x^j) = 0$$

This of course implies that $T(\sum_{j=s+1}^{n} c_j x^j) = 0$; hence $(\sum_{j=1}^{n} c_j x^j) \in N_T$, and there are scalars b_1 , ..., b_s such that

$$\sum_{j=s+1}^{n} c_j x^j = \sum_{k=1}^{s} b_k x^k$$

Since the set $\{x^1, ..., x^s, x^{s+1}, ..., x^n\}$ is a basis for X, $c_i = 0$, $i = s + 1,..., n$, $b_k = 0, k = 1,..., s$, and the vectors $T(x^i), i = s + 1,..., n$, are independent. Since they span the range of T, they form a basis for this set, which therefore has dimension $r = n - s$. The theorem follows.

We are now in a position to determine conditions for the invertibility of a linear operator.[†] Let T map an n-dimensional vector space X into itself. A preliminary theorem is

THEOREM 2.7.   If the linear operator $T: X \rightarrow X$ is invertible (one-one and onto), then its inverse function, T^{-1}, is a linear operator.

Proof.   Let x^1, x^2 be in X, and let $T(x^1) = y^1$, $T(x^2) = y^2$; y^1 and y^2 are also in X. Then $T^{-1}(y^1) = x^1$, $T^{-1}(y^2) = x^2$. The operator T itself is linear; therefore, for c_1 , c_2 in F,

$$T(c_1 x^1 + c_2 x^2) = c_1 T(x^1) + c_2 T(x^2) = c_1 y^1 + c_2 y^2$$

The only vector mapped into $c_1 y^1 + c_2 y^2$ is $c_1 x_1 + c_2 x^2$. It follows that T^{-1} is linear

$$T^{-1}(c_1 y_1 + c_2 y_2) = c_1 x^1 + c_2 x^2 = c_1 T^{-1}(y^1) + c_2 T^{-1}(y^2)$$

The following theorems give the basic conditions for the invertibility of a linear operator T.

THEOREM 2.8.   A linear operator T is one-one if and only if its nullity is zero.

Proof.   Let the nullity of T be zero. The null space of T consists then only of the zero vector. We show that, if $T(x) = T(y)$, then $x = y$. Indeed, if $T(x) = T(y)$, $T(x - y) = 0$, since T is linear. It follows

† Remember that an operator is, for us, a transformation that maps a space into itself.

that $x - y$ is in the null space of T. Since this consists only of the zero vector, $x - y = 0$, or $x = y$; T is therefore one-one. If T is one-one, that is, if $T(x - y) = 0$ implies that $x - y = 0$, then the null space of T consists only of the zero vector, and its nullity is zero.

THEOREM 2.9. A linear operator T mapping X, an n-dimensional linear space, into itself is onto if and only if its rank is n.

Proof. Let T be onto. Then its range is all of the space X, which has dimension n; the rank of T is therefore n. Suppose now that the rank of T is n. Since the range of T is a subspace of X, and since any proper subspace of X has rank less than n, the range of T coincides with the whole space X. T is onto.

It is interesting to note that if the nullity of the operator T is zero, then its rank is n, as follows from Theorem 2.6. An operator T is onto if and only if it is one-one.

We have, therefore, completely characterized an invertible operator: An operator such as T is invertible if and only if its rank is n (or its nullity is zero). This condition will be used often in the following chapters. On many ocassions, though, the operator whose invertibility is to be investigated will be defined by its matrix relative to some basis. It is convenient, therefore, to restate the above condition on corresponding characteristics of the matrix. With this purpose in mind, we will at this point translate some of the definitions concerning transformations into definitions concerning the matrices of these transformations relative to some bases.

Let T be a linear transformation of X, an n-dimensional space into Y, an m-dimensional space. The matrix $T = (a_{ij})$ of T relative to the bases $\{x^1, ..., x^n\}$ of X and $\{y^1, ..., y^m\}$ of Y is then an $m \times n$ matrix, that is, it has m rows and n columns. The entries of this matrix are, of course, elements of the field F. The columns of the matrix can then be considered as m-vectors; they are called the *column vectors* of the matrix; they are in either R^m or C^m. The rows of the matrix will be considered as n-vectors, the *row vectors* of the matrix, and they will be in R^n or C^n; note that to construct these vectors it is necessary simply to transpose each row of the matrix, transforming them into columns.

Definition. The *row space* of a matrix T is the subspace of R^n spanned by its row vectors. The *column space* of T is the subspace of R^m spanned by its column vectors.

Definition. The *row rank* of a matrix T is the dimension of its row space; its *column rank* is the dimension of its column space.

The rank of a transformation and the row and column ranks of its matrix relative to some basis are very simply related, as proved in the following theorem.

THEOREM 2.10. Let T be a linear transformation $X \to Y$, where X and Y have the basis $\alpha = \{x^1, ..., x^n\}$ and $\beta = \{y^1, ..., y^m\}$ respectively. Let $T = (a_{ij})$ be the matrix of T relative to the bases α and β. Then the rank of the transformation T = row rank of T = column rank of T.

Proof. We will only prove the equality between the rank of T and the column rank of T because the proof of the equality concerning the row rank of T requires concepts which have not been introduced. [3, p. 105].

Let $x \in X$ be given by $x = \sum_{j=1}^{n} c_j x^j$, then $T(x) = \sum_{j=1}^{n} c_j T(x^j)$. It follows that every vector in the range of the transformation T is a linear combination of the vectors $T(x^j)$, $j = 1, ..., n$. Assume that r of these vectors are independent, and for the sake of simplicity of notation assume further that these vectors are the ones corresponding to $1 \leqslant j \leqslant r$. Then the set $\{T^j(x), j = 1, 2, ..., r\}$ is a basis for the range of the transformation T. Consider now the fact that vectors in this basis have coordinates relative to the basis β; $T(x^j)$ has as coordinates the entries corresponding to the jth column of the matrix T. We can construct a set with the n-tuples corresponding to the vectors $T^j(x)$, $j = 1, ..., r$. These n-tuples are vectors in R^n, which span the column space of the matrix T. Since they are independent, the column rank of the matrix T equals r, the rank of the transformation T.

As a consequence of this theorem, we can make a definition.

Definition. The *rank* of a matrix T equals the rank of its transformation T. The rank of a matrix is therefore independent of the particular bases with respect to which its entries have been computed.

The following example will provide the reader with an illustration of the definitions and relationships given above.

EXAMPLE

Let $T: R^3 \to R^3$ be a linear operator whose matrix relative to the standard basis in R^3 is

$$T = \begin{bmatrix} 1 & 0 & 1 \\ 0 & -1 & -1 \\ -1 & 0 & -1 \end{bmatrix}$$

The column space of T is the one spanned by the vectors

$$\begin{bmatrix} 1 \\ 0 \\ -1 \end{bmatrix}, \quad \begin{bmatrix} 0 \\ -1 \\ 0 \end{bmatrix}, \quad \begin{bmatrix} 1 \\ -1 \\ -1 \end{bmatrix}$$

Since the third vector equals the sum of the other two, the column space of T is a 2-dimensional subspace of R^3; its column rank is 2. Its row rank is of course also 2; the row space is spanned by

$$\begin{bmatrix} 1 \\ 0 \\ 1 \end{bmatrix}, \quad \begin{bmatrix} 0 \\ -1 \\ -1 \end{bmatrix}, \quad \begin{bmatrix} -1 \\ 0 \\ -1 \end{bmatrix}$$

These vectors are not independent, since the third equals minus the first one. The operator T with matrix T given above is therefore not invertible.

Note that the determinant of T in the previous example is zero, while there are 2×2 determinants in the matrix which are not zero. This is no accident, and there is indeed a relationship between the rank of a matrix and the maximal nonzero determinant which can be picked from its rows and columns. This relationship is not, however, very important from the standpoint of actually determining numerically the rank of a matrix; determinants are seldom nice things to work with computationally. The relationship is given in the following theorem, which we present without proof.

THEOREM 2.11.   Let T be an $m \times n$ matrix. A *submatrix* of T is any $r \times r$ matrix obtained by deleting the appropriate number of rows and columns of T; r is an integer not larger than the smallest of the two numbers m and n. The *determinant rank* of T is the largest positive integer d such that some $d \times d$ submatrix has a nonzero determinant. The determinant rank of T equals its rank.

It follows easily that an operator T is invertible if and only if the determinant of its matrix T with respect to a basis, det T, is not zero. The matrix T is said to be *nonsingular* or *invertible*.

If T is invertible and has a matrix T relative to a basis, the matrix of its inverse operator T^{-1} with respect to the same basis will be denoted by T^{-1}. Clearly, $T^{-1}T = I$; remember that I is the identity operator. It follows that $T^{-1}T = I_n$; we assume that T maps an n-dimensional space into itself.

An explicit expression for the entries of T^{-1} in terms of those of T can be derived easily, even if it has little numerical value. Consider the equation $TS = I_n$, where T is nonsingular. Clearly the solution S is T^{-1}. Call s_{ij} the (unknown) entries of S. Then,

$$\sum_{k=1}^{n} a_{ik}s_{kj} = \delta_{ij}, \qquad i, j = 1, ..., n$$

where δ_{ij} is the Kronecker symbol defined by $\delta_{ii} = 1$, $\delta_{ij} = 0$ if $i \neq j$. If we fix the value of j we obtain a system of n linear equations for the quantities s_{kj}, with determinant $\det T \neq 0$. The solutions are given by

$$s_{ij} = \frac{\hat{a}_{ji}}{\det T}$$

where $\hat{a}_{ij}$ is the cofactor of a_{ij} in the expansion of the determinant of T.

EXAMPLE

If

$$A = \begin{bmatrix} 1 & 3 \\ 4 & 2 \end{bmatrix}$$

$\det A = 2 - 12 = -10$, and then

$$A^{-1} = \frac{1}{-10} \begin{bmatrix} 2 & -3 \\ -4 & 1 \end{bmatrix}$$

Exercises

*1. Show that if V and W are subspaces of a finite-dimensional vector space X and (i) they have the same dimension; (ii) $W \supseteq V$, then $W = V$.

*2. If T is an invertible linear operator which maps a linear space X into itself, show that if $\{x^1, ..., x^n\}$ is a basis for X then $\{T(x^1), ..., T(x^n)\}$ is also a basis for it.

*3. Let W_1 and W_2 be subspaces of a linear space X such that $W_1 \cap W_2$ is the zero vector. If every vector $x \in X$ can be written as $x = x^1 + x^2$, $x^1 \in W_1$, $x^2 \in W_2$, show that these vectors x^1 and x^2 are unique.

4. Let T be a linear transformation from R^3 into R^2 and U a linear transformation from R^2 into R^3. Prove that the transformation UT is not invertible.

5. If the $n \times n$ matrix A is nonsingular, show that for every matrix B the matrices B, AB, and BA have the same rank.

6. If A satisfies $A^2 - A + 3I_n = 0_n$, prove that A is invertible. (Hint: Determine the vectors in the null space of A.)

7. Suppose that A, an $n \times n$ matrix, is such that $A^2 = 0_n$. Show that A is not invertible.

8. Find the range and null space of the transformations defined below.

    i.

$$T\left(\begin{bmatrix} x_1 \\ x_2 \end{bmatrix}\right) = \begin{bmatrix} 2x_1 + x_2 \\ x_1 - x_2 \\ 2x_2 \end{bmatrix}$$

    ii.

$$T\left(\begin{bmatrix} x_1 \\ x_2 \end{bmatrix}\right) = \begin{bmatrix} x_1 - 2x_2 \\ x_2 - x_1 \\ -2x_1 \end{bmatrix}$$

2.8. Metric Properties of Linear Spaces. Norms

We have reached a stage in our development of the fundamentals of linear spaces in which it is necessary to give these spaces a new type of structure by means of the definition of metric properties. These will be of two kinds. The first is simply modeled on the concept of length of a vector or magnitude of a real number; the second, to be studied in the following section, is richer and gives rise to more interesting structures; it makes possible the definition in general linear spaces of something akin to the concept of angle between two lines on a plane.

The reader should be aware that many of the properties of real numbers have their origin in the fact that real numbers have magnitude, and thus one can say, for instance, which of two numbers is closer to a third. This leads us to define a magnitude, or *norm*, of the elements of vector spaces.

Let X be a linear space over the field F. With every $x \in X$ we associate a real number, denoted by $\| x \|$ and called the *norm* of x, such that for all $x, y \in X$, all $\lambda \in F$,

i. $\| x \| \geqslant 0$. $\| x \| = 0$ only if $x = 0$,

ii. $\| x + y \| \leqslant \| x \| + \| y \|$,   (2.31)

iii. $\| \lambda x \| = | \lambda | \| x \|$.

It should be recognized that on a linear space many such norms can be defined. We will consider first the problem of defining norms for the finite-dimensional spaces R^n and C^n. If a vector $x \in R^n$ or C^n has coordinates $x_1, x_2, ..., x_n$ with respect to a basis, then the following are norms for this vector:

$$\| x \|_1 = \left(\sum_{i=1}^{n} | x_i |^2 \right)^{1/2} \tag{2.32}$$

$$\| x \|_2 = \max_i | x_i | \tag{2.33}$$

$$\| x \|_3 = \sum_{i=1}^{n} | x_i | \tag{2.34}$$

$$\| x \|_4 = \left(\sum_{i=1}^{n} | x_i |^p \right)^{1/p}, \quad 1 \leqslant p < \infty \tag{2.35}$$

Clearly, the norm (2.32) is a special case of (2.35). Note that when $x \in C^n$, the absolute value of the complex numbers x_i appears in the expressions for the norms.

It is not difficult to prove that all the functions defined above are actually norms, i.e., that they satisfy the conditions for a norm given above. We show this for the norm (2.33).

Condition i is satisfied; if $\| x \|_2 = 0$, $\max(| x_1 |, | x_2 |, ..., | x_n |) = 0$, that is, $x_1 = x_2 = \cdots = x_n = 0$, which of course implies $x = 0$. Since

$$\max_i | x_i | + \max_i | y_i | \geqslant \max_i | x_i + y_i |$$

condition ii is also satisfied. Condition iii is obviously satisfied; we conclude that (2.33) is indeed a norm.

The selection of a particular norm for a specific problem is, in the case of finite-dimensional vector spaces, a matter mainly of algebraic convenience. No new spectacular results can be attained merely by changing the norm used in a given problem We will not prove this (rather vague) assertion; the interested reader can consult the literature [4] concerning the fact that in finite-dimensional spaces all norms are said to be *equivalent*.

Norms can also be defined in spaces which are not finite-dimensional. We present several examples.

i. Let X be the space of all continuous, real-valued functions x defined on the interval [0, 1]. This is a linear space over the field R or C. Let the norm of $x \in X$ be

$$\| x \| = \max_{t \in [0,1]} | x(t) | \qquad (2.36)$$

This norm is well defined, since a continuous function does attain a maximum on a closed interval. It also satisfies the axioms which define a norm; axioms i and iii are satisfied trivially, while ii follows from

$$\| f + g \| = \max_{t \in [0,1]} | f(t) + g(t) | \leqslant \max_{t \in [0,1]} \{| f(t) | + | g(t) |\}$$

$$\leqslant \max_{t \in [0,1]} | f(t) | + \max_{t \in [0,1]} | g(t) | = \| f \| + \| g \|$$

where $f, g \in X$.

This normed linear space will be denoted by $C[0, 1]$. A space $C[a, b]$ can be defined in a similar way.

ii. Let X be the space of bounded sequences of numbers; that is,

$$x = \{x_1, x_2 ..., x_n, ...\}$$

and $| x_j | \leqslant M_x$ for all j. Then, we define

$$\| x \| = \sup_j | x_j | \qquad (2.37)$$

the norm is well defined here since a set of bounded numbers has a least upper bound. It can be shown that the norm axioms are satisfied in this case.

iii. Consider now the space X of all bounded, real-valued functions f defined on $[0, 1]$. Here we put[†]

$$\|f\| = \sup_{t \in [0,1]} |f(t)| \tag{2.38}$$

This norm is well defined because bounded functions defined on $[0, 1]$ have a least upper bound.

iv. Let $X = L_p[0, 1]$, the space of real-valued functions f defined on $[0, 1]$ such that they have an integrable pth power, that is, for which the following integral

$$\int_0^1 |f(t)|^p \, dt$$

exists and is finite; p is assumed to satisfy $p \geqslant 1$. If $f \in L_p[0, 1]$, we define

$$\|f\| = \left[\int_0^1 |f(t)|^p \, dt \right]^{1/p} \tag{2.39}$$

A word of caution is necessary here because the integrals above are generally taken in the more general, Lebesgue, sense rather than in the well-known Riemann sense.

Once a norm has been selected, one can define the *distance* $d(x, y)$ of two vectors x and y in X as follows:

$$d(x, y) = \| x - y \| \tag{2.40}$$

It can be easily proved that the function $d(x, y)$ *generated* by the norm $\| \cdot \|$ satisfies the following conditions:

i. $d(x, y) \geqslant 0$, all x and y in X, and $d(x, y) = 0$ implies $x = y$; of course, if $x = y$, $d(x, y) = 0$.

ii. $d(x, y) = d(y, x)$

iii. $d(x, y) \leqslant d(x, z) + d(z, y)$ (triangle inequality).

The reader should be aware that distance functions can be defined which are not generated by a norm. We will not be concerned with these functions, however.

[†] Remember that, if S is a set of real numbers which is not empty and which has an upper bound, then sup S is the *least upper bound* of S, that is, an upper bound of S such that there is no smaller number which is also an upper bound of S.

EXAMPLE

The usefulness of the concept of distance can be illustrated by referring briefly to the discipline known as approximation theory. In many engineering and mathematical situations, it is necessary or convenient to approximate a function by a linear combination of other functions. Let x be a real-valued, continuous function defined on $[0, 1]$; define

$$y^i(t) = \sin it, \qquad i = 1, ..., m, \quad t \in [0, 1]$$

We would like to find a set of m numbers, $\lambda_1, ..., \lambda_m$, so that the function defined by

$$\lambda_1 y^1(t) + \cdots + \lambda_m y^m(t), \qquad t \in [0, 1]$$

is, in some sense, close to the given function x. The sense in which this closeness is to be measured is of course specified by the norm defined for the space of real-valued, continuous functions; if we choose (2.36), the extent of the closeness of the two functions, x and the one defined above, is measured by

$$\| x - (\lambda_1 y^1 + \cdots + \lambda_m y^m)\| = \max_{t \in (0,1)} | x(t) - (\lambda_1 y_1(t) + \cdots + \lambda_m y^m(t))|$$

The distance between the two functions can be measured in a so-called "mean-square sense" by the use of another norm:

$$\| x - (\lambda_1 y^1 + \cdots + \lambda_m y^m)\| = \left\{ \int_0^1 | x(t) - (\lambda_1 y^1(t) + \cdots + \lambda_m y^m(t))|^2 \, dt \right\}^{1/2}$$

The main problem in approximation theory is concerned with finding a set of numbers $\lambda_1, ..., \lambda_m$ so as to make the norm chosen as small as possible. The usual Fourier series concept can be put in this framework.

With the help of the notion of *distance* between two vectors, linear spaces acquire a geometrical structure. It is not surprising, then, that the following definitions have a strongly geometrical flavor.

Definition. The set $S_r(x^0) = \{x : x \in X \text{ and } \| x - x^0 \| < r\}$ is called an *open sphere* with center at the vector (or *point*) x^0 in a linear space X. If the equality sign is included in the inequality inside the braces in the definition of $S_r(x^0)$, then the corresponding set is called a *closed sphere*. An open (closed) *neighborhood* of $x^0 \in X$ is an open (closed) sphere with center x^0.

EXAMPLE

In $C[0, 1]$, the set of all real-valued continuous function with maximum absolute value less than a constant r is an open sphere with center at the point $x^0 = 0$.

Definition. A set $A \subset X$ is said to be *bounded* if it lies in a sphere, that is, if there is a vector $x^0 \in X$ and a positive number r such that $A \subset \{x: \| x^0 - x \| < r\}$.

Definition. Let a set $M \subset X$. A vector $x \in X$ is said to be a *limit point* of M if every open neighborhood of x contains at least a point of M different from x.

A limit point of a set is not necessarily in the set. For instance, consider the open interval $(0, 1)$, and the set of open neighborhoods of zero $\{(-\epsilon, \epsilon), 0 < \epsilon < \infty\}$. Every one of these neighborhoods contains at least one point of $(0, 1)$; we conclude that the point 0 is a limit point of this open interval. This point, however, is not in the interval. These considerations lead us to make the following:

Definition. A set $M \subset X$ is said to be *closed* if it contains all of its limit points. A set $N \subset X$ is said to be *open* if its complement $X - N$ is closed.

The interval $[0, 1]$, for instance, is closed. The interval $(0, 1)$, which has as complement with respect to the whole of R the closed set $(-\infty, 0] \cup [1, \infty)$, is open.

At last, we introduce the concept of the norm of a matrix. Let us consider an operator $T: R^n \to R^n$, with $T = (a_{ij})$ the matrix of the operator T relative to a basis $\{x^1, ..., x^n\}$ of R^n. We shall first use the norm in R^n defined in (2.32). Then, if $x = \sum_{j=1}^n c_j x^j$,

$$\| Tx \|_1^2 = \sum_{i=1}^n \left(\sum_{j=1}^n a_{ij} c_j \right)^2 \leqslant \sum_{i=1}^n \left(\sum_{j=1}^n a_{ij}^2 \right) \left(\sum_{j=1}^n c_j^2 \right)$$

$$= \left(\sum_{i,j=1}^n a_{ij}^2 \right) \left(\sum_{j=1}^n c_j^2 \right) = \left(\sum_{i,j=1}^n a_{ij}^2 \right) \| x \|_1^2 \tag{2.41}$$

define

$$\| T \|_1 = \left(\sum_{i,j=1}^n a_{ij}^2 \right)^{1/2} \tag{2.42}$$

it follows that

$$\| Tx \|_1 \leqslant \| T \|_1 \| x \|_1 \tag{2.43}$$

The expression $\| T \|_1$ in (2.42) is said to be a *norm* of the matrix T.

It is convenient to be able to associate with any vector norm in R^n a norm function for matrices, such that an inequality such as (2.43) holds. The algebraic development in (2.41) is not in general possible for an arbitrary vector norm, so that we must have recourse to other, more powerful methods. Let $\| \cdot \|$ denote any vector norm in R^n, and define

$$\| T \| = \max_{x \neq 0} \frac{\| Tx \|}{\| x \|} \tag{2.44}$$

We show first that the maximum in (2.44) exists. Put $\bar{x} = x/\| x \|$; then,

$$\| T \| = \max_{\|\bar{x}\|=1} \| A\bar{x} \| \tag{2.45}$$

the function with values $\| A\bar{x} \|$ in R and domain $\{\bar{x} : \| \bar{x} \| = 1\}$ is a continuous function of $\bar{x}$ defined on a closed, bounded set; there is a point in this set for which the function takes a maximum value. The norm (2.44) is well defined. It satisfies

$$\| Tx \| \leqslant \| T \| \| x \| \tag{2.46}$$

since, if $x \in R^n$, $x \neq 0$,

$$\frac{\| Tx \|}{\| x \|} \leqslant \max_{x \neq 0} \frac{\| Tx \|}{\| x \|} = \| T \| \tag{2.47}$$

It can be proved that, if $\| x \| = \max_i | x_i |$,

$$\| T \| = \max_k \sum_j | a_{kj} | \tag{2.48}$$

We remark here that the norm (2.44) equals the smallest number k such that, for a fixed vector norm, $\| Tx \| \leqslant k \| x \|$. The norm (2.44) is said to be *induced* by the corresponding vector norm. We note that the norm (2.42) is *not* induced by the vector norm (2.32).

Exercises

*1. Let X be a linear normed space and let A be a subset of it. For any $x \in X$, the distance from x to A, denoted by $d(x, A)$, is defined as

$$d(x, A) = \inf_{y \in A} \| x - y \|$$

Show that $d(x, A) = 0$ if and only if x is either in A or is a limit point of A. (The set composed of A and its limit points, which is denoted by $\bar{A}$, is said to be the *closure* of A. This theorem could be stated: Show that $d(x, A) = 0$ if and only if $x \in \bar{A}$).

*2. Let X be a linear normed space. Show that the closure of an open sphere, $\bar{S}_r(x_0)$, is a subset of the corresponding closed sphere. Note that there are points in the boundary of the closed sphere which are not limit points of the open sphere.

3. A point in $M \subset X$ is said to be an *interior point* of M if it is the center of an open sphere contained in M. Prove that an open set consists only of interior points.

*4. Show that the norm (2.33) can be considered as the limit of the norm (2.35) when p tends to infinity.

5. Prove expression (2.46).

*6. Show by means of an example that the norm (2.42) and the norm induced by the vector norm (2.32) are not equal.

*7. Prove that the norm (2.44) equals the smallest number k such that, for a fixed vector norm, $\| Tx \| \leqslant k \| x \|$.

*8. Derive an expression such as (2.42) when $T: C^n \to C^n$.

2.9. Inner Products. Adjoint Operators

We have seen in the previous section how, by defining a norm in a vector space (and thus making it a *normed space*), we managed to give to this space a geometrical structure which would otherwise be lacking. It so happens, however, that sometimes this structure is not rich enough for many problems; we are forced in such cases to define even more operations of the type we have called metric on the elements of a vector space. We proceed to do just this by defining the *inner product* of two vectors in a vector space, which becomes then an *inner product space*. We will consider only the n-tuple spaces R^n and C^n.

Definition. If x and y are in R^n, the *inner product* of x and y, to be denoted by (x, y) is defined by:

$$(x, y) = \sum_{i=1}^{n} x_i y_i \tag{2.49}$$

If x and y are in C^n, the inner product is defined by

$$(x, y) = \sum_{i=1}^{n} x_i \bar{y}_i \tag{2.50}$$

where $\bar{y}_i$ denotes the complex conjugate of y_i. Note that the inner product operation in C^n is not symmetrical in the two vectors x and y, and that, as occurs with all expressions in which vectors in C^n appear, (2.50) is transformed into the equivalent expression for R^n, (2.49), when the vectors x and y are purely real. The space R^n is called a *euclidean space* when the inner product (2.49) is defined on its elements.

If in either expressions (2.49) or (2.50) we put $y = x$, we obtain

$$(x, x) = \sum_{i=1}^{n} | x_i |^2 \tag{2.51}$$

that is, we obtain the norm (2.32), which is said to be *generated* by the inner product. Other properties of the inner product are listed below

$$(x, y) = \overline{(y, x)}$$
$$(\alpha_1 x_1 + \alpha_2 x_2 , y) = \alpha_1(x_1 , y) + \alpha_2(x_2 , y)$$
$$(x, \alpha y) = \bar{\alpha}(x, y)$$

Note that these properties refer to the inner product (2.50); in the case of real vectors, the conjugation may be ignored.

The inequality (2.52) is the Cauchy–Schwarz inequality, of importance in many problems.

THEOREM 2.12.   If x and y are vectors in R^n or C^n, then

$$|(x, y)| \leqslant \| x \| \cdot \| y \| \tag{2.52}$$

where the norm is the one generated by the inner product.

Proof.   Let x and y be in C^n with $x \neq 0$, and form the function of a real variable λ:

$$\psi(\lambda) = \sum_{i=1}^{n} | \lambda x_i + y_i |^2 = \sum_{i=1}^{n} (\lambda x_i + y_i)(\overline{\lambda x_i + y_i})$$

$$= \lambda^2 \sum_{i=1}^{n} | x_i |^2 + | \lambda | \left\{ \sum_{i=1}^{n} (x_i \bar{y}_i + \bar{x}_i y_i) \right\} + \sum_{i=1}^{n} | y_i |^2$$

$$= \lambda^2 \| x \|^2 + | \lambda | \{ (x, y) + \overline{(x, y)} \} + \| y \|^2$$

$$= \lambda^2 \| x \|^2 + 2 | \lambda | \operatorname{Re}(x, y) + \| y \|^2$$

$$\leqslant \lambda^2 \| x \|^2 + 2 | \lambda | |(x, y)| + \| y \|^2$$

$$= \left\{ | \lambda | \| x \| + \frac{|(x, y)|^2}{\| x \|^2} \right\}^2 + \left\{ \| y \|^2 - \frac{|(x, y)|^2}{\| x \|^2} \right\}$$

If the above is to remain nonnegative for all λ, it is clear that

$$\| y \|^2 - \frac{|(x, y)|^2}{\| x \|^2} \geqslant 0$$

which is equivalent to the Cauchy–Schwarz inequality.

An important concept in an inner-product space is that of *orthogonality*.

Definition.   Two vectors x and y in R^n (or in C^n) are said to be *orthogonal* if $(x, y) = 0$.

Definition.   A basis $\{x^1, x^2, ..., x^n\}$ in R^n or C^n will be said to be *orthogonal* if $(x^i, x^j) = 0$, $i \neq j$. If in addition to the above $\| x^i \| = 1$ for all i, then the basis is *orthonormal*.

For instance, the standard basis in R^n is orthonormal.

It can be proved that in a finite-dimensional vector space an orthonormal basis can always be constructed [3, p. 230]. The process to do so is called the Gram–Schmit orthogonalization procedure. The importance of orthonormal basis stems mainly from the following result.

THEOREM 2.13.   If $\{x^1, ..., x^n\}$ is an orthonormal basis for R^n or C^n, and if

$$x = \sum_{i=1}^{n} c_i x^i$$

then

$$c_i = (x, x^i) \tag{2.53}$$

Suppose now that W is a subspace of R^n (or of C^n), that is, a subset of R^n which satisfies all axioms of a linear space. Let W be k-dimensional. We now select an orthonormal basis $\{x^1, ..., x^n\}$ in R^n so that the set $\{x^1, ..., x^k\}$ spans W. The elements $\{x^{k+1}, ..., x^n\}$ span a second subspace of dimension $n - k$, to be denoted by $W^\perp$ (read "W perp"). If $x \in W$ and $y \in W^\perp$, $(x, y) = 0$.

The importance of this concept stems from the fact that any vector $z \in R^n$ (or C^n) can be decomposed in the form

$$z = x + y$$

with $x \in W$, $y \in W^\perp$, W being an arbitrary subspace of R^n (or C^n); this follows simply by partitioning an appropriate basis. The uniqueness of the decomposition follows from the fact that, if $z = x^1 + y^1$ is assumed to be another such decomposition, then

$$(x - x^1) + (y - y^1) = 0$$

which implies $\| x - x^1 \| = 0$, since $(x - x^1, y - y^1) = 0$; then $x = x^1$. Similarly, $y = y^1$.

It is said that x is the *projection* of z on W, and y the projection of z on $W^\perp$.

At last, we consider the important concept of *adjoint* operator. We will deal with the real case first. Let $T: R^n \to R^n$, and let $T = (a_{ij})$ be the matrix of T with respect to a basis in R^n. If the coordinates of x with respect to the same basis are $x_1, ..., x_n$, and those of y are $y_1, ..., y_n$, we can compute the following inner product:

$$(T(x), y) = \sum_{i=1}^{n} \left(\sum_{j=1}^{n} a_{ij} x^j \right) y^i = \sum_{j=1}^{n} x^j \left(\sum_{i=1}^{n} a_{ij} y^i \right)$$

$$= x_1 \left(\sum_{i=1}^{n} a_{i1} y^i \right) + x_2 \left(\sum_{i=1}^{n} a_{i2} y^i \right) + \cdots + x_n \left(\sum_{i=1}^{n} a_{in} y^i \right) \tag{2.54}$$

It should be recognized that the expressions in the brackets are the components of a vector obtained by applying the *transpose* matrix

$$T' = (a_{ji}) \tag{2.55}$$

to the vector y. We have therefore

$$(T(x), y) = (Tx, y) = (x, T'y) \qquad (2.56)$$

The operator whose matrix is T' with respect to the chosen basis in R^n is called the *adjoint* operator of T. Note that the rows of T' are the columns of T and the rows of T are the columns of T'.

If $T = T'$, the matrix T is said to be *symmetric*. Symmetric matrices have many interesting properties which will be studied in later chapters.

Note that if x is a column vector, x' is a row vector. The inner product of two vectors (of the same dimensions, of course) can be written $x'y$, where both x and y are column vectors.

When the space C^n is considered, the counterpart of the matrix T' is played by the complex conjugate of the transpose of T, to be designated by T^*. If $T = T^*$, we say that T is *Hermitian*.

Exercises

1. Show that there is no nonzero linear operator in C^2 such that $(x, T(x)) = 0$ for every x in C^2.

2. Show that AB is not necessarily symmetric if A and B are.

3. Show that if A has real entries and $AA' = 0$, then A is zero.

4. If $TT' = I_n$, then T is said to be *orthogonal*. Show that the determinant of an orthogonal matrix is either 1 or -1.

*5. Show that two vectors are orthogonal if and only if $\| x + y \| = \| x \| + \| y \|$, where the norm is the one generated by the inner product.

2.10. An Introduction to Matrix Algebra

Many of the concepts presented in the previous pages are of great importance, and will be used throughout the book and indeed during the whole scientific career of the student. The technical aspect of this subject, matrix algebra, forms a bridge between the conceptual content introduced above and the numerical, computational methods of the Appendix. It is of importance, however, in developments quite remote from any numerical aspects; for instance we will use matrix algebra later on to obtain explicit solutions for systems of linear differential equations; these solutions will be used later to treat, for instance, problems concerning the controllability of such systems.

This section develops and summarizes the results related to matrix algebra which have been already obtained.

Let A, B, and C be matrices of the same dimensions with entries in a field F. The following formulas are quite simple to derive:

$$A + B = B + A \tag{2.57}$$

$$c(A + B) = cA + cB, \quad c \in F \tag{2.58}$$

$$(A + B) + C = A + (B + C) = (A + C) + B \tag{2.59}$$

Let now A and B be two square matrices of the same dimension. Then the product AB is well defined. Assume further that A and B are nonsingular. What is the inverse of the product AB? If we call this inverse X, clearly

$$ABX = I$$

If we multiply both sides of this equality by A^{-1} on the left, $BX = A^{-1}$ and if we multiply this by B^{-1} on the left, we obtain

$$X = (AB)^{-1} = B^{-1}A^{-1} \tag{2.60}$$

If A is $m \times n$ and B is $n \times m$, both AB and BA exist and are square matrices. The *trace* of a square matrix is defined as the sum of its diagonal elements; then

$$\operatorname{tr}(AB) = \sum_{i,k} a_{ik}b_{ki} = \sum_{k,i} a_{ki}b_{ik} = \operatorname{tr}(BA) \tag{2.61}$$

The following two formulas are of importance, even if we will not prove them:

$$\det(AB) = \det A \cdot \det B \tag{2.62}$$

$$\| AB \| \leqslant \| A \| \| B \| \tag{2.63}$$

If a matrix A is square, its *powers* are defined as follows:

$$A^2 = AA$$

$$A^n = \underset{n \text{ times}}{AA \cdots A} \tag{2.64}$$

$$A^{-2} = (A^2)^{-1} = A^{-1}A^{-1} \tag{2.65}$$

$$A^{-n} = \underset{n \text{ times}}{A^{-1} \cdots A^{-1}} \tag{2.66}$$

Clearly,

$$A^{n+m} = A^n A^m \tag{2.67}$$

We present now some properties of the *transpose* matrix. Of course,

$$(A + B)' = A' + B' \tag{2.68}$$

$$(A')' = A \tag{2.69}$$

What is the transpose of AB? By definition,

$$(ABx, y) = (x, (AB)'y) = (Bx, A'y) = (x, B'A'y),$$

so that

$$(AB)' = B'A' \tag{2.70}$$

If A is square,

$$\det A = \det A' \tag{2.71}$$

it follows that the operations of transposing and inverting do commute, so that

$$(A^{-1})' = (A')^{-1} \tag{2.72}$$

We will often *partition* matrices into submatrices. For instance, consider an $n \times m$ matrix A with complex entries. Each of the m columns can be used to form a column vector with n complex entries; in this way we construct the vectors $a^1, ..., a^m$. We will write often

$$A = [a^1\, a^2 \cdots a^m]$$

For instance, if B is a $p \times n$ matrix, the product BA is equal to

$$BA = [Ba^1\, Ba^2 \cdots Ba^m] \tag{2.73}$$

that is, the m columns of BA are obtained from those of A by acting on them by the matrix B. This of course follows from the definition of matrix multiplication; it is, however, often useful to employ such a rule directly.

In a similar way, one can partition matrices as follows:

$$A = \begin{bmatrix} A_1 & A_2 \\ A_3 & A_4 \end{bmatrix}$$

The matrices A_1, A_2, A_3, A_4 can be of arbitrary dimensions provided that they cover the whole of the matrix A. If a matrix B is partitioned similarly

$$B = \begin{bmatrix} B_1 & B_2 \\ B_3 & B_4 \end{bmatrix}$$

then, supposing AB exists, it is given by

$$AB = \begin{bmatrix} A_1B_1 + A_3B_3 & A_1B_2 + A_2B_4 \\ A_1B_3 + A_3B_4 & A_2B_2 + A_4B_4 \end{bmatrix} \tag{2.74}$$

provided that the partitions of A and B are such that all the products which appear in the matrix AB are well defined. The reader can convince himself of the validity of this rule by trying out a few simple examples. The interest of (2.74) lies in the fact that matrices there are treated formally as scalar entries, so that the usual rules of multiplication for matrices are conserved provided of course that all products are well defined.

Exercises

*1. Prove that $\| AB \| \leqslant \| A \| \| B \|$. A and B are $n \times n$ matrices.

2. Partition the following matrix

$$\begin{bmatrix} 1 & -1 & 2 & 3 \\ 2 & 2 & 0 & 2 \\ 4 & 1 & -1 & -1 \\ 1 & 2 & 3 & 0 \end{bmatrix}$$

so that you can compute its inverse matrix with ease.

*3. A *skew-symmetric* matrix is one for which $A = -A'$. Show that any square matrix can be written as the sum of a symmetric and an skew symmetric matrix.

4. Show that the determinant of a skew-symmetric matrix of odd order is zero.

SUPPLEMENTARY NOTES AND REFERENCES

The student interested in improving his mathematical background beyond the elements given in this book should study

W. A. Porter, "Modern Foundations of Systems Engineering." Macmillan, New York, 1966.

A reference on finite-dimensional vector spaces other than the book by Hoffman and Kunze is the classic

P. R. Halmos, "Finite-Dimensional Vector Spaces." Van Nostrand, Princeton, New Jersey, 1958.

As a preparation for the study of Chapter 9, the reader should start about now to learn about elementary functional analysis. The writer found the following book by Vulikh a great help in this task.

B. Z. Vulikh, "Introduction to Functional Analysis for Scientists and Technologists." Pergamon Press, New York, 1963.

REFERENCES

1. P. R. Halmos, "Naive Set Theory." Van Nostrand, Princeton, New Jersey, 1960.
2. G. Birkhoff and S. MacLane, "A Survey of Modern Algebra." Macmillan, New York, 1965.
3. K. Hoffman and R. Kunze, "Linear Algebra." Prentice-Hall, Englewood Cliffs, New Jersey, 1961, pp. 316–317.
4. G. Bachman and L. Narici, "Functional Analysis." Academic Press, New York, 1966.

3

Differential Systems I

3.1. Introduction

We will be concerned in this chapter with the development of some properties of simple differential systems such as those examined in Chapter 1. The features to be treated here are mostly of a nonalgebraic nature; we will be interested in the general form of the solutions of the system equations, as well as in some characteristics of the systems themselves which can be derived from these forms.

The *homogeneous* systems, systems without inputs, will be our first subject of study. Once these are taken care of, the general solution and properties of the systems with inputs will follow readily. We will then summarize the properties found so as to put them in a form suitable for the inductive process of Chapter 9; after all, these simple systems will serve as models there for those more general entities, the dynamical systems.

It should be clear from the examples presented in Chapter 1 that the equations which describe simple differential systems are of the form:

$$\dot{x}_i = \sum_{j=1}^{n} a_{ij}x_j + \sum_{k=1}^{r} b_{ik}u_k, \qquad i = 1, ..., n$$

$$y_j = \sum_{k=1}^{n} c_{jk}x_k + \sum_{l=1}^{r} d_{jl}u_l, \qquad j = 1, ..., m \tag{3.1}$$

Note that these equations are given in the form of equality of *functions*. There are two ways to denote the equality of two functions f and g defined

71

on a set S of the real numbers, say. One way is by means of the *values* of the functions; we write

$$f(t) = g(t), \qquad t \in S$$

The other way, which is equivalent, is to write that the *functions* are equal:

$$f = g$$

In the case of Equation (3.11), the set S will always be a closed interval such as $I = [t_a, t_b]$. The equation (3.1) can therefore be written also as

$$\dot{x}_i(t) = \sum_{j=1}^{n} a_{ij}(t) x_j(t) + \sum_{k=1}^{r} b_{ik}(t) u_k(t), \qquad i = 1, ..., n$$

$$\tag{3.2}$$

$$y_j(t) = \sum_{k=1}^{n} c_{jk}(t) x_k(t) + \sum_{l=1}^{r} d_{jl}(t) u_l(t), \qquad j = 1, ..., m$$

for all $t \in [t_a, t_b]$; to these equations we add a set of n numbers, $x_1^0, x_2^0, ..., x_n^0$, which are the values of $x_1, x_2, ..., x_n$ respectively at any instant of time $t_0 \in [t_a, t_b]$. The inputs u_k, $k = 1, ..., r$, are real-valued functions defined on $[t_a, t_b]$. The quantities $y_j, j = 1, ..., m$, are the outputs of the system, the quantities $x_i, i = 1, ..., n$, will turn out to be the state variables.

The systems with which we will be concerned are generally real, in the sense that the coefficients of the Equation (3.2) represent combinations of real parameters such as resistances, capacitances, etc., and are therefore real numbers. It happens sometimes, however, that in order to simplify a system of equations such as (3.2), it is transformed into another system by the operation known as a change of basis; with respect to the new basis the system may have complex coefficients. We will for the moment ignore this, waiting until Chapter 4, where the procedure consisting of the change of basis will be discussed in detail, for a comment on the properties of systems such as (3.2) with complex coefficients.

The set of equations (3.2) can be interpreted as defining several transformations in various spaces. The first equation, for instance, says that, at each time $t \in [t_a, t_b]$, $\dot{x}_i$ equals a linear combination of the x_j's plus another linear combination of the u_k's. The functions defined by the linear combinations can be described in matrix form by choosing bases in the corresponding spaces; for instance, the operation of taking linear combinations of the x_j's can be described with respect to the standard basis in R^n. We will also describe the other linear operations with respect to the standard bases in the appropriate spaces, so that the set (3.2) can be written as

$$\dot{x}(t) = A(t) x(t) + B(t) u(t),$$
$$\qquad\qquad\qquad\qquad t \in [t_a, t_b] \qquad\qquad (3.3)$$
$$y(t) = C(t) x(t) + D(t) u(t),$$

here $x(t)$ is an n-vector, $A(t) = (a_{ij}(t))$ is an $n \times n$ matrix, $u(t)$ is an r-vector, $B(t) = (b_{ik}(t))$ is an $n \times r$ matrix, $y(t)$ is an m-vector, $C(t) = (c_{jk}(t))$ is an $m \times n$ matrix, and $D(t) = (d_{jl}(t))$ is an $m \times r$ matrix. All vector are column vectors in R^n, R^r, or R^m. The conditions on x at t_0 are written $x(t_0) = x^0$. Of course, $\dot{x}(t)$ is the vector with components $\dot{x}_1(t),..., \dot{x}_n(t)$.

We should emphasize that any other bases could have been chosen in R^n, R^r, or R^m to describe in matrix form the linear operations involved in the set of Equations (3.2). Indeed, in Chapter 4 we will study in detail the process consisting of the change of basis for a system such as (3.3); this change is performed with the purpose of simplifying the matrices or otherwise making the system more amenable to treatment and study.

The matrix equations (3.3) can be described also by equality of vector- and matrix-valued functions of t with domain $[t_a, t_b]$:

$$\dot{x} = Ax + Bu$$
$$\qquad\qquad\qquad\qquad\qquad (3.4)$$
$$y = Cx + Du, \qquad x(t_0) = x^0$$

In most of this chapter the interest will be centered on the first of the equations in (3.4) or (3.3). Clearly, since y does not appear in this equation, once a set of initial conditions on x, $x(t_0)$, and an input vector u are given, the first step has to consist in solving for x in the first equation and studying the properties of the solutions; once this is done, y is determined, and its properties follow readily from those of x and u.

Generally speaking, the input u in the first of Equations (3.4) will be specified on a half-open interval, $[t_0, t_1)$, $t_1 > t_0$, which of course is to be contained in the basic interval of definition $I = [t_a, t_b]$. Why define the input in such an interval? It happens that sometimes it is necessary to define an input composed of two successive, previously defined inputs

$$u(t) = \begin{matrix} u^1(t), & t_0 \leqslant t < t_1 \\ u^2(t), & t_1 \leqslant t < t_2 \end{matrix}$$

It would be impossible to make this definition if u^1 were specified at t_1 unless the unlikely situation prevailed that $u^1(t_1) = u^2(t_1)$.

Note that we have not, so far, defined the class of functions u which will be taken as admissible inputs. This will be done in Section 3.6.

3.2. The Homogeneous System

The *homogeneous system* associated with the first equation in (3.4) is defined to be:

$$\dot{x} = Ax, \quad x(t_0) = x^0, \quad t_0 \in I = [t_a, t_b] \tag{3.5}$$

This equation is said to describe the *free* (that is, without inputs) motion of the state vector.

Let the entries of A be continuous functions of time on the interval $I = [t_a, t_b]$. We will prove that under this condition, the system (3.3) has one and only one solution on I [1, 2].

Before becoming immersed in the first phase of this program, the one dealing with the proof that Equation (3.5) has at least one solution, let us define the *integral* of a vector- or matrix-valued function. It will be understood that the integral of such a function is a vector, or a matrix, as the case may be, whose entries are the integrals of the entries of the original vector or matrix. We write

$$\int_{t_0}^{t_1} x(t)\, dt \qquad \text{or} \qquad \int_{t_0}^{t_1} A(t)\, dt$$

It can be proved that

$$\left\| \int_{t_0}^{t_1} x(t)\, dt \right\| \leqslant \int_{t_0}^{t_1} \| x(t) \|\, dt$$

and

$$\left\| \int_{t_0}^{t_1} A(t)\, dt \right\| \leqslant \int_{t_0}^{t_1} \| A(t) \|\, dt$$

The symbol which denotes norm behaves with respect to vectors and matrices much as the sign which indicates absolute value does with respect to real numbers.

To show the existence of a solution of Equation (3.5), it will prove convenient to consider instead the following integral equation

$$x(t) = x^0 + \int_{t_0}^{t} A(\lambda)\, x(\lambda)\, d\lambda, \quad t \in I \tag{3.6}$$

We prove that, under some conditions, a solution of (3.6) is also a solution of (3.5). Indeed, let x be a differentiable (with respect to time, for all $t \in I$) solution of (3.6). Of course, $x(t_0) = x^0$. Let us differentiate both sides of (3.6) at an instant $t \in I$. The derivative of the left-hand side is of course $\dot{x}(t)$. This should be equal to the derivative of the right-hand side, if this exists. Since A is continuous and x is differentiable (and thus continuous), the integrand is continuous for all $t \in I$; the derivative of the integral with respect to its upper limit is then the integrand function evaluated at the value of the upper limit, that is, $A(t)\, x(t)$ [3, p. 520]. Since this is true for any $t \in I$, a differentiable solution of (3.6) also satisfies (3.5).

The advantage of considering (3.6) consists in the possibility of constructing a solution of this equation by means of successive approximations. Take an arbitrary but fixed value of $t \in I$ and define a sequence of n-vectors $x^0(t)$, $x^1(t)$, ..., $x^k(t)$, ... such that

$$x^0(t) = x^0$$

$$x^1(t) = x^0 + \int_{t_0}^{t} A(s)\, x^0(s)\, ds$$

$$\vdots \tag{3.7}$$

$$x^{k+1}(t) = x^0 + \int_{t_0}^{t} A(s)\, x^k(s)\, ds, \qquad t \in I$$

$$\vdots$$

Note that t is *not* taken to be a higher value than t_0; rather, it can assume either higher or lower values than t_0.

Let us consider first the question whether this sequence has a limit. From (3.7) it follows that

$$x^{k+1}(t) - x^k(t) = \int_{t_0}^{t} A(s)\, [x^k(s) - x^{k-1}(s)]\, ds, \qquad t \in I$$

and therefore

$$\| x^{k+1}(t) - x^k(t) \| \leqslant \left| \int_{t_0}^{t} \| A(s) \|\, \| x^k(s) - x^{k-1}(s) \|\, ds \right|, \qquad t \in I \tag{3.8}$$

note that, since t may be less than t_0, it is necessary to use the absolute value of the right-hand side of (3.8). Here we have used any norm for the vectors in R^n and the matrix norm generated by it. Since the entries of $A(t)$ are continuous, the norm of this matrix is bounded on I, so that there exists a number M such that $\| A(t) \| < M$ for $t \in I$. Therefore,

$$\| x^{k+1}(t) - x^k(t) \| \leqslant M \left| \int_{t_0}^{t} \| x^k(s) - x^{k-1}(s) \|\, ds \right|, \qquad t \in I$$

Since, for $t \in I$,

$$\| x^1(t) - x^0(t) \| \leqslant \left| \int_{t_0}^{t} \| A(s) \| \| x^0 \| \, ds \right| \leqslant M \| x^0 \| \, | t - t_0 |$$

it should be clear from these two expressions that

$$\| x^2(t) - x^1(t) \| \leqslant M \left| \int_{t_0}^{t} \| x^1(s) - x^0(s) \| \, ds \right| \leqslant \frac{1}{2!} M^2 \| x^0 \| \, | t - t^0 |^2$$

By a simple process of induction, it follows that

$$\| x^{k+1}(t) - x^k(t) \| \leqslant \frac{M^{k+1} \| x^0 \| \, | t - t_0 |^{k+1}}{(k+1)!} \tag{3.9}$$

We revert now to work with the components of the vectors $x^k(t)$; after all, we know a great deal about convergence of sequences of real numbers. We choose a norm

$$\| x \| = \max_i | x_i | \tag{3.10}$$

noting that our proof can be carried out using any other norm in R^n; the norm above happens to be very convenient, since from (3.9) it follows now that

$$| x_i^{k+1}(t) - x_i^k(t) | \leqslant \frac{M^{k+1} \| x^0 \| \, | t - t_0 |^{k+1}}{(k+1)!} \tag{3.11}$$

for $i = 1, \ldots, n$, $t \in I$. We use now the identity

$$x_i^k(t) = x_i^0(t) + (x_i^1(t) - x_i^0(t))$$

$$+ (x_i^2(t) - x_i^1(t)) + \cdots + (x_i^k(t) - x_i^{k-1}(t)) \tag{3.12}$$

It is our intention to prove that the left-hand side of (3.12) converges as k tends to infinity. The right-hand side of this equation can be considered as a partial sum of an infinite series. We will prove that the series formed with the absolute values of the terms converges. Indeed,

$$| x_i^0(t)| + | x_i^1(t) - x_i^0(t) | + | x_i^2(t) - x_i^1(t) | + \cdots + | x_i^k(t) - x_i^{k-1}(t) |$$

$$\leqslant \| x^0 \| \left[1 + M \frac{| t - t_0 |}{1!} + \cdots + M^k \frac{| t - t_0 |^k}{k!} \right]$$

that is, this series is majorized by a convergent series and thus the series itself converges. It follows that the series defined by the right-hand side of (3.12) is absolutely convergent, and thus convergent. Hence, there exist numbers $x_i(t)$, for each $t \in I$, such that

$$\lim_{k \to \infty} | x_i{}^k(t) - x_i(t) | = 0, \qquad i = 1, ..., n \qquad (3.13)$$

The sequence of vectors $x^k(t)$ can be proved to converge for all $t \in I$. Indeed, since Equation (3.13) is valid for all i,

$$\lim_{k \to \infty} \| x^k(t) - x(t) \| = \lim_{k \to \infty} \max_i | x_i{}^k(t) - x_i(t) | = 0 \qquad (3.14)$$

The vector $x(t)$ has components $x_i(t)$, $i = 1, ..., n$.

The limit $x(t)$ just obtained will be shown to satisfy Equation (3.6), and thus (3.5). Consider the following inequality; the norm is, as above, the one defined by Equation (3.10):

$$\left\| x(t) - \left(x^0 + \int_{t_0}^t A(\lambda) \, x(\lambda) \, d\lambda \right) \right\|$$

$$= \left\| (x(t) - x^{k+1}(t)) + \left(x^{k+1}(t) - x^0 - \int_{t_0}^t A(\lambda) \, x(\lambda) \, d\lambda \right) \right\|$$

$$\leqslant \| x(t) - x^{k+1}(t) \| + \left| \int_{t_0}^t \| A(\lambda)(x^k(\lambda) - x(\lambda)) \| \, d\lambda \right|, \qquad t \in I \quad (3.15)$$

Our intention is, of course, to show that the first norm in zero. For this purpose, we must prove that the two norms in the right-hand side of the inequality can be made arbitrarily small at time t. The first norm presents no problem, since, by definition of a limit, given a number $\epsilon > 0$, an integer $K(t)$ which, as indicated, depends in general on the particular value of the time, can be found such that

$$\| x(t) - x^{k+1}(t) \| < \epsilon \qquad (3.16)$$

for $k > K(t)$. The second norm requires, however, a more sophisticated treatment. Since $\| A(\lambda) \| < M$ on this interval, we have

$$\left| \int_{t_0}^t \| A(\lambda)(x^k(\lambda) - x(\lambda)) \| \, d\lambda \right| \leqslant M \left| \int_{t_0}^t \| x^k(\lambda) - x(\lambda) \| \, d\lambda \right| \qquad (3.17)$$

Suppose that we try to make the right-hand side of this expression arbitrarily small by choosing k large enough. Given an $\epsilon > 0$, the norm inside the integral can be made smaller than ϵ at a given value of λ,

$$\| x^k(\lambda) - x(\lambda) \| < \epsilon \qquad (3.18)$$

by choosing $k > K_1(\lambda)$. However, to write

$$M \left| \int_{t_0}^{t} \| x^k(\lambda) - x(\lambda) \| \, d\lambda \right| \leqslant M \, | \, t - t_0 \, | \, \epsilon \qquad (3.19)$$

it is necessary to consider *all* values of $\lambda \in [t_0, t]$, that is, to find a number K_1, independent of λ, such that the inequality (3.18) is achieved *for all* λ by making $k > K_1$. The reader may argue that it is sufficient to make

$$K_1 = \max_{\lambda \in [t_0, t]} K_1(\lambda)$$

it is not known, however, if this maximum exists.

It so happens that in this case there is a simple way out of this difficulty because the number $K_1(\lambda)$ does not, after all, depend on λ; it is said that the convergence of $x^k(t)$ to $x(t)$ is *uniform* in t [3, Chapter XVIII] for $t \in I$. (Remember that any symbol can be used for the argument of a function without altering the meaning.)

The uniformity of the convergence of $x^k(t)$ to $x(t)$ follows very simply from Equation (3.11). Since $t, t_0 \in I = [t_a, t_b]$, $| \, t - t_0 \, | < (t_b - t_a)$; Equation (3.11) becomes

$$| \, x_i^{k+1}(t) - x_i^{k}(t) \, | \leqslant \frac{M^{k+1} \| x^0 \| (t_b - t_a)^{k+1}}{(k+1)!} \qquad (3.20)$$

Any particular value of the time t does not appear any more, and the integers $K(t)$ and $K_1(t)$ can indeed be chosen independently of the time. Therefore, given an $\epsilon > 0$, k can be chosen so that both Equations (3.16) and (3.18) are satisfied by making it larger than the largest of the numbers K and K_1; it follows that

$$\left\| x(t) - \left[x^0 + \int_{t_0}^{t} A(\lambda) \, x(\lambda) \, d\lambda \right] \right\| \leqslant \epsilon + M(t - t_0)\epsilon$$

Since ϵ is arbitrary, the norm has to be zero since, if it were not zero, there would be values of ϵ, perhaps very small, for which the previous inequality would not be satisfied. It follows that x, the limit of the iterative process, is a solution of Equation (3.6). In order to show it to be a solution of the differential equation (3.5), we must prove its differentiability with respect to t for $t \in I$. Indeed, this property is a consequence of the following facts: Each of the iterates x^k is a differentiable function of t with derivative function $\dot{x}^k$ defined by

$$\dot{x}^k(t) = A(t) \, x^{k-1}(t), \qquad t \in I$$

Since the functions x^k tend uniformly to a limit x, and A is bounded over I, the functions $\dot{x}^k$ tend (uniformly) to a limit which [3] is the derivative of the limit function x. This is therefore differentiable, and thus a solution of the differential equation (3.5).

We prove at last the uniqueness of this solution. It should be evident that every solution of the differential equation (3.5) is also a solution of the integral equation (3.6), so that if this equation is shown to have a unique solution, then the differential equation also has a unique solution.

Assume that (3.6) has a second solution y. We shall prove that $\| x(t) - y(t) \| = 0$ for all $t \in I$. We find that

$$\| x(t) - y(t) \| \leqslant \left| \int_{t_0}^{t} \| A(\lambda) \| \, \| x(\lambda) - y(\lambda) \| \, d\lambda \right|$$

$$\leqslant M \left| \int_{t_0}^{t} \| x(\lambda) - y(\lambda) \| \, d\lambda \right| \tag{3.21}$$

Since x and y are differentiable, and thus continuous, we can define

$$m = \max_{\tau \in [t_0, t]} \| x(\tau) - y(\tau) \| \tag{3.22}$$

Then, from (3.21) and (3.22),

$$\| x(t) - y(t) \| \leqslant mM \, | t - t_0 |$$

We can use this bound again in (3.21) to obtain

$$\| x(t) - y(t) \| \leqslant mM^2 \frac{1}{2!} \, | t - t_0 |^2$$

again we use this new bound

$$\| x(t) - y(t) \| \leqslant mM^3 \frac{1}{3!} \, | t - t_0 |^3$$

It follows by iteration that

$$\| x(t) - y(t) \| \leqslant mM^j \frac{1}{j!} \, | t - t_0 |^j \tag{3.23}$$

for all j.

If $\| x(t) - y(t) \|$ were not zero, the inequality (3.23) would be contradicted for some sufficiently high values of j. Then

$$x(t) = y(t), \qquad t \in I$$

The solution found by the iterative method is indeed unique.

We summarize our findings in the following theorem.

THEOREM 3.1.   The homogeneous system

$$\dot{x} = Ax, \quad x(t_0) = x^0, \qquad t_0 \in [t_a, t_b]$$

has a unique solution for all $t \in [t_a, t_b]$, provided the entries of the matrix A are continuous over this interval.

The fact that the solution of this system is unique has the following rather startling consequence.

THEOREM 3.2.   The solution of the homogeneous system (3.5) never equals the zero vector, unless it is identically equal to it for all $t \in I$.

Proof.   We prove simply that if a solution of (3.5) equals zero at some $t_0 \in I$, then it must be identically equal to zero for all $t \in I$. Indeed, a solution of

$$\dot{x} = Ax, \qquad x(t_0) = 0$$

is certainly $x(t) = 0$, $t \in I$; and, by the previous theorem, this is the only such solution. It follows, of course, that a solution which is not zero at some $t_0 \in I$ is not zero for all $t \in I$.

From the iterative construction given above, we can obtain a number of properties of the solution of (3.5). For instance, we see that, since the coefficients of A are continuous on I, and since therefore all of the iterates are well defined on I, then x is well defined on I. Consider now a slightly different problem: Suppose that a coefficient of A is not continuous on I, but that it has a singularity at a given point. Then the iterates (and also the solution x) might have a singularity at the same point. One example of this situation is the well-known Bessel equation

$$\ddot{y} + \frac{1}{t}\dot{y} + y = 0 \tag{3.24}$$

or

$$\dot{x}_1 = x_2, \qquad \dot{x}_2 = -x_1 - \frac{1}{t}x_2$$

which has solutions which are singular at $t = 0$.

So far, we have considered the interval $I = [t_a, t_b]$ to be finite. What happens when the matrix A is defined (and has continuous entries) on an unbounded set, such as the whole of the real line, or the positive half-line ? This is the case, for instance, when the entries of A do not depend on the time. (Such a system is said to be a *constant* or *time-invariant* homogeneous system.) Suppose that A is defined for all t. The corresponding homogeneous system has a solution on any finite interval. Since the real line is the union of a set of finite intervals, we conclude that in such a case the homogeneous system has a unique solution which is determined by specifying its value $x(t_0)$ at an arbitrary value of the time t_0. This solution is defined for all time.

Exercises

*1. Show that

$$\lim_{k \to \infty} \| x^k(t) \| \leqslant \| x^0 \| e^{M(t-t_0)}, \qquad t \in I$$

here x^k is defined by (3.7).

2. Consider the equation

$$\ddot{y}(t) + a(t)\,\dot{y}(t) + b(t)\,y(t) = 0, \qquad t \in I$$

Let y_1 and y_2 be solutions of this equation such that $y_1(t_0) = \alpha_1$, $\dot{y}_1(t_0) = \beta_1$, $y_2(t_0) = \alpha_2$, $\dot{y}_2(t_0) = \beta_2$, $t_0 \in I$. Show that

$$| y_1(t) - y_2(t)| \leqslant |\alpha| + \frac{|\alpha| + |\beta|}{M+1} [e^{(M+1)(t-t_0)} - 1]$$

$t \in I$, where $\alpha = \alpha_1 - \alpha_2$, $\beta = \beta_1 - \beta_2$,

$$M = \max[\max_{t \in I} | a(t)|, \max_{t \in I} | b(t)|]$$

We assume, of course, that a and b are continuous on I.

3. Obtain a bound on $| \dot{y}_1(t) - \dot{y}_2(t)|$, $t \in I$, for the system of Problem 2.
4. Generalize the result of Problem 2, that is, consider two solutions, x^1 and x^2, of $\dot{x} = Ax$ for two different initial conditions, and estimate $\| x^1(t) - x^2(t) \|$ for $t \in I$. Assume that the $n \times n$ matrix A has continuous entries over I.

3.3. Fundamental Matrices

In this section we will consider the set of all possible solutions of the homogeneous system (3.5). The whole of this set can be generated simply by specifying all possible initial conditions at an arbitrary value of the time $t_0 \in I$. It will be shown that the set so constructed can be endowed with an algebraic structure, it is indeed a finite-dimensional linear space.

The main ideas leading to the development of such a structure will be shown by means of an example. Consider the constant homogeneous system

$$\dot{x}_1 = x_2$$
$$\dot{x}_2 = -2x_1 - 3x_2 \tag{3.25}$$

the set I here is the whole real line R. Let t_0 be an arbitrary value of the time. The vector-valued function of the time x^1 defined by

$$x^1(t) = \begin{bmatrix} 2e^{-(t-t_0)} - e^{-2(t-t_0)} \\ -2e^{-(t-t_0)} + 2e^{-2(t-t_0)} \end{bmatrix} \tag{3.26}$$

can be shown to be the solution of (3.25) when $x_1(t_0) = 1$ and $x_2(t_0) = 0$. (Indeed, the reader can easily derive this solution using his knowledge of elementary transform theory; this will show him that to know how to "solve" systems such as (3.25) is just a beginning, and never the end, for the actual understanding of their properties.) In the same way, the vector-valued function x^2 defined by

$$x^2(t) = \begin{bmatrix} e^{-(t-t_0)} - e^{-2(t-t_0)} \\ -e^{-(t-t_0)} + 2e^{-2(t-t_0)} \end{bmatrix} \tag{3.27}$$

is the solution of (3.25) when $x_1(t_0) = 0$ and $x_2(t_0) = 1$.

Consider now the solution of (3.25) with $x_1(t_0) = c_1$ and $x_2(t_0) = c_2$; c_1 and c_2 are two arbitrary real numbers. We claim that this solution is simply $x = c_1 x^1 + c_2 x^2$. Clearly $x(t_0) = c_1 x^1(t_0) + c_2 x^2(t_0) = [c_1 \ c_2]'$; x satisfies the conditions at t_0. Since $\dot{x}^1 = Ax^1$ and $\dot{x}^2 = Ax^2$,

$$\dot{x} = \frac{d}{dt}(c_1 x^1 + c_2 x^2) = c_1 \dot{x}^1 + c^2 \dot{x}^2 = c_1 Ax^1 + c_2 Ax^2$$

$$= A(c_1 x^1 + c_2 x^2) = Ax$$

the function x satisfies the homogeneous system and is therefore the solution of it which satisfies the conditions at t_0.

It should be clear, then, that every solution of the system (3.25) is a linear combination of the vector-valued functions x^1 and x^2; these span the set of all solutions, which consists therefore of a 2-dimensional linear space; its vectors are vector-valued functions of the time defined on all of the real line. The operations of addition and scalar multiplication are defined in this vector space as follows. Given x^1 and x^2, two vectors in it, and a_1 and a_2, two real numbers, the vector $x = a_1 x^1 + a_2 x^2$ is defined by

$$x(t) = a_1 x^1(t) + a_2 x^2(t), \qquad t \text{ real}$$

This is of course a vector-valued function of time.

This situation can be generalized to cover the set of all solutions of the homogeneous system (3.5)

THEOREM 3.3.   The set of all solutions of (3.5) forms an n-dimensional linear space over R.

Proof. If x^1 and x^2 are solutions of (3.5), then $a_1x^1 + a_2x^2$ is also a solution for any a_1, a_2 in R; hence, solutions form a linear space. Take now the standard basis in R^n, e^1, ..., e^n, and form n solutions of (3.5), x^1, ..., x^n, such that $x^i(t_0) = e^i$, $i = 1$, ..., n. Assume that these solutions are dependent, that is, that there are numbers b_1, ..., b_n in R, not all of which are zero, such that $b_1x^1 + \cdots + b_nx^n = 0$. This implies, according to the meaning given to this equality, that

$$b_1x^1(t) + \cdots + b_nx^n(t) = 0$$

for all $t \in I$. This, however, implies that

$$b_1x^1(t_0) + \cdots + b_nx^n(t_0) = b_1e^1 + \cdots + b_ne^n = 0$$

which contradicts the fact that the vectors e^i are independent. Therefore, the vectors x^1, ..., x^n are linearly independent.

We show now that the set x^1, ..., x^n spans the vector space of all solutions of (3.5). Let x be a solution such that $x(t_0) = x^0$. Since e^1, ..., e^n is a basis in R^n, we can write

$$x^0 = \sum_{i=1}^{n} c_ie^i \tag{3.28}$$

it follows that x is given by

$$x(t) = \sum_{i=1}^{n} c_ix^i(t), \qquad t \in I$$

since

i. $\sum_{i=1}^{n} c_ix^i(t_0) = \sum_{i=1}^{n} c_ie^i = x^0$;
ii. $\sum_{i=1}^{n} c_ix^i$ satisfies (3.5);
iii. The solution of (3.5) with $x(t_0) = x^0$ is unique.

We have proved, therefore, that the set $\{x^1, ..., x^n\}$ is a basis for the linear space of all solutions of (3.5). Hence, this space has dimension n.

There are, of course, sets of linearly independent vectors in this linear space other than the one described above. As a matter of fact, any such set can be constructed simply by taking any set of n linearly independent vectors in R^n as the conditions at t_0 for the n corresponding solutions of (3.5). A set of n linearly independent solutions of this system is said to be a *fundamental* set of solutions.

The set $\{x^1, x^2\}$ is a fundamental set for the system (3.25). It is very convenient to construct a matrix, to be called a *fundamental matrix* of the system (3.25), using as columns the vectors which form this fundamental set

$$X(t) = [x^1(t)\ x^2(t)]$$

$$= \begin{bmatrix} 2e^{-(t-t_0)} - e^{-2(t-t_0)} & e^{-(t-t_0)} - e^{-2(t-t_0)} \\ -2e^{-(t-t_0)} + 2e^{-2(t-t_0)} & -e^{-(t-t_0)} + 2e^{-2(t-t_0)} \end{bmatrix} \quad (3.29)$$

This matrix has several interesting properties:

i. Its determinant can be readily computed to be

$$\det X(t) = e^{-3(t-t_0)} \tag{3.30}$$

which is nonzero for all $t \in R$, so that the columns of the matrix are independent for all $t \in R$.

ii. In a sense to be defined below, this matrix satisfies an equation similar to (3.25). Indeed, if A is the matrix of coefficients of this equation, we know that

$$\dot{x}^1(t) = Ax^1(t), \qquad \dot{x}^2(t) = Ax^2(t), \qquad t \in R \tag{3.31}$$

Let us compute the product of matrices $AX(t)$.

$$AX(t) = A[x^1(t)\ x^2(t)]$$

$$= [Ax^1(t)\ Ax^2(t)] = [\dot{x}^1(t)\ \dot{x}^2(t)]$$

$$\equiv \dot{X}(t), \qquad t \in R$$

the matrix X satisfies the matrix equation $\dot{X} = AX$.

The properties of this fundamental matrix can be generalized and extended.

Definition. The matrix equation

$$\dot{X} = AX \tag{3.32}$$

is said to be the *associated matrix equation* of the system (3.5)

Let X be a fundamental matrix of (3.5), that is, a matrix whose columns are a fundamental set of solutions of this system. This matrix satisfies (3.32), since

$$AX = A[x^1 \cdots x^n] = [Ax^1 \cdots Ax^n]$$

$$= [\dot{x}^1 \cdots \dot{x}^n] = \dot{X} \tag{3.33}$$

However, not all solutions of (3.33) are fundamental matrices of (3.5).

THEOREM 3.4.   A solution X of (3.32) is a fundamental matrix of (3.5) if and only if the determinant of this matrix, det $X(t)$, is nonzero for some $t_0 \in I$. If this is the case, det $X(t)$ is nonzero for all $t \in I$.

Proof.   Let $x^1, ..., x^n$ be the columns of X. If this is a fundamental matrix, a nonzero solution x of (3.5) is a linear combination of these vectors; that is, there are *unique* scalars, not all of which are zero, such that

$$x = \sum_{i=1}^{n} a_i x^i$$

or, in matrix form

$$x = Xa \tag{3.34}$$

with a the column vector with entries $a_1, ..., a_n$. At $t = t_0$, $t_0 \in I$, this equality becomes $x(t_0) = X(t_0)\, a$. Note that this expression can be considered as a system of n linear algebraic equations in the entries of the vector a; since we know that there is a unique, nonzero solution for this vector, then the matrix of coefficients is nonsingular at t_0, that is, det $X(t_0)$ is nonzero. Since t_0 is any arbitrary element of I, we conclude that the matrix $X(t)$ is nonsingular for all $t \in I$, det $X(t) \neq 0$, $t \in I$.

Let now det $X(t_0) \neq 0$ for some $t_0 \in I$. Then, if x is a solution, $x(t_0) = X(t_0)a$ has a nontrivial solution for the vector a, and x, the solution we started with, equals Xa because of uniqueness. The columns of X are therefore a basis for the space of all solutions and the matrix is a fundamental matrix. By the necessity of the condition, det $X(t)$ is nonzero for all $t \in I$. The theorem follows.

It is worth noting that this theorem establishes that the vectors $x^1(t), ..., x^n(t)$, are independent, as elements of the space R^n, for all $t \in I$. In Theorem 3.3 we proved only that there is at least one value of $t \in I$ for which these vectors are independent. A clear picture emerges from this theorem. In order to characterize the space of all solutions of the homogeneous system (3.5), it is necessary to know only n linearly independent solutions; these are the columns of a fundamental matrix of the system, which is a solution of the associated matrix equation (3.32) such that, for some arbitrary $t_0 \in I$, $X(t_0)$ is nonsingular. The solution of the associated matrix equation under this condition characterizes completely the solutions of (3.5).

There are advantages in making $X(t_0)$ the $n \times n$ identity matrix. As in the case of the system (3.25), the solution of the system (3.5) with $x(t_0) = x^0$ can be written explicitly as

$$x(t) = X(t)\, x(t_0) \tag{3.35}$$

where $X(t)$ is here the solution of Equation (3.32) with $X(t_0) = I_n$; I_n is as usual the $n \times n$ identity matrix. This matrix solution of (3.32) is so important that it deserves a special name—*transition matrix*—and a special section for its study, the next one.

Exercises

*1. Show that the determinant of any solution of the associated matrix equation (3.32) is

$$\det X(t) = [\det X(t_1)] \exp \int_{t_1}^{t} \operatorname{tr} A(\lambda) \, d\lambda$$

(Hint: Develop first a formula for the derivative of a determinant with entries which are functions of the time, and then show that

$$\frac{d}{dt} [\det X(t)] = [\det X(t)] \operatorname{tr} A(t)$$

2. Relate the result of Problem 1 with Theorem 3.4. Note that the entries of A are continuous functions of the time.

*3. Consider an nth order differential equation

$$y^{(n)} + a_1 y^{(n-1)} + \cdots + a_n y = 0$$

By transforming this equation to a system of n first-order equations, show that $\det X(t)$ is in this case given by

$$\det X(t) = [\det X(t_1)] \exp \int_{t_1}^{t} a_1(\lambda) \, d\lambda$$

4. For a differential equation of the form

$$a_0(t) y^{(n)}(t) + a_2(t) y^{(n-2)}(t) + \cdots + a_n(t) y(t) = 0$$

in which there is no derivative term of order $(n - 1)$, show that $\det X(t)$ is constant.

5. Show that a system such as (3.5) cannot have more than n independent solutions.

6. Show that if X is a fundamental matrix for the system (3.5), XC, where C is a constant nonsingular $n \times n$ matrix, is also a fundamental matrix for the system (3.5).

3.4. The Transition Matrix

The transition matrix has the property of generating any solution of the system (3.5) by means of expressions such as (3.35). A word now concerning notation. This matrix depends on both t_0 and t, as in (3.29). In our terminology, it is a matrix-valued function with domain on $R \times R$. However, following the notation used in most textbooks, we will ignore for once the difference between a function and its value, and write $\Phi(t, t_0)$ for the transition matrix of a system such as (3.5). Note that this is a function of two real variables, t and t_0 .

We can write (3.35) as

$$x(t) = \Phi(t, t_0) x(t_0), \qquad t, t_0 \in I \tag{3.36}$$

This is an explicit expression for the solution of the system (3.5). It is not intended for computation. Rather it expresses in an explicit form the structure of the solution; it indicates that the value of the vector-valued function x, the solution of (3.5), is obtained from its value at another instant of time by operating on it with a matrix which depends on both values of the time. This matrix governs the transition of the values of x from one instant of time to another.

The importance of this concept is that by the knowledge of properties of the transition matrix we can find new characteristics of the solution of the homogeneous system (and, as we shall see later, of the inhomogeneous system as well). We derive now the properties of the transition matrix which are useful in this endeavor.

Consider a situation such as the one shown in Figure 3.1. The state at

FIGURE 3.1.   Two-dimensional representation of a trajectory of a homogeneous system, showing the progression of the state vector, $x(t_0) \to x(t_1) \to x(t_2)$.

t_1 is related to the state at t_0 by

$$x(t_1) = \Phi(t_1, t_0) x(t_0) \tag{3.37}$$

while the state at t_2 is related to the state at t_1 by

$$x(t_2) = \Phi(t_2, t_1) x(t_1) \tag{3.38}$$

We can, however, ignore the passage of the state through the intermediate point t_1, and write simply

$$x(t_2) = \Phi(t_2, t_0) x(t_0) \tag{3.39}$$

It follows, hence, from (3.37), (3.38), and (3.39) that, since $x(t_0)$ is arbitrary,

$$\Phi(t_2, t_0) = \Phi(t_2, t_1) \Phi(t_1, t_0) \tag{3.40}$$

This is a fundamental property of the transition matrix, known as the *group property*. We will not go into the reasons for this name at this time, since they will be explored at length in Chapter 9. It so happens that this property is so crucial for the understanding of the behavior of the solutions of linear systems such as (3.3) and for the derivation of new properties of these systems, that it will be taken as a model, suitably modified, for the behavior of more complex systems.

Note that there is no reason why t_0, t_1, and t_2 should be such that $t_0 \leqslant t_1 \leqslant t_2$, as suggested by Figure 3.1. For instance, t_1 could be less than t_0, and this less than t_2. We have, so far, no preferred direction for the flow of time. In this connection, we must note an important consequence of Equation (3.40). Put $t_2 = t_0$ there. Then

$$\Phi(t_2, t_2) = \Phi(t_2, t_1)\, \Phi(t_1, t_2)$$

It follows from (3.36) that $\Phi(t, t) = I_n$, the $n \times n$ identity matrix, so that we have, for all t and τ in I,

$$\Phi(t, \tau)\, \Phi(\tau, t) = I_n \qquad (3.41)$$

This equation says that a transition matrix is invertible, and that the inverse of $\Phi(t, \tau)$ is $\Phi(\tau, t)$. Of importance is not only the fact that the inverse of the transition matrix can be computed by switching the roles of t and τ, but simply that this inverse exists: A transition matrix is nonsingular for all t and τ in I. This follows also from Theorem 3.4.

Of course, we could have derived the property (3.41) directly from (3.36); indeed,

$$x(t_0) = \Phi(t_0, t)\, x(t)$$

remember that (3.36) does not presuppose any ordering in the numbers t_0 and t. It is clear that from an instant t it is possible "to go back in time," in the sense that knowledge of the value of x at t, $x(t)$, and of the transition matrix, are sufficient for the determination of the state at any time prior to t.

A crucial point in the development of these ideas is that it is possible to obtain many results and properties concerning the solution of the homogeneous system (3.5) just by knowing the form of Equation (3.36) and some properties of the transition matrix, such as the group property, without the need for computing this matrix. We illustrate this now; many other examples of this situation will appear in the following chapters.

Let us consider an important engineering aspect of the solutions of the homogeneous system. Suppose that a solution of this system depends on various parameters. For instance, the coefficients of the matrix A may depend on a parameter such as a resistance or inductance, as in Section 1.2, which would of course make the solutions dependent on these same parameters. The vector $x(t_0)$, specified as the value of the solution at t_0, can be considered as a set of n parameters. What happens if this vector changes slightly? Do momentous things happen to the solution? Or does it just experience a slight change? If a resistor is varied slightly in a network such as the one in Section 1.2, what happens to the solution of the corresponding homogeneous system?

We will answer here the question involving the behavior of the solution of (3.5) when the condition at t_0 experiences a slight change. Assume that we examine at $t \in I$ two different solutions of (3.5), x and y, resulting from two different values at t^0, x^0 and y^0. Of course,

$$x(t) = \Phi(t, t_0) \, x^0, \qquad y(t) = \Phi(t, t_0) \, y^0 \qquad (3.42)$$

so that

$$x(t) - y(t) = \Phi(t, t_0)(x^0 - y^0) \qquad (3.43)$$

and therefore

$$\| x(t) - y(t) \| \leqslant \| \Phi(t, t_0) \| \, \| x^0 - y^0 \| \qquad (3.44)$$

This expression is sufficient to provide an answer to our query. The norm of the difference between the values of the solutions at a time $t \in I$ can be kept as small as desired by choosing x^0 and y^0 sufficiently close, that is, by making $\| x^0 - y^0 \|$ small enough; that is, given $\epsilon > 0$, a number $\delta(\epsilon, t)$ can be found such that $\| x(t) - y(t) \| < \epsilon$ if $\| x^0 - y^0 \| < \delta(\epsilon, t)$. We say that the solution of the homogeneous system is a *continuous* function of the conditions at some time t_0. Nothing much happens to the solution if the conditions at t_0, $x(t_0)$, are changed slightly.

This result can be improved by invoking a new property of the transition matrix. Remember that its columns are actually solutions of the system (3.5); the entries of the matrix are thus differentiable in t, and hence continuous in t. Over the finite interval I, the norm of $\Phi(t, t_0)$ is therefore bounded, by M say, so that (3.44) becomes

$$\| x(t) - y(t) \| \leqslant M \| x^0 - y^0 \|, \qquad t \in I \qquad (3.45)$$

the number $\delta(\epsilon, t)$ can be chosen therefore independently of t. The solution of the system (3.5) is said to be *uniformly* continuous with respect to t over the interval I.

We pass now to the derivation of an explicit form for the transition matrix of a constant system. It should be pointed out again that an explicit form such as this is not generally suitable to compute with; its main usefulness resides in its structure, by means of which new properties can be proved concerning, in this case, the solution of the homogeneous system. Other explicit forms will be derived in Chapter 4.

As we already know, the associated matrix equation of system (3.5) is

$$\dot{X} = AX \tag{3.46}$$

Here A is a constant $n \times n$ matrix. We are interested in a solution of this equation which equals I_n, the $n \times n$ identity matrix, at a fixed but arbitrary t_0.

It is well known that the solution of the scalar equation $\dot{x} = ax$, $x(t_0) = 1$, is

$$x(t, t_0) = e^{(t-t_0)a}, \qquad t \in R \tag{3.47}$$

By analogy, we might expect the solution of (3.46) to be

$$\Phi(t, t_0) = e^{(t-t_0)A}, \qquad t \in R \tag{3.48}$$

the matrix exponential would have to be suitably defined.

Indeed, let us apply the iterative method studied above to each column of the matrix X. A sequence of matrices is defined in this way:

$$X^0(t) = I_n$$

$$X^1(t) = I_n + \int_{t_0}^{t} AI_n \, ds = I_n + (t - t_0)A$$

$$X^2(t) = I_n + \int_{t_0}^{t} A(I_n + (t - t_0)A) \, ds$$

$$= I_n + \frac{1}{1!} (t - t_0)A + \frac{1}{2!} (t - t_0)^2 A^2 \tag{3.49}$$

$$\vdots$$

$$X^k(t) = I_n + \frac{1}{1!} (t - t_0)A + \frac{1}{2!} (t - t_0)^2 A^2 + \cdots + \frac{1}{k!} (t - t_0)^k A^k$$

It was proved above that this sequence [or rather, the sequences formed by the n columns of the matrices $X^0(t)$, ..., $X^k(t)$, ...] converges, in this case for all $t \in R$, and that its limit, to be denoted by $e^{(t-t_0)A}$, satisfies Equation (3.46) and the condition $X(t_0) = I_n$. This limit is, therefore, the transition matrix of the homogeneous, constant system.

An interesting consequence is that the matrix exponential satisfies the relation

$$\frac{d}{dt} (e^{tA}) = Ae^{tA}$$

Observe that when A is a constant matrix the transition matrix depends only on the difference $(t - t_0)$. We shall write $T(t - t_0)$ for this matrix; that is,

$$T(t) = e^{tA} \tag{3.50}$$

The matrix exponential has several properties which are of importance. The group property can be put as follows:

$$T(t - \tau) = T(t - \lambda) \, T(\lambda - \tau) \tag{3.51}$$

or

$$e^{(t-\tau)A} = e^{(t-\lambda)A} e^{(\lambda-\tau)A}, \qquad t, \lambda \in R$$

from which it follows by putting $\lambda = 0$ and replacing τ by $-\tau$, that

$$e^{(t+\tau)A} = e^{tA} e^{\tau A} \tag{3.52}$$

In a similar way,

$$T^{-1}(t - \tau) = T(\tau - t)$$

$$T^{-1}(t) = T(-t)$$

that is,

$$(e^{At})^{-1} = e^{-At} \tag{3.53}$$

An interesting formula exists for the exponential of a matrix whose components are *square matrices* placed in the diagonal. If A is

$$A = \begin{bmatrix} A_1 & & & \\ & A_2 & & \\ & & \ddots & \\ & & & A_m \end{bmatrix}$$

with A_1, ..., A_m, square matrices of various sizes, then (note that all entries in the matrix above which have not been filled are assumed to be equal to zero)

$$e^A = \begin{bmatrix} e^{A_1} & & & \\ & e^{A_2} & & \\ & & \ddots & \\ & & & e^{A_m} \end{bmatrix} \tag{3.54}$$

as can easily seen by applying the rules learned in Chapter 2 for the manipulation of matrices whose entries are other matrices and the definition of the exponential matrix derived above.

If the coefficients of the matrix A are explicit functions of the time, one may expect that the transition matrix would be given by

$$\Phi(t, t_0) = \exp \int_{t_0}^{t} A(\lambda) \, d\lambda \qquad (*)$$

this is not the case, however; the same method as above can be applied here, and the resulting series

$$\Phi(t, t_0) = I_n + \int_{t_0}^{t} A(\lambda) \, d\lambda + \int_{t_0}^{t} A(\lambda) \left[\int_{t_0}^{\lambda} A(\mu) \, d\mu \right] d\lambda + \cdots \qquad (3.55)$$

becomes identical with $(*)$ if and only if $A(t)$ and $\int_{t_0}^{t} A(\lambda) \, d\lambda$ commute. Of course, if A is a constant, this series becomes identical with the series for $e^{(t-t_0)A}$.

Exercises

*1. Show that if X is any fundamental matrix of the homogeneous system (3.5), then the transition matrix of this system is

$$\Phi(t, \lambda) = X(t) \, X^{-1}(\lambda)$$

2. If $\Phi(t, \lambda)$ is the transition matrix of $\dot{x}(t) = A(t) \, x(t)$, show that

$$A(t) = \frac{\partial \Phi(t, \lambda)}{\partial t} \bigg|_{\lambda=t}$$

3. Show that

$$e^{(A+B)t} = e^{At} e^{Bt}$$

for all t if and only if A and B commute.

4. Find a solution for

$$\dot{X} = AX + XB$$

X, A, and B are $n \times n$ matrices, A and B are constant.

5. Let the matrix A in (3.5) be skew-symmetric, that is, assume that it satisfies $A' = -A$. Show that under this condition the norm of the solution of (3.5) is constant for all t.

3.5. The Adjoint System

The adjoint system of the homogeneous system $\dot{x} = Ax$ is defined to be the homogeneous system

$$\dot{x} = -A'x \qquad (3.56)$$

We prove the following theorem.

THEOREM 3.5. If $\Phi(t, t_0)$ is the transition matrix for $\dot{x} = Ax$, then $(\Phi'(t, t_0))^{-1}$ is the transition matrix for its adjoint.

Proof. We prove that $(\Phi'(t, t_0))^{-1}$ satisfies the associated matrix equation of $\dot{x} = -A'x$, that is $\dot{X} = -A'X$. Indeed,

$$\frac{d}{dt}(\Phi\Phi^{-1}) = 0 = \left(\frac{d}{dt}\Phi\right)\Phi^{-1} + \Phi\left(\frac{d}{dt}\Phi^{-1}\right)$$

$$= (A\Phi)\Phi^{-1} + \Phi\frac{d}{dt}\Phi^{-1} = A + \Phi\frac{d}{dt}\Phi^{-1}$$

that is,

$$-A = \Phi\frac{d}{dt}\Phi^{-1}$$

or

$$\frac{d}{dt}\Phi^{-1} = -\Phi^{-1}A$$

Transposing both sides of this equation, we obtain

$$\frac{d}{dt}(\Phi')^{-1} = -(\Phi^{-1}A)' = -A'(\Phi')^{-1}$$

where we have used the fact that

$$\left(\frac{d}{dt}\Phi^{-1}\right)' = \frac{d}{dt}(\Phi')^{-1}$$

Adjoint systems are useful for reducing the order of homogeneous systems of which some solutions are known [2]. They also occur quite naturally in connection with many control problems. Our interest, however, will be centered in the use of this concept for deriving an explicit expression for the solution of nonhomogeneous systems of equations.

Exercises

1. Let A be a constant matrix, b and c two n-vectors. Show that if the solution of $\dot{x} = Ax$ is $e^{\lambda t}b$, and if the solution of $\dot{x} = -A'x$ is $e^{-\mu t}c$, then if $\lambda \neq \mu$, $c'b = 0$. What can you say about the n-vectors b and c?

2. Find the adjoint system of the system corresponding to the nth order differential equation

$$y^{(n)}(t) + a_1(t)y^{(n-1)}(t) + \cdots + a_n(t)y(t) = 0$$

and reduce this adjoint system to an nth order differential equation.

3. Carry on the process found in Problem 2 for the system

$$\frac{d}{dt}\left(P(t)\frac{d}{dt}y(t)\right) + Q(t)\,y(t) = 0$$

Do you notice anything special in the resulting second-order equation? This equation is said to be *self-adjoint*.

3.6. The General Solution of Nonhomogeneous Systems

We consider now the whole of the first of equations (3.4), that is,

$$\dot{x} = Ax + Bu \qquad\qquad (3.57)$$

The entries of A and B are continuous functions of the time for $t \in I = [t_a, t_b]$. The control function u in (3.57) is defined on a half-open interval $[t_0, t_1) \subset I$, $t_0 < t_1$. The objective of this section is to derive an explicit expression for the solution x of (3.57) which results from a specified value of $x(t_0)$ and a control function u defined on $[t_0, t_1)$. Of course, the solution will be found for all $t \in [t_0, t_1)$.

It is interesting to note that in this problem time has taken at last a preferred direction; we want to determine the effect of $x(t_0)$ and u on the value of x *at later times*. The reason for this choice is not mathematical, since we could as well have made, from a strictly mathematical standpoint, the opposite one, but deeply physical: A later input cannot influence the value of the solution at an earlier time. This will be clarified below, and a rather interesting conclusion obtained.

Let $\varPhi(t, t_0)$ be the transition matrix of the system $\dot{x} = Ax$, and denote the transition matrix of (3.56) by $\varPsi(t, t_0)$; by Theorem 3.5,

$$\varPsi(t, t_0) = (\varPhi'(t, t_0))^{-1}$$

it follows that $\varPsi'(t, t_0) = \varPhi(t, t_0)^{-1}$. Since $\varPsi(t, t_0)$ satisfies the equation

$$\dot{X} = -A'X$$

$\varPsi'(t, t_0)$ satisfies

$$\varPsi' = -\varPsi'A \qquad\qquad (3.58)$$

as can easily be proved by taking tansposes of both sides of the equation $\dot{\varPsi} = -A'\varPsi$.

Consider now the expression

$$\frac{d}{dt}(\varPsi'x) = \dot{\varPsi}'x + \varPsi'\dot{x} = -\varPsi'Ax + \varPsi'(Ax + Bu)$$
$$= -\varPsi'Ax + \varPsi'Ax + \varPsi'Bu = \varPsi'Bu$$

Hence, integrating from t_0 to t

$$\Psi'(t, t_0)\, x(t) = \Psi'(t_0, t_0)\, x(t_0) + \int_{t_0}^{t} \Psi'(\lambda, t_0)\, B(\lambda)\, u(\lambda)\, d\lambda$$

Clearly,

$$x(t) = (\Psi'(t, t_0))^{-1} \Psi'(t_0, t_0)\, x(t_0)$$

$$+ (\Psi'(t, t_0))^{-1} \int_{t_0}^{t} \Psi'(\lambda, t_0)\, B(\lambda)\, u(\lambda)\, d\lambda$$

and since $(\Psi'(t, t_0))^{-1} = \Phi(t, t_0)$

$$x(t) = \Phi(t, t_0)\, \Phi^{-1}(t_0, t_0)\, x(t_0)$$

$$+ \Phi(t, t_0) \int_{t_0}^{t} \Phi^{-1}(\lambda, t_0)\, B(\lambda)\, u(\lambda)\, d\lambda$$

At last, since $\Phi^{-1}(t, \tau) = \Phi(\tau, t)$, and since $\Phi(t_0, t_0) = I_n$,

$$x(t) = \Phi(t, t_0)\, x(t_0) + \int_{t_0}^{t} \Phi(t, \lambda)\, B(\lambda)\, u(\lambda)\, d\lambda, \qquad t \in [t_0, t_1) \qquad (3.59)$$

Note that the derivation of the expression (3.59) has been purely formal, since no conditions have as yet been put on the problem so as to ensure the existence of the integral in its right-hand side. As a matter of fact, this existence condition will serve as a defining postulate for the class of admissible control functions u. Before proceeding any further, therefore, we should define the sense in which the existence of integrals is to be taken. All integrals in this text, other than in some cases in Chapter 9, are to be taken as Riemann integrals. The class of admissible controls for the system (3.4) is therefore the class of vector-valued functions defined on $[t_0, t_1)$, such that the integral in (3.59) exists in a Riemann sense. Since Φ and B are continuous over the interval $[t_0, t_1)$, this class includes vector-valued functions with components which are:

i. Continuous functions over $[t_0, t_1)$.

ii. Piecewise continuous functions over $[t_0, t_1)$, that is, functions which are continuous over open subintervals of $[t_0, t_1)$ but which may be discontinuous at the boundaries between such subintervals; only a finite number of such discontinuities is allowed, and the magnitudes of the jumps are to stay finite. This class includes of course all continuous functions.

iii. Functions which are continuous "almost everywhere" on $[t_0, t_1)$. The most important subclass of functions which are continuous almost everywhere is that of piecewise continuous functions.

The whole subject of how adequate the Riemann integral actually is for supplying explicit expressions for the solutions of linear systems depends really on how large the class of allowable inputs is under this type of integral; a more general class of integrals would allow a wider class of inputs. It is unfortunate that we cannot go deeper into these matters.

By examining Equation (3.59), we can see that time has indeed taken a preferred direction. The value of x at t, $t \in [t_0, t_1)$, depends on the value of x at t_0, $x(t_0)$, and on all the values of u over $[t_0, t)$. If we were to take $t < t_0$, the value of x at t would depend on the values of the input at the future values of the time from t to t_0. This is not a situation which we can admit physically. Of course, Equation (3.59) can be used for calculating backward the value of $x(t_0)$ which gives rise to $x(t)$, $t \in [t_0, t_1)$, given u over this interval.

Note that a consequence of this choice is that in the integral of Equation (3.59), λ, which lies in the set $[t_0, t)$, is of course always *less than* t; it follows that the only values of $\Phi(t, \lambda)$ which are of any importance in this integral are those defined on that subset of its domain R^2 for which $t \geqslant \lambda$. It can be concluded that two systems whose transition matrices coincide for $t \geqslant \lambda$ have identical responses when $x(t_0) = 0$; they are said to be *zero-state equivalent*. This fact is of importance in the *realization problem*, which is the subject matter of Chapter 6.

A warning now: Expression (3.59) is not to compute with; no competent numerical analyst would disgrace his computer program by using it to estimate values of the solution of (3.57); such a computation would take much too long and it would be very wasteful of memory space. The value of the expression (3.59) lies as in the case of the corresponding expression for the homogeneous system, in the structure it gives to the state vector in terms of functions, such as the transition matrix, whose properties, such as the group property, are helpful in establishing new properties of the solution of the nonhomogeneous system. We will do exactly this in Chapters 5, 6, and indeed in most of the text.

The following example clarifies the meaning of Equation (3.59).

EXAMPLE

Consider the system of equations

$$\dot{x}_1 = x_2 + u_1, \qquad \dot{x}_2 = -2x_1 - 3x_2 + u_2, \qquad x(t_0) = x^0$$

The transition matrix of this system is given by (3.29). If u_1 is identically zero for all $t \in I$ and u_2 is defined by

$$u_2(t) \begin{array}{ll} = 1 & t_0 \leqslant t < 1 + t_0 \\ = -1 & t \geqslant 1 + t_0, \end{array} \qquad t_0 \in R$$

then

$$\begin{bmatrix} x_1(t) \\ x_2(t) \end{bmatrix} = \begin{bmatrix} 2e^{-(t-t_0)} - e^{-2(t-t_0)} & e^{-(t-t_0)} - e^{-2(t-t_0)} \\ -2e^{-(t-t_0)} + 2e^{-2(t-t_0)} & -e^{-(t-t_0)} + 2e^{-2(t-t_0)} \end{bmatrix} \begin{bmatrix} x_1^0 \\ x_2^0 \end{bmatrix}$$

$$+ \int_{t_0}^{t} \begin{bmatrix} 2e^{-(t-\lambda)} - e^{-2(t-\lambda)} & e^{-(t-\lambda)} - e^{-2(t-\lambda)} \\ -2e^{-(t-\lambda)} + 2e^{-2(t-\lambda)} & -e^{-(t-\lambda)} + 2e^{-2(t-\lambda)} \end{bmatrix} \begin{bmatrix} 1 & 0 \\ 0 & 1 \end{bmatrix} \begin{bmatrix} u_1(\lambda) \\ u_2(\lambda) \end{bmatrix} d\lambda$$

that is,

$$x_1(t) = (2e^{-(t-t_0)} - e^{-2(t-t_0)}) x_1^0 + (e^{-(t-t_0)} - e^{-2(t-t_0)}) x_2^0$$

$$+ \int_{t_0}^{t} (e^{-(t-\lambda)} - e^{-2(t-\lambda)}) u_2(\lambda) d\lambda$$

$$x_2(t) = (-2e^{-(t-t_0)} + 2e^{-2(t-t_0)}) x_1^0 + (-e^{-(t-t_0)} + 2e^{-2(t-t_0)}) x_2^0$$

$$+ \int_{t_0}^{t} (-e^{-(t-\lambda)} + 2e^{-2(t-\lambda)}) u_2(\lambda) d\lambda$$

If $t \leqslant t_0 + 1$,

$$x_1(t) = (2e^{-(t-t_0)} - e^{-2(t-t_0)}) x_1^0 + (e^{-(t-t_0)} - e^{-2(t-t_0)}) x_2^0$$

$$+ \tfrac{1}{2}(-e^{-(t-t_0)}) + \tfrac{1}{2}e^{-2(t-t_0)}$$

with a similar expression holding for $x_2(t)$. The best way to compute $x_1(t)$ for $t > 1 + t_0$ is:

a. To compute $x_1(t_0 + 1)$ and $x_2(t_0 + 1)$

b. To use the formula (3.59) with $t_0 + 1$ instead of t_0, and $x_1(t_0 + 1)$ and $x_2(t_0 + 1)$ instead of x_1^0 and x_2^0 respectively.

The algebra is left to the reader. This is an illustration of the fact that t_0 is actually any instant of time at which the value of x, $x(t_0)$, is known.

The solution (3.59) of system (3.57) can be proved to be unique in a simple way. Let x^1 and x^2 be two solutions of (3.57) with the same input u and satisfying $x(t_0) = x^0$. Clearly,

$$x^1 - x^2 = A(x^1 - x^2)$$

and $x^1(t_0) - x^2(t_0) = 0$. Hence, $x^1(t) - x^2(t) = 0$ for all $t \in I$, since the only solution of $\dot{z} = Az$, $z(t_0) = 0$, is the zero vector.

It will be necessary sometimes to write the solution (3.59) as

$$x(t; t_0, x^0; u)$$

When there is no possibility of misunderstanding the input function, we will write

$$x(t; t_0, x^0)$$

The interval of definition of the control u, and then of the solution x, will not be included in this notation. In the same way, we write an expression for the values of the output y of the system as

$$y(t; t_0, x^0; u) = C(t) \, \Phi(t, t_0) \, x^0$$

$$+ \, C(t) \int_{t_0}^{t} \Phi(t, \lambda) \, B(\lambda) \, u(\lambda) \, d\lambda + D(t) \, u(t)$$

$$t \in [t_0, t_1) \subset I \qquad\qquad (3.60)$$

Exercises

1. Derive the explicit solution (3.59) by the method of *undetermined coefficients*, that is, assume that Φc is a solution of the nonhomogeneous system (3.57), and derive which properties c, if it exists, is to have.

2. Suppose $x(t_0) = 0$. It is known that the components of the input are bounded above and below by the numbers α and β, $\alpha > \beta$. Can you obtain bounds on the state vector at $t > t_0$?

3. Suppose $x(t_1; t_0, x^0; u^1) = x^1$. Does this imply $x(t_1; t_0, x^1; (-u^1)) = x^0$?

4. Consider the system

$$\dot{x}_1 = x_2 + u, \qquad \dot{x}_2 = x_1 + x_2$$

Find a control u^1 of the form

$$u^1(t) = a \sin 2t, \qquad t \geqslant 0$$

such that $x(t_1; 0, x^0; u^1) = x^1$, where x^0 is a fixed vector in R^2, t_1 is an unspecified value of the time, x^1 is a vector in R^2 specified only by the condition that its entry $x_2^1 = 0$. Under what conditions on a, x^0, and t_1 is this possible?

3.7. A Summary and Some Conclusions

It is our intention to summarize in this section the properties of the solutions of the system (3.4), in such a way that these properties can be used directly as models for the definition of the more general entities known as dynamical systems.

The following elements are of importance in the construction of solutions of the system (3.4):

i. The *state space* R^n, where the solutions of the first of Equation (3.4) take values.

ii. A subset $I = [t_a, t_b]$ of the space of the reals whose elements are referred to as the values of *time*.

iii. A space U of functions with domains consisting of half-open intervals lying in I and range in the space R^r. These are r-vector-valued control functions with entries which are either continuous, piecewise continuous or continuous almost everywhere.

iv. A space Y of outputs of the system.

v. A *transition function* $x(\cdot, \cdot, \cdot, \cdot)^\dagger$ with domain the product space $I \times I \times R^n \times U$ and with range in R^n, which defines the state $x(t; t_0, x^0; u)$ for any initial time $t_0 \in I$, any initial state $x^0 \in R^n$ and any input u defined on $[t_0, t)$, $t \in I$. This function is defined by:

$$x(t; t_0, x^0; u) = \Phi(t, t_0) x^0 + \int_{t_0}^t \Phi(t, \lambda) B(\lambda) u(\lambda) \, d\lambda \qquad (3.61)$$

vi. An *output function* $y(\cdot, \cdot, \cdot)$ with domain the product space $I \times R^n \times U$ which assigns to each $t \in I$, each $x \in R^n$ and each $u \in U$ an output

$$y(t, x, u) = C(t)x + D(t) u(t)$$

The transition function has the following properties:

i. $x(t_0; t_0, x^0; u) = x^0$ for all $u \in U$, $t_0 \in I$, $x^0 \in R^n$.

ii. Consider now three instants of time, t_0, t_1, and t_2 such that $t_2 \geq t_1 \geq t_0$. If two inputs u^1 and u^2 are defined on $[t_0, t_1)$ and $[t_1, t_2)$ respectively, we define u as

$$u(t) = \begin{array}{ll} u^1(t), & t \in [t_0, t_1) \\ u^2(t), & t \in [t_1, t_2) \end{array}$$

Then,

$$x(t_2; t_0, x^0; u) = \Phi(t_2, t_0) x^0 + \int_{t_0}^{t_2} \Phi(t_2, \lambda) B(\lambda) u(\lambda) \, d\lambda$$

$$= \Phi(t_2, t_1) \Phi(t_1, t_0) x^0 + \int_{t_0}^{t_1} \Phi(t_2, \lambda) B(\lambda) u^1(\lambda) \, d\lambda$$

$$+ \int_{t_1}^{t_2} \Phi(t_2, \lambda) B(\lambda) u^2(\lambda) \, d\lambda$$

† The notation $x(\cdot, \cdot, \cdot, \cdot)$ for the function $x: I \times I \times R^n \times U \to R^n$ is used so as to make clear that this is a function of four variables. This notation was introduced in Section 2.3.

$$= \varPhi(t_2, t_1) \left\{ \varPhi(t_1, t_0) x^0 + \int_{t_0}^{t_1} \varPhi(t_1, \lambda) B(\lambda) u^1(\lambda) d\lambda \right\}$$

$$+ \int_{t_1}^{t_2} \varPhi(t_2, \lambda) B(\lambda) u^2(\lambda) d\lambda$$

$$= x(t_2; t_1, x(t_1; t_0, x^0; u^1); u^2) \tag{3.62}$$

The reader should go carefully through this development, which, though simple, illustrates well several points concerning the algebra of these expressions; we will use this algebra very often in Chapter 5.

Note that Equation (3.62) is a consequence of the group property of the transition matrix as well as of the additivity of the integral.

iii. The following property will be mentioned here for the sake of completeness; the main ideas involved, which are important, will be developed fully in Chapter 9. It happens that the transition function $x(\cdot, \cdot, \cdot, \cdot)$ is a continuous function of its arguments. In order to follow this up properly, we must define continuity in a suitable way; this will be done in Chapter 9.

The output function $y(\cdot, \cdot, \cdot, \cdot)$ is also a continuous function of its arguments.

Another property of interest of the transition function is that it is linear with respect to x^0 when u is zero and linear with respect to u when x^0 is zero. This function satisfies, for all α, β in R,

$$x(t; t_0, \alpha x^{01} + \beta x^{02}; \alpha u^1 + \beta u^2)$$

$$= \alpha x(t; t_0, x^{01}; u^1) + \beta x(t; t_0, x^{02}; u^2) \tag{3.63}$$

the proof is left to the reader. Because of this linearity, the function $x(\cdot, t_0, x^0; u)$ can be separated into the sum of two well-identified parts, the *state zero-input response* and the *state zero-state response*, equal to

$$\phi(\cdot, t_0) x^0 \qquad \text{and} \qquad \int_{t_0}^{\cdot} \phi(\cdot, \lambda) B(\lambda) u(\lambda) d\lambda$$

respectively. The output $y(\cdot, t_0, x^0; u)$, defined by Equation (3.60), can be separated also in the sum of a term which is zero when x^0 is zero, and a term which is zero when u is zero (that is, when $u(\lambda) = 0$, $\lambda \in [t_0, t)$). The first term, $C(\cdot) \phi(\cdot, t_0) x^0$, is said to be the *zero-input* response of the system, while the second is said to be the *zero-state* response.

At last, we note that if the matrices A, B, C, and D are constant, then a shift in the time origin changes neither the values of the state nor of the output, provided that the input is shifted accordingly. The system in this case said to be *time-invariant* or *constant*.

SUPPLEMENTARY NOTES AND REFERENCES

§3.1. The student desiring to learn about differential equations should read, besides the books listed in the references, the following:

S. Lefschetz, "Differential Equations: Geometric Theory." Wiley (Interscience), New York, 1963.
E. A. Coddington and N. Levinson, "Theory of Ordinary Differential Equations." McGraw-Hill, New York, 1955.

An interesting monograph on linear differential equations is:

N. P. Erugin, "Linear Systems of Ordinary Differential Equations." Academic Press, New York, 1966.

§3.2. The homogeneous system $\dot{x}(t) = A(t)\,x(t)$ has a solution even if the entries of A are not continuous; the condition given in the text is only sufficient. If the components of A are of *integrable square* (that is, the integrals over I of their squares are finite), then the system has a solution, even though it may not actually satisfy the homogeneous equation at all times, but only at most of them, that is, "almost everywhere" on I.

§3.6. Of course, impulse functions are also acceptable as inputs. The resulting solutions, however, are not continuous in t, a property which we want our solutions to have. Otherwise, how could they be differentiable, and then satisfy (3.4) for all t? The answer, of course, lies in the theory of distributions, with which we will not be concerned in this text. The reader is referred to the text by Zemanian mentioned in Chapter I.

We would like to repeat that a solution such as (3.59) of the Equation (3.57) exists for a wider class of input functions than the ones described in the text, provided that the concept of integral is broadened correspondingly.

The reader should be aware that this is not simply an academic matter. It so happens that in many control problems one is forced to include in the set of allowable inputs vectors with components which are not Riemann integrable (but are Lebesgue integrable). For an example of this situation, see

L. S. Pontryagin, V. G. Boltyanskii, R. V. Gamkrelizde, and E. F. Mishchenko, "The Mathematical Theory of Optimal Processes." Wiley (Interscience), New York, 1962, Chapter III.

REFERENCES

1. R. Bellman, "Introduction to Matrix Analysis." McGraw-Hill, New York, 1970.
2. H. Hochstadt, "Differential Equations: A Modern Approach." Holt, New York, 1964.
3. A. E. Taylor, "Advanced Calculus." Ginn, Boston, Massachusetts, 1955.

4

Differential Systems II

4.1. Introduction

The matrix representation of the system of equations (3.2) used in the last chapter is just one of the possible descriptions of this system; it should be remembered that it was obtained by choosing the standard basis as a basis in R^n. It is conceivable that a matrix description of the operations involved in (3.2) relative to other bases may prove more amenable to treatment. After all, the standard basis, in spite of its simple form, is in no way related to the structure of the operations; perhaps a basis which depends on the system itself would prove of help. This is indeed the case.

We will consider in this chapter only constant systems. Let S be a nonsingular constant, $n \times n$ matrix which, as we will see later, in general has complex entries. Define a new state vector z by

$$z = S^{-1}x \qquad (4.1)$$

Note that, since S is complex, S^{-1} is also complex and z may have complex entries, so the new state space is C^n, the space of n-tuples with complex entries. The transformation (4.1) is equivalent to changing the basis with respect to which the operations on (3.2) are described; this aspect of the transformation will be treated in the next section.

The system (3.4) is transformed by this change of basis into the following

$$\dot{z} = S^{-1}ASz + S^{-1}Bu$$

$$y = CSz + Du \qquad (4.2)$$

The matrices A and $S^{-1}AS$ are said to be *similar*. The new system (4.2) is *equivalent* to the system (3.4). Since they are different descriptions of the same basic set of relationships (3.2), it is to be expected that many properties of (3.4) are shared by (4.2); this situation will be illustrated many times in the chapters to follow. On the other hand, S can be chosen so that the system (4.2) is more amenable to treatment than (3.4); since they have the same properties, we choose to study the simpler (4.2) rather than the more complex (3.4).

Systems such as (4.2), which result from a change of basis, can have complex coefficients; this, and this alone, is the reason why we have to study the properties of systems with C^n as the state space, and for which the matrices have coefficients in C. As a matter of fact, these systems are no different from those studied in Chapter 3; all the proofs and theorems can painlessly (but rather boringly) be shown to apply in this case as well. Of course, the input and output vectors in (4.2) are the same as in (3.4), therefore real. The new state vector z, however, is not necessarily real, even though Sz has to be real for y to be real; since S has complex components, it may well be that z has complex components as well. Examples of this situation will be given later on in this chapter. Note that there appear to be preferred bases for which the state vector is real, and therefore, in the terminology of quantum mechanics, an observable, while there may be other bases for which the state vector does not have this property.

The following section shows, among other things, a way of simplifying matrices. Instead of presenting the bare procedure, we will found it on solid, fundamental concepts belonging to the theory of linear spaces.

4.2. Similarity, Eigenvalues, and Eigenvectors

Let T be an operator which maps X, an n-dimensional linear space over the field C, into itself. We say that a number λ in C is an *eigenvalue* of T if there is at least one nonzero vector in x such that

$$T(x) = \lambda x \qquad (4.3)$$

If λ is an eigenvalue of T, any vector $x \in X$ such that $T(x) = \lambda x$ is said to be an *eigenvector* of T; more precisely, it is said to be an eigenvector *associated with the eigenvalue* λ. Each eigenvector is associated with an eigenvalue and each eigenvalue is associated at least with one eigenvector. The names characteristic value and characteristic vector are also used.

EXAMPLE

Let $T: R^3 \rightarrow R^3$ be defined as

$$T(x) = \begin{bmatrix} x_1 + 3x_2 \\ 3x_1 - 2x_2 - x_3 \\ -x_2 + x_3 \end{bmatrix}$$

The number $\lambda = 1$ is an eigenvalue of T because the equation $T(x) = x$ can be solved for the components of x, which is then an eigenvector:

$$x_1 + 3x_2 = x_1$$
$$3x_1 - 2x_2 - x_3 = x_2$$
$$-x_2 + x_3 = x_3$$

The first equation implies $x_2 = 0$, while the third becomes then an identity. The vectors which satisfy the second equation are such that $3x_1 = x_3$; the eigenvectors of T associated with the eigenvalue $\lambda = 1$ are all of the form

$$\begin{bmatrix} x_1 \\ 0 \\ 3x_1 \end{bmatrix}$$

for some number x_1. Note that the set of all these vector is a one-dimensional subspace of R^3; many eigenvectors are associated with the eigenvalue $\lambda = 1$, and they are all multiples of each other. Observe that $\lambda = 1$ happens to have a set of eigenvectors associated with it (that is, it happens to be an eigenvalue) because the homogeneous system of equations $T(x) = \lambda x$ has solutions for $\lambda = 1$, which is due to the fact that the determinant of this system of equations is zero.

Let T be the matrix of T relative to a basis. We say that the eigenvalues of the matrix T are the eigenvalues of T, while its eigenvectors are the n-tuples whose (complex) entries are the components of the eigenvectors of T with respect to the basis. Of course, there are many bases in X, and to each basis corresponds a different matrix associated with the operator T. In order to gain insight into the effect of a change of basis in the eigenvalues and eigenvectors of a matrix, we show first the relationship which exist between two different matrices which are descriptions of the same operator relative to two different bases.

Let $\alpha = \{x^1, ..., x^n\}$ be a basis for the linear space X, and let an $n \times n$ invertible matrix $S = (s_{ij})$, with complex entries, be used in describing a new basis for the space, $\beta = \{z^1, ..., z^n\}$, as follows:

$$z^i = \sum_{j=1}^{n} s_{ji} x^j, \qquad i = 1, ..., n \qquad (4.4)$$

Note that the matrix of coefficients in (4.4) is S', the transpose of S. Before proceeding any further, we must find out if the set β actually constitutes a basis, that is, if the vectors $z^1,..., z^n$ are independent. Assume they are not independent. Then there are numbers $c_1,..., c_n$, not all of which are zero, such that

$$\sum_{i=1}^{n} c_i z^i = 0$$

However, this implies that

$$\sum_{i=1}^{n} c_i z^i = \sum_{i=1}^{n} c_i \left(\sum_{j=1}^{n} s_{ji} x^j \right) = \sum_{j=1}^{n} \left(\sum_{i=1}^{n} c_i s_{ji} \right) x^j = 0 \qquad (4.5)$$

Since the vectors x^j, $j = 1,..., n$, are independent, the coefficients of these vectors in (4.5) are zero. This, however, contradicts the facts that S is a nonsingular matrix and that not all of the numbers c_i are zero. Indeed, consider the following homogeneous system of equations for the numbers c_i

$$\sum_{i=1}^{n} c_i s_{ji} = 0, \qquad j = 1, ..., n \qquad (4.6)$$

The determinant of this system is not zero, since S is nonsingular [and then S', the matrix of coefficients in (4.6), is nonsingular as well]. Therefore the only numbers which satisfy (4.6) are $c_i = 0$, $i = 1,..., n$. The set β is a basis.

LEMMA.   Let a vector in X have coordinates $x_1 ,..., x_n$ and $z_1 , ..., z_n$ relative to the bases α and β defined above. If x and z are the vectors in C^n with entries equal to these coordinates, then these vectors are related by

$$x = Sz \qquad (4.7)$$

Proof.   Indeed,

$$\sum_{i=1}^{n} z_i z^i = \sum_{i=1}^{n} z_i \left(\sum_{j=1}^{n} s_{ji} x^j \right) = \sum_{j=1}^{n} \left(\sum_{i=1}^{n} s_{ji} z_i \right) x^j$$

then

$$x_j = \sum_{i=1}^{n} s_{ji} z_i \qquad \text{or} \qquad x = Sz$$

We are in a position now to relate the matrices of an operator relative to the bases α and β.

THEOREM 4.1.   Let T and T_1 be the matrices of $T\colon X \to X$ relative to the basis α and β respectively. Then

$$T_1 = S^{-1}TS \tag{4.8}$$

where S is the matrix defined in (4.4). The matrices of an operator relative to different bases are similar.

Proof.   Let x, w be the vectors in C^n corresponding to the coordinates relative to α of a vector in X and its transform by T respectively. Then of course $w = Tx$.

If the coordinates of the vector and its transform with respect to β form the n-tuples z and v, then $v = T_1z$; T_1 is the matrix of T relative to β. But of course $w = Sv$, $x = Sz$, so that

$$Sv = TSz \qquad \text{or} \qquad v = S^{-1}TSz$$

which of course implies $T_1 = S^{-1}TS$.

Since all bases in X can be described with respect to the basis α, say, by a set of relationships such as (4.4), then for every basis a matrix S exists such that the expression (4.8) holds. Therefore, all matrices of the operator T relative to some basis are similar to T. We show now that these matrices are similar to each other, that is, that given T_1 and T_2, matrices of T relative to some bases, a nonsingular matrix S_a exists such that $T_1 = S_a^{-1}T_2S_a$.

Indeed, matrices S_1 and S_2 exist such that

$$T_1 = S_1^{-1}TS_1, \qquad T_2 = S_2^{-1}TS_2$$

so that $S_1 T_1 S_1^{-1} = S_2 T_2 S_2^{-1}$; that is,

$$T_1 = S_1^{-1}S_2T_2S_2^{-1}S_1$$

the contention follows by making $S_a = S_2^{-1}S_1$.

Conversely, any matrix similar to a matrix T which is the matrix of an operator relative to a basis is a matrix of the operator relative to some basis. Remember that the eigenvalues of a matrix T which is the matrix of an operator T relative to a basis are defined as being the eigenvalues of the operator T; it follows that all matrices of T relative to some basis have the same eigenvalues; Theorem 4.2 also follows.

THEOREM 4.2.   All similar matrices have the same eigenvalues.

The eigenvectors of matrices of an operator T are related by an expression such as (4.7); let x be an eigenvector of the matrix T, that is, the

n-tuple composed of the coordinates of an eigenvector of the operator T. Then, by the lemma, the eigenvector z of another matrix of T is given by an expression such as (4.7).

How many eigenvalues does an operator T that maps an n-dimensional space X into itself have? The answer is contained in the following theorem. We write $\lambda - T$ for the operator $\lambda I - T$, I is the identity operator.

THEOREM 4.3. If λ is an eigenvalue of an operator T which maps X, an n-dimensional space, into itself, then the operator $\lambda - T$ is not invertible. If T is the matrix of the operator T with respect to any basis in X, then $\det(\lambda I_n - T) = 0$.

Proof. Let λ be an eigenvalue of T. Then, for a nonzero eigenvector x of T,

$$\lambda x = T(x) \qquad \text{or} \qquad (\lambda - T)(x) = 0$$

Since x is not zero, it follows that the null space of $\lambda - T$ is not composed of the zero vector alone, so that the operator is neither onto nor one–one, and thus not invertible. The matrix of $\lambda - T$ is $\lambda I_n - T$, so that $\det(\lambda I_n - T) = 0$.

The converse of this theorem is rather trivial. Let T be an $n \times n$ matrix, the matrix of an operator $T: X \to X$ with respect to some basis. If the scalar μ satisfies the equation $\det(\mu I_n - T) = 0$, then μ is an eigenvalue of the matrix T (and of the operator T).

Thus , a scalar μ is an eigenvalue of a matrix T if and only if it is a root of the equation

$$\det(\lambda I_n - T) = 0 \tag{4.9}$$

This equation is a polynomial equation of nth order in λ, said to be the *characteristic equation* of the operator T (and of the matrix T). It has n roots, generally complex, not necessarily distinct, to be called $\lambda_1 , ..., \lambda_n$. The polynomial in the left-hand side of (4.9) is said to be the *characteristic polynomial* of the operator T (and of the matrix T).

EXAMPLE

We show several matrices, their characteristic polynomials and eigenvalues.

$$A = \begin{bmatrix} 1 & 3 & 0 \\ 3 & -2 & -1 \\ 0 & -1 & 1 \end{bmatrix}$$

$$\det(\lambda I_3 - A) = \begin{vmatrix} \lambda - 1 & -3 & 0 \\ -3 & \lambda + 2 & 1 \\ 0 & 1 & \lambda - 1 \end{vmatrix} = \lambda^3 - 13\lambda + 12 = (\lambda - 1)(\lambda - 4)(\lambda + 3)$$

The eigenvalues are $\lambda_1 = 1$, $\lambda_2 = -4$, $\lambda_3 = 3$.

$$A = \begin{bmatrix} 3 & 0 \\ 5 & 3 \end{bmatrix}$$

$$\det(\lambda I_2 - A) = \begin{vmatrix} \lambda - 3 & 0 \\ -5 & \lambda - 3 \end{vmatrix} = (\lambda - 3)^2$$

The eigenvalues of this matrix are identical, $\lambda_1 = \lambda_2 = 3$.

$$A = \begin{bmatrix} 3 & -6 \\ 2 & -3 \end{bmatrix}$$

$$\det(\lambda I_2 - A) = \begin{vmatrix} \lambda - 3 & 6 \\ -2 & \lambda + 3 \end{vmatrix} = \lambda^2 + 3$$

Here the eigenvalues are purely imaginary, $\lambda_1 = i\sqrt{3}$, $\lambda_2 = -i\sqrt{3}$.

The characteristic polynomials of similar matrices are equal. Indeed, the coefficients of the polynomial in the left-hand side are uniquely determined by the roots of (4.9), since

$$\det(\lambda I_n - T) = (\lambda - \lambda_1) \cdots (\lambda - \lambda_n)$$

and, as proved in Theorem 4.2, similar matrices have identical eigenvalues. The coefficients of λ^{n-1}, λ^{n-2}, ..., λ^0, in (4.9) are said to be *invariant* under a change of basis.

Exercises

*1. Show that if A, a square matrix with real entries, is symmetric, then all its eigenvalues are real.

*2. Let A be as in Problem 1. Show that two eigenvectors of A which are associated with different eigenvalues are orthogonal.

*3. Show that if an operator has only nonzero eigenvalues, then it is invertible. Show that invertible operators have only nonzero eigenvalues.

4. Let T be an $n \times n$ matrix with real entries. If $V = TT'$ and $V^n = I_n$, show that the eigenvalues of V are all equal to 1.

5. Let A, B be $n \times n$ matrices. If $A^2 = B^2 = 0$, show that A and B are similar if and only if they have the same rank.

6. Can the eigenvalues of $A + B$ obtained in a simple way from those of A and B?

*7. Show that an eigenvector cannot be associated with two different eigenvalues.

8. Show that A and A' have the same characteristic equation.

*9. Show that the trace and the determinant of a matrix are invariant, that is, if S is nonsingular,

$$\mathrm{tr}(SAS^{-1}) = \mathrm{tr}\,A, \qquad \det(SAS^{-1}) = \det A$$

10. If A is skew-symmetric, show that its eigenvalues are either zero or purely imaginary.

4.3. Reduction

It is possible sometimes to find a basis relative to which the matrix of an operator $T: X \to X$ is very simple; for instance, in some cases it is possible to find a basis relative to which the matrix is diagonal, that is, one in which only the diagonal elements are not zero. In other cases a basis can be found with respect to which the matrix is slightly less simple, but still very useful to work with. We devote this and the following sections to the study of these phenomena.

The eigenvectors of T associated with a given eigenvalue λ constitute the null space of $T - \lambda$. In the example following Equation (4.3), this subspace is spanned by the vector $[1, 0, 3]'$; it is, therefore, a 1-dimensional subspace. With each eigenvalue one can associate one subspace of the original space X (and, of course, the corresponding subspace of the space C^n), each of which has, of course, a basis. Suppose that we take the n bases, one from each subspace, and form a set consisting of all possible linear combinations of the vectors in these bases. The set of vectors thus obtained may or may not be the whole space X, as shown in the following examples. When it is the whole space, striking things happen.

EXAMPLE

Let $T: R^2 \to R^2$ have the following matrix with respect to the standard basis in R^2

$$T = \begin{bmatrix} 1 & 0 \\ 5 & -1 \end{bmatrix}$$

The characteristic polynomial is

$$\det(\lambda I_2 - T) = \begin{vmatrix} \lambda - 1 & 0 \\ -5 & \lambda + 1 \end{vmatrix} = (\lambda - 1)(\lambda + 1)$$

The eigenvectors associated with $\lambda_1 = 1$ are characterized by

$$(\lambda_1 I_2 - T)x = \begin{bmatrix} 0 & 0 \\ -5 & 2 \end{bmatrix}\begin{bmatrix} x_1 \\ x_2 \end{bmatrix} = 0$$

so that they are defined by the equality $-5x_1 + 2x_2 = 0$; the corresponding subspace in R^2 is spanned by

$$x^1 = \begin{bmatrix} 1 \\ \frac{5}{2} \end{bmatrix}$$

The second subspace can be determined in a similar way:

$$(\lambda_2 I_2 - T)x = \begin{bmatrix} -2 & 0 \\ -5 & 0 \end{bmatrix}\begin{bmatrix} x_1 \\ x_2 \end{bmatrix} = 0$$

This subspace is spanned by the vector

$$x^2 = \begin{bmatrix} 0 \\ 1 \end{bmatrix}$$

The set spanned by the vectors x^1 and x^2 equals the whole of R^2, since x^1 and x^2 are independent. It so happens that the matrix of T relative to the basis $\{x^1, x^2\}$ is quite simple. The matrix S which describes the change of basis in this case is

$$S = \begin{bmatrix} 1 & 0 \\ \frac{5}{2} & 1 \end{bmatrix}$$

with inverse

$$S^{-1} = \begin{bmatrix} 1 & 0 \\ -\frac{5}{2} & 1 \end{bmatrix}$$

so that the matrix of T relative to the new basis is

$$S^{-1}TS = \begin{bmatrix} 1 & 0 \\ -\frac{5}{2} & 1 \end{bmatrix}\begin{bmatrix} 1 & 0 \\ 5 & -1 \end{bmatrix}\begin{bmatrix} 1 & 0 \\ \frac{5}{2} & 1 \end{bmatrix} = \begin{bmatrix} 1 & 0 \\ 0 & -1 \end{bmatrix}$$

It is said that T has been *diagonalized* by the change of basis.

If the vectors x^1 and x^2 had not spanned the whole of R^2, this construction would not have been possible, as show in the following example.

EXAMPLE

Let $T: R^2 \to R^2$ have the matrix T relative to the standard basis

$$T = \begin{bmatrix} 1 & 3 \\ 0 & 1 \end{bmatrix}$$

Then,

$$\det(\lambda I_2 - T) = \begin{vmatrix} \lambda - 1 & -3 \\ 0 & \lambda - 1 \end{vmatrix} = (\lambda - 1)^2$$

The first subspace is spanned by the vector

$$\begin{bmatrix} 1 \\ 0 \end{bmatrix}$$

The subspace associated with the second eigenvalue is identical to the one associated with the first; the eigenvalues of T do not span the whole of R^2. The construction shown in the previous example is not possible in this case.

The situations depicted in these examples are fairly typical; some matrices can be diagonalized by making a suitable change of basis, while some other matrices are like the one in the second example, which turns out to be not diagonalizable by means of a change of basis.

In order to pursue these ideas in a more general context, we need some more powerful tools.

Definition. Let $X_1, ..., X_p$ be subspaces of a vector space X. If each vector in X can be expressed uniquely as

$$x = x^1 + x^2 + \cdots + x^p$$

with $x^i \in X_i$, $i = 1, ..., p$, then it is said that the space X is the *direct sum* of the subspaces X_i, $i = 1, ..., p$; we write

$$X = X_1 \oplus X_2 \oplus \cdots \oplus X_p \tag{4.10}$$

There are several consequences of this definition:

i. If α_i is a basis for X_i, $i = 1, ..., p$, then the union of all these bases is a basis for X.

ii. The dimension of X equals the sum of the dimensions of the subspaces X_i, $i = 1, ..., p$.

An example of a direct sum is provided by the first of the two examples above, in which the space R^2 is the direct sum of two 1-dimensional subspaces, the null spaces of $T - \lambda_1$ and $T - \lambda_2$.

Definition. A subspace W of a vector space X is said to be *invariant* under an operator $T: X \to X$ if $x \in W$ implies $T(x) \in W$.

Consider, for instance, the null space of the operator $T - \lambda$ where λ is an eigenvalue of T. If x is in this null space, $(T - \lambda)(x) = 0$; then $T(x) = \lambda x$, and $T(x)$ is also in this null space because λx is in it if x is in it. Note that λ has to be an eigenvalue of T because otherwise the null space of $T - \lambda$ is composed only of the zero vector.

Another example of an invariant subspace is provided by the null space of $(T - \lambda)^m$, m being an arbitrary positive integer and λ an eigenvalue of T. If x is in this null space, $(T - \lambda)^m (x) = 0$. Let us compute $(T - \lambda)^m (T(x))$. Since $(T - \lambda)^m$ is a polynomial in T,

$$(T - \lambda)^m(T(x)) = T((T - \lambda)^m(x)) = 0$$

which of course implies that $T(x)$ is in the null space of $(T - \lambda)^m$. This subspace is invariant under T.

Definition. Let W be an invariant subspace of X under T. With each $x \in W$ we associate a unique element $T(x)$ also in W. The operator thus defined is said to be the operator *induced* on W by T. This operator maps W into itself.

EXAMPLE

Let $T: R^3 \to R^3$ be defined by the following matrix relative to the standard basis

$$T = \begin{bmatrix} 0 & 1 & 0 \\ 1 & 0 & 1 \\ 0 & 1 & 0 \end{bmatrix}$$

The eigenvalues can be readily computed to be $\lambda_1 = \sqrt{2}$, $\lambda_2 = 0$, and $\lambda_3 = -\sqrt{2}$. The null spaces of $T - \lambda_1$, $T - \lambda_2$, and $T - \lambda_3$ are spanned by the following vectors, relative to the standard basis in R^3

$$T - \lambda_1: \qquad x^1 = \begin{bmatrix} 1 \\ \sqrt{2} \\ 1 \end{bmatrix}$$

$$T - \lambda_2: \qquad x^2 = \begin{bmatrix} 1 \\ 0 \\ -1 \end{bmatrix}$$

$$T - \lambda_3: \qquad x^3 = \begin{bmatrix} 1 \\ -\sqrt{2} \\ 1 \end{bmatrix}$$

The subspaces spanned respectively by x^1, x^2, and x^3 are invariant under T. What are the operators induced on these subspaces by T? Clearly, if x is in the null space of $T - \lambda_i$, $Tx = \lambda_i x$, so the effect of T on a member of the null space of $T - \lambda_i$ is simply one of multiplication by λ_i.

Of course, these induced operators can be characterized by matrices relatives to bases in the corresponding subspace. Bases for the null spaces of $T - \lambda_i$, $i = 1, 2, 3$, are the vectors x^i above, relative to which the operators induced by T on these subspaces, to be denoted by T_i, have matrices T_i, which in this case are

$$T_i = [\lambda_i], \qquad i = 1, 2, 3$$

Note that R^3 is the direct sum of the subspaces spanned by x^1, x^2, and x^3. We can compute the matrix of T relative to the basis $\{x^1, x^2, x^3\}$ in R^3; this happens to be

$$\begin{bmatrix} \lambda_1 & & \\ & \lambda_2 & \\ & & \lambda_3 \end{bmatrix} = \begin{bmatrix} T_1 & & \\ & T_2 & \\ & & T_3 \end{bmatrix}$$

Here we follow the standard practice of omitting the entries which are zero. Observe as well that we are using the notation, introduced in Chapter 2, which consists in using matrices as elements of other matrices.

The situation depicted in the previous example can be easily generalized. Let X, an n-dimensional space, be the direct sum of q subspaces, S_1, ..., S_q, which are invariant under T, a linear operator defined on X. If these subspaces have bases α_1, ..., α_q, then the union of these bases is a basis α for X. Let T_1, ..., T_q be the matrices of the operators induced on

S_1, ..., S_q respectively by T, and let T be the matrix of T relative to α. Then,

$$
T = \begin{bmatrix} T_1 & & & \\ & T_2 & & \\ & & \ddots & \\ & & & T_q \end{bmatrix}
$$

where again we have omitted the zero entries. Note that the matrices T_1, ..., T_q are square; the dimension of T_i equals the dimension of the subspace S_i, $i = 1$, ..., q. The matrix T is said to be the *direct sum* of the matrices T_1, ..., T_q. We will write sometimes

$$
T = T_1 \oplus T_2 \oplus \cdots \oplus T_q
$$

A crucial role will be played in what follows by polynomial functions of an operator T which equals the zero operator, that is, by expressions of the type $\sum_{i=0}^{n} a_i T^i = 0$; T^0 is the identity operator. The coefficients a_i will in general be complex numbers. We prove first an important result.

THEOREM 4.4 (Hamilton–Cayley). Let a linear operator T map X, an n-dimensional space, into itself. Let $\sum_{i=0}^{n} a_i \lambda^i$ be the characteristic polynomial of T, $a_n = 1$. Then the operator

$$
\sum_{i=0}^{n} a_i T^i
$$

equals the zero operator; it is also said that T satisfies its own characteristic equation. If T is the matrix of T relative to a basis, then

$$
\sum_{i=0}^{n} a_i T^i = 0_n
$$

$T^0 = I_n$, the $n \times n$ identity matrix.

Proof. Consider the inverse of the matrix $\lambda I_n - T$; λ is not, of course, an eigenvalue of T. As shown in Chapter 2, there is an $n \times n$ matrix $U(\lambda)$ such that

$$
(\lambda I_n - T)^{-1} = \frac{U(\lambda)}{\det(\lambda I_n - T)}
$$

The entries of $U(\lambda)$ are polynomials in λ. We have for all complex λ which are not eigenvalues of T,

$$
(\lambda I_n - T) U(\lambda) = f(\lambda) I_n \tag{4.11}
$$

where $f(\lambda) = \det(\lambda I_n - T)$ is the characteristic polynomial of T.

We compute now

$$f(\lambda) I_n - f(T) = \sum_{i=0}^{n} a_i(\lambda^i I_n - T^i)$$

Since the polynomial in $\lambda^i I_n - T^i$ can be written as $(\lambda I_n - T)$ times a matrix dependent on λ, there is a matrix $V(\lambda)$ such that

$$f(\lambda) I_n - f(T) = (\lambda I_n - T) V(\lambda) \qquad (4.12)$$

Note that the entries of $V(\lambda)$ are polynomials in λ. It follows from (4.11) and (4.12) that

$$f(T) = (\lambda I_n - T)(U(\lambda) - V(\lambda))$$

We prove that the polynomial $W(\lambda) = U(\lambda) - V(\lambda)$ has to be identically zero. Suppose first that $W(\lambda) = B_0$, a constant matrix. Then $f(T) = \lambda B_0 - T B_0$ for all λ which are not eigenvalues of T, so the coefficient of λ in this expression has to be zero since $f(T)$ does not depend on λ, that is, $B_0 = W(\lambda) = 0$. Suppose now $W(\lambda) = B_0 + \lambda B_1$. Then

$$f(T) = \lambda^2 B_1 + \lambda(B_0 - T B_1) - T B_0$$

which implies that the coefficients of λ^2 and of λ have to be zero, which in turn implies $W(\lambda) = 0$. Suppose now that

$$W(\lambda) = B_0 + \lambda B_1 + \cdots + \lambda^{n-1} B_{n-1}$$

so that

$$\begin{aligned} f(T) &= (\lambda I_n - T)(B_0 + \lambda B_1 + \cdots + \lambda^{n-1} B_{n-1}) \\ &= \lambda^n B_{n-1} + \lambda^{n-1}(B_{n-2} - T B_{n-1}) + \cdots + \lambda(B_0 - T B_1) - T B_0 \end{aligned}$$

so that, since the coefficients of $f(T)$ do not depend on λ,

$$B_{n-1} = 0, \qquad B_{n-2} = T B_{n-1} = 0, \quad \ldots, \quad B_0 = T B_1 = 0$$

It follows that $f(T) = 0$. The matrix T indeed satisfies its own characteristic equation; the operator T satisfies this equation as well.

There are many implications of this useful theorem, some of which we will meet in the following chapters. Most of its success is due to the fact that the theorem establishes that, in effect, higher powers of T are linear combinations of the first $(n - 1)$ powers, so that for instance an expression such as e^T depends only on a finite number of powers of T.

EXAMPLE

Suppose T, a 3×3 matrix, satisfies

$$T^3 + 3T^2 - T + I_3 = 0_3$$

Let us multiply both sides of this equality by T^k. Then,

$$T^{k+3} + 3T^{k+2} - T^{k+1} + T^k = 0_3$$

so that

$$T^{k+3} = -3T^{k+2} + T^{k+1} - T^k$$

It follows that

$$T^3 = -3T^2 + T - I_3$$
$$T^4 = -3T^3 + T^2 - T = 10T^2 - 4T + 3I_3 , \ldots .$$

The matrix T is invertible. We multiply the first equality by T^{-1}; then

$$T^2 + 3T - I_3 + T^{-1} = 0_3$$

so that

$$T^{-1} = -T^2 - 3T + I_3$$

a rather surprising result.

Our present interest on the Hamilton–Cayley equation is focused on its relevance to the subject of reduction of operators. In this context, the following example is of importance.

EXAMPLE

Let

$$T = \begin{bmatrix} 3 & 0 & 0 \\ 5 & 3 & 0 \\ 0 & 0 & 4 \end{bmatrix}$$

then,

$$\det(\lambda I_3 - T) = \begin{vmatrix} \lambda - 3 & 0 & 0 \\ -5 & \lambda - 3 & 0 \\ 0 & 0 & \lambda - 4 \end{vmatrix} = (\lambda - 3)^2 (\lambda - 4)$$

It follows that

$$(T - 3I_3)^2 (T - 4I_3) = \begin{bmatrix} 0 & 0 & 0 \\ 5 & 0 & 0 \\ 0 & 0 & 1 \end{bmatrix} \begin{bmatrix} 0 & 0 & 0 \\ 5 & 0 & 0 \\ 0 & 0 & 1 \end{bmatrix} \begin{bmatrix} -1 & 0 & 0 \\ 5 & -1 & 0 \\ 0 & 0 & 0 \end{bmatrix} = 0_3$$

Note that $T - 3I_3 \neq 0$, $T - 4I_3 \neq 0$, $(T - 3I_3)^2 \neq 0$ and $(T - 3I_3)(T - 4I_3) \neq 0$. The polynomial $(T - 3I_3)^2 (T - 4I_3)$ happens to be the polynomial in T of lowest degree which equals the zero 3×3 matrix, 0_3.

EXAMPLE

A rather different situation obtains for the matrix

$$T = \begin{bmatrix} 3 & 2 & 0 \\ -3 & -2 & 0 \\ 0 & 0 & 1 \end{bmatrix}$$

whose characteristic polynomial is $(\lambda - 1)^2\lambda$. Of course, T satisfies $(T - I_3)^2\, T = 0_3$; but also

$$(T - I_3)T = \begin{bmatrix} 2 & 2 & 0 \\ -3 & -3 & 0 \\ 0 & 0 & 0 \end{bmatrix}\begin{bmatrix} 3 & 2 & 0 \\ -3 & -2 & 0 \\ 0 & 0 & 1 \end{bmatrix} = 0_3$$

there is therefore a polynomial $m(\lambda)$ of lower order than the characteristic polynomial such that $m(T) = 0_3$.

The situation depicted in the previous two examples is representative of a general phenomenon: Given an operator T (or its matrix T relative to some basis), all polynomials $g(\lambda)$ with the property that $g(T) = 0$ (or that $g(T) = 0_n$) are the multiples of one polynomial, $m(\lambda)$, to be called the *minimal polynomial* of the operator T (or of the matrix T). The name *minimal* is due to the fact, which follows readily from the previous assertion, that $m(\lambda)$ is the smallest (in the sense of least degree) polynomial with the property that $m(T) = 0$. In the first example above, it coincides with the characteristic polynomial, while in the second it does not. We will not give the proof of the existence of a minimal polynomial, since this would entail previous knowledge of some topics in advanced algebra (See [3]).

In the second example above, the characteristic polynomial is $(\lambda - 1)^2\lambda$, while the minimal polynomial is $(\lambda - 1)\lambda$; we see that in this case every root of the minimal polynomial is an eigenvalue. This is a general property.

THEOREM 4.5   Every eigenvalue of an operator $T: X \to X$ is a root of the minimal polynomial associated with T. Every root of the minimal polynomial is an eigenvalue of T.

Proof. Let μ be an eigenvalue of T, that is, $T(x) = \mu x$, $x \neq 0$. If g is any polynomial, $g(T)(x) = g(\mu)\, x$; for instance,

$$T^2(x) = T(T(x)) = T(\mu x) = \mu T(x) = \mu^2 x$$

Therefore, $g(\mu)$ is an eigenvalue of $g(T)$. If $g(T)$ is the zero operator, then $g(\mu)$ is an eigenvalue of it, *and must then be zero*, $g(\mu) = 0$. Since $m(T) = 0$, $m(\mu) = 0$; every eigenvalue of T is a root of the minimal polynomial associated with this operator.

Since the minimal polynomial is a factor of the characteristic polynomial, every root of it is also an eigenvalue of the operator.

We have characterized very clearly the structure of the minimal polynomial. It has the same roots as the characteristic polynomial, but these roots may occur with lower multiplicity than in the characteristic polynomial. For instance, in the second example above, these polynomials are $(\lambda - 1)^2 \lambda$ and $(\lambda - 1)\lambda$ respectively. In general, if the eigenvalues of T are $\lambda_1, ..., \lambda_n$, only some, say q, of them, will be distinct. Without loss of generality, we can assume that these are the first q eigenvalues, so that the minimal polynomial can be factored as follows

$$m(\lambda) = (\lambda - \lambda_1)^{k_1}(\lambda - \lambda_2)^{k_2} \cdots (\lambda - \lambda_q)^{k_q} \qquad (4.13)$$

Here $k_1, ..., k_q$ are positive integers. The numbers $\lambda_1, ..., \lambda_q$ are distinct. The characteristic polynomial of the operator has a structure similar to (4.13) but in general with different exponents.

At last we can develop the fundamental decomposition theorems. The main ideas will be presented by means of an example.

EXAMPLE

If $T: R^3 \to R^3$ has as matrix with respect to the standard basis

$$T = \begin{bmatrix} 1 & 1 & 1 \\ 0 & 2 & 0 \\ 0 & -3 & 2 \end{bmatrix}$$

the minimal polynomial, which in this case equals the characteristic polynomial, is

$$m(\lambda) = (\lambda - 1)(\lambda - 2)^2$$

The null spaces of the operators $T - 1$ and $(T - 2)^2$ play an important role in what follows. They will be denoted by S_1 and S_2 respectively. S_1 is spanned by

$$x^1 = \begin{bmatrix} 1 \\ 0 \\ 0 \end{bmatrix}$$

since if $x \in S_1$, $(T - 1)(x) = 0$, that is, $x_2 + x_3 = 0$, $x_2 = 0$, $-3x_2 + x_3 = 0$, which implies $x_2 = x_3 = 0$.

If $x \in S_2$, $(T - 2)^2 (x) = 0$, that is, $x_1 - 4x_2 - x_3 = 0$, which implies that S_2 is spanned by

$$x^2 = \begin{bmatrix} 4 \\ 1 \\ 0 \end{bmatrix}, \qquad x^3 = \begin{bmatrix} 1 \\ 0 \\ 1 \end{bmatrix}$$

The space R^3 is the direct sum of S_1 and S_2; the set $\{x^1, x^2, x^3\}$ is a basis for it. It is of interest to characterize the operators induced on these subspaces by T. If $x \in S_1$,

$T_1(x) = x$; T_1 is the operator induced on S_1 by T. Its matrix relative to the basis $\{x^1\}$ in S_1 is simply $[1]$. The matrix of T_2, the operator induced by T on S_2, with respect to the basis $\{x^2, x^3\}$ in S_2, is obtained as follows:

$$T_2(x^2) = Tx^2 = \begin{bmatrix} 5 \\ 2 \\ -3 \end{bmatrix} = 2x^2 - 3x^3$$

$$T_2(x^3) = Tx^3 = \begin{bmatrix} 2 \\ 0 \\ 2 \end{bmatrix} = 0x^2 + 2x^3$$

so the matrix of T_2 relative to $\{x^2, x^3\}$ is

$$\begin{bmatrix} 2 & 0 \\ -3 & 2 \end{bmatrix}$$

The matrix of T relative to $\{x^1, x^2, x^3\}$ can be written by inspection:

$$\begin{bmatrix} 1 & 0 & 0 \\ 0 & 2 & 0 \\ 0 & -3 & 2 \end{bmatrix}$$

Note that this matrix depends on the particular vectors chosen as basis in the subspaces; if x^3 is chosen instead as

$$\begin{bmatrix} 3 \\ 0 \\ 3 \end{bmatrix}$$

the matrix of T relative to $\{x^1, x^2, x^3\}$ is

$$\begin{bmatrix} 1 & 0 & 0 \\ 0 & 2 & 0 \\ 0 & -1 & 2 \end{bmatrix}$$

Again, this example was chosen so as to show a very general situation.

THEOREM 4.6. Let the minimal polynomial of an operator $T: X \to X$, X being an n-dimensional space, be given by Equation (4.13). Let S_i be the null space of $(T - \lambda_i)^{k_i}$, for $i = 1, 2, ..., q$. Then X is the direct sum of the invariant subspaces $S_1, S_2, ..., S_q$.

Proof. We show first that any $x \in X$ can be written uniquely as $x = y + z$, with y in S_1, while z is in the null space of the operator

$$m_1(T) = (T - \lambda_2)^{k_2}(T - \lambda_3)^{k_3} \cdots (T - \lambda_q)^{k_q}$$

Because the polynomials $(\lambda - \lambda_1)^{k_1}$ and $m_1(\lambda)$ have no common factors, it is well known [3] that there are polynomials $h(\lambda)$ and $g(\lambda)$ such that

$1 = h(\lambda)(\lambda - \lambda_1)^{k_1} + g(\lambda)\, m_1(\lambda)$ for all $\lambda \in C$. Then, by substituting T for λ we get

$$I = h(T)(T - \lambda_1)^{k_1} + g(T)\, m_1(T) \qquad (4.14)$$

so that, if $x \in X$,

$$x = h(T)(T - \lambda_1)^{k_1}x + g(T)\, m_1(T)x$$

this is the decomposition sought with

$$z = h(T)(T - \lambda_1)^{k_1}x, \qquad y = g(T)\, m_1(T)x$$

Indeed,

$$(T - \lambda_1)^{k_1}y = (T - \lambda_1)^{k_1}g(T)\, m_1(T)x$$

$$= g(T)(T - \lambda_1)^{k_1}m_1(T)x = g(T)\, m(T)x$$

$$= 0$$

since $m(T)$ is the zero operator; $m(\lambda)$ is, as usual, the minimal polynomial. Therefore, $y \in S_1$. In a similar way, it can be shown that z is in the null space of $m_1(T)$. This decomposition is unique; assume there is another decomposition of the same type,

$$x = y^1 + z^1$$

Then $y^1 - y = z - z^1$ is in both S_1 and the null space of $m_1(T)$; (4.14) implies, however, that in this case $y^1 - y = z - z^1 = 0$. The space X is the direct sum of S_1 and the null space of $m_1(T)$.

It is possible to apply the previous argument to show that the null space of $m_1(T)$ is the direct sum of S_2 and the null space of $(T - \lambda_3)^{k_3} \cdots (T - \lambda_q)^{k_q}$; the argument is applied then to show that this null space is the direct sum of S_3 and the null space of $(T - \lambda_4)^{k_4} \cdots (T - \lambda_q)^{k_q}$, etc; the theorem follows after applying the argument a sufficiently large number of times.

Of course, the matrix of T relative to the basis formed by the union of basis of $S_1, ..., S_q$ is the direct sum of the matrices of the operators induced in these invariant subspaces by T. As a matter of fact, the bases for the subspaces can be chosen so that the matrix T has a particularly simple structure, as shown in the example above; we will pursue the study of the most general case in a later section, while concentrating in the next one on operators with matrices which can be diagonalized.

Exercises

*1. Show that if a linear space X is the direct sum of the subspaces X_i, $i = 1, ..., p$, then if α_i is a basis for X_i, the union of all these bases is a basis for X.

2. Let X be a finite-dimensional linear space and let W be any subspace of this space. Show that there is a subspace V of X such that $X = W \oplus V$.

3. Find the characteristic polynomial of A^{-1} in terms of the coefficients of the characteristic polynomial of A.

*4. Find a general expression for e^A, in terms of the coefficients of the Hamilton–Cayley equation and a finite number of powers of A.

5. Show that similar matrices have the same minimal polynomial.

*6. Consider R^n as an inner-product space, and let S be any set of vectors in R^n. Then $S^\perp$ (read "S perp"), the *orthogonal complement* of S, is the set of all vectors in R^n which are orthogonal to every vector in S. Show that if W is a subspace of R^n,

$$R^n = W \oplus W^\perp$$

4.4. Diagonalization

We start this discussion with the fundamental theorem on diagonalization. It is a simple consequence of Theorem 4.6.

THEOREM 4.7. Let an operator T be as in Theorem 4.6. A basis in X can be found such that the matrix of T relative to this basis is diagonal if and only if the minimal polynomial of T is a product of distinct linear factors. If such is the case, the operator T is said to be *diagonalizable*.

Proof. If

$$m(\lambda) = (\lambda - \lambda_1) \cdots (\lambda - \lambda_q) \tag{4.15}$$

with $\lambda_1, ..., \lambda_q$ distinct complex numbers, Theorem 4.6 indicates that x is the direct sum of subspaces S_i, $i = 1, ..., q$; each subspace S_i is the null space of $T - \lambda_i$, so that it consists of all eigenvectors associated with the eigenvalue λ_i.

Let the subspace S_i have dimension n_i. Since all vectors in S_i are eigenvectors of the operator T, a basis for this subspace can be found so that all elements of the basis are eigenvectors of T. Let this basis be $\{x_i^1, ..., x_i^{n_i}\}$. Then,

$$T(x_i^k) = \lambda_i x_i^k, \qquad k = 1, ..., n_i$$

so that the matrix of the operator T_i induced by T on S_i with respect to this basis in S_i is simply an $n_i \times n_i$ diagonal matrix with all diagonal elements equal to the eigenvalue λ_i. The matrix of T with respect to the union of all bases of the subspaces S_i is the direct sum of the individual matrices, and therefore a diagonal matrix itself.

Suppose now that there is a basis relative to which the matrix of T is diagonal with distinct diagonal entries $\lambda_1, ..., \lambda_q$. The minimal polynomial of this matrix (which equals the minimal polynomial of T) can be computed to be exactly equal to (4.15). (The algebra takes much space, so we omit it; the reader can try an example of, say, a 6×6 matrix with three eigenvalues of multiplicities 3, 2, and 1 respectively.) The theorem follows.

The structure of the diagonal matrix of a diagonalizable operator with eigenvalues λ_1, ..., λ_q is clear from the previous theorem:

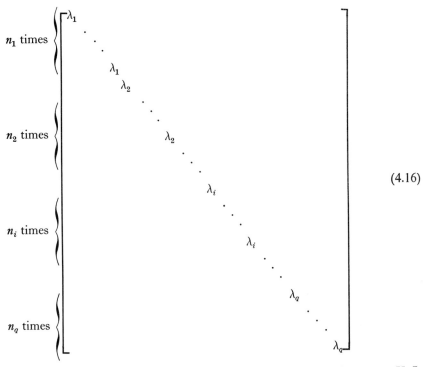

$$(4.16)$$

Of course, $n_1 + n_2 + \cdots + n_q = n$, the dimension of the space X. It so happens that the numbers n_i can be characterized in a very simple manner. The characteristic polynomial of the matrix (4.16) is

$$f(\lambda) = (\lambda - \lambda_1)^{n_1}(\lambda - \lambda_2)^{n_2} \cdots (\lambda - \lambda_q)^{n_q}$$

This is of course the characteristic polynomial of the operator T and indeed of every matrix of T relative to any basis. It follows that the numbers n_1, n_2, ..., n_q are simply the multiplicities of the eigenvalues λ_1, λ_2, ..., λ_q respectively. Each eigenvalue λ_i appears in the diagonal of (4.16) as many times as the value of its multiplicity, that is, as the value of the exponent of $(\lambda - \lambda_i)$ in the characteristic equation of the operator T.

A matrix which is similar to a diagonal matrix is said to be *diagonalizable*. It follows from the previous theorem that every matrix of a diagonalizable operator relative to any basis whatsoever is diagonalizable.

EXAMPLE

Let $T: X^3 \rightarrow X^3$ be defined by the following matrix relative to the standard basis

$$T = \begin{bmatrix} 1 & 1 & 0 \\ 0 & 2 & 0 \\ 0 & 0 & 2 \end{bmatrix}$$

The characteristic polynomial of this operator is $(\lambda - 1)(\lambda - 2)^2$; its minimal polynomial, however, is $(\lambda - 1)(\lambda - 2)$. It should be, therefore, a diagonalizable operator. Indeed, S_1, the null space of $T - 1$, is spanned by

$$x^1 = \begin{bmatrix} 1 \\ 0 \\ 0 \end{bmatrix}$$

its dimension is $n_1 = 1$, and the matrix of T_1 relative to the basis $\{x^1\}$ is the matrix [1]. The null space of $T - 2$, S_2, is spanned by

$$x^2 = \begin{bmatrix} 1 \\ 1 \\ 0 \end{bmatrix}, \qquad x^3 = \begin{bmatrix} 0 \\ 0 \\ 1 \end{bmatrix}$$

its dimension is $n_2 = 2$; the matrix of T_2 relative to the basis $\{x^2, x^3\}$ is the 2×2 matrix

$$\begin{bmatrix} 2 & 0 \\ 0 & 2 \end{bmatrix}$$

The matrix of T relative to the basis $\{x^1, x^2, x^3\}$ is

$$\begin{bmatrix} 1 & 0 & 0 \\ 0 & 2 & 0 \\ 0 & 0 & 2 \end{bmatrix}$$

This matrix is, of course, identical to $S^{-1}TS$, where S describes the change from the standard basis to $\{x^1, x^2, x^3\}$:

$$S = \begin{bmatrix} 1 & 1 & 0 \\ 0 & 1 & 0 \\ 0 & 0 & 1 \end{bmatrix}, \qquad S^{-1} = \begin{bmatrix} 1 & -1 & 0 \\ 0 & 1 & 0 \\ 0 & 0 & 1 \end{bmatrix}$$

$$S^{-1}TS = \begin{bmatrix} 1 & -1 & 0 \\ 0 & 1 & 0 \\ 0 & 0 & 1 \end{bmatrix} \begin{bmatrix} 1 & 1 & 0 \\ 0 & 2 & 0 \\ 0 & 0 & 2 \end{bmatrix} \begin{bmatrix} 1 & 1 & 0 \\ 0 & 1 & 0 \\ 0 & 0 & 1 \end{bmatrix}$$

$$= \begin{bmatrix} 1 & -1 & 0 \\ 0 & 2 & 0 \\ 0 & 0 & 2 \end{bmatrix} \begin{bmatrix} 1 & 1 & 0 \\ 0 & 1 & 0 \\ 0 & 0 & 1 \end{bmatrix} = \begin{bmatrix} 1 & 0 & 0 \\ 0 & 2 & 0 \\ 0 & 0 & 2 \end{bmatrix}$$

The following corollary to Theorem 4.7 affords a very simple way to identify some diagonalizable operators.

COROLLARY. If all the eigenvalues of an operator $T: X \to X$, X a finite dimensional space, are distinct, then the operator is diagonalizable.

Diagonalizable operators can be characterized still in another way.

THEOREM 4.8. An operator T which maps X, a finite-dimensional space, into itself is diagonalizable if and only if its eigenvectors span the whole of the space X.

Proof. Let the eigenvectors span the whole space X. Then enough vectors can be chosen from them to form a basis $\{x^1, ..., x^n\}$ for x. Since $T(x^i) = \lambda_i x^i$, $i = 1, ..., n$, the matrix of T relative to this basis is diagonal, and its diagonal entries are the eigenvalues of T.

Suppose now that the operator T is diagonalizable, and let T be its diagonal matrix. Eigenvectors of this matrix can be chosen so as to form a basis for C^n; indeed, the standard basis vectors $e^i, ..., e^n$ are eigenvectors of T. The corresponding vectors in X, which are eigenvectors of the operator T, span then the whole of X.

An operator T is generally specified by its $n \times n$ matrix T relative to a basis. If the operator is diagonalizable, the task in hand consists in obtaining the basis with respect to which the matrix of T is diagonal. This is done by working in the space C^n; the n eigenvalues of the matrix T are computed first, corresponding n eigenvectors which span C^n are then constructed, and a matrix S is formed with these eigenvectors as columns. Of course, the diagonal matrix is $S^{-1}TS$. Several examples of this procedure have been shown above.

We can at long last obtain a representation of the transition matrix of the system (3.4) when the matrix A which appears in this system is diagonalizable. Consider the associated matrix equation of the homogeneous system $\dot{x} = Ax$,

$$\dot{X} = AX, \qquad X(t_0) = I_n \qquad (4.17)$$

Put $Z = S^{-1}X$, where S is the matrix formed with eigenvectors of A as described above. Then

$$\dot{Z} = (S^{-1}AS)Z, \qquad Z(t_0) = S^{-1}$$

is a matrix differential equation with solution

$$Z(t) = e^{(t-t_0)S^{-1}AS}S^{-1} \qquad (4.18)$$

It is very simple to obtain an explicit representation for the matrix exponential now because $S^{-1}AS$ is a diagonal matrix. Indeed, if the diagonal entries of $S^{-1}AS$ are the complex numbers $c_1, c_2, ..., c_n$, (which are not necessarily distinct), then we obtain, by means of the series for the matrix exponential,

$$e^{tS^{-1}AS} = \begin{bmatrix} e^{tc_1} & & & \\ & e^{tc_2} & & \\ & & \ddots & \\ & & & e^{tc_n} \end{bmatrix} \tag{4.19}$$

again we have omitted the zero entries.

It follows, of course, that the solution of (4.17) is

$$X(t) = T(t - t_0) = S e^{(t-t_0)S^{-1}AS} S^{-1} \tag{4.20}$$

where the exponential matrix is given by the simple expression (4.19).

EXAMPLE

Let the homogeneous system be

$$\dot{x}_1 = x_2, \qquad \dot{x}_2 = -2x_1 - 3x_2$$

the same used in the main example of Section 3.3. Then,

$$A = \begin{bmatrix} 0 & 1 \\ -2 & -3 \end{bmatrix}$$

$\det(\lambda I_2 - A) = (\lambda + 1)(\lambda + 2)$, $\lambda_1 = -1$, $\lambda_2 = -2$. The subspace S_1 (the null space of $A + I_2$) is spanned by

$$x^1 = \begin{bmatrix} 1 \\ -1 \end{bmatrix}$$

while S_2, the null space of $A + 2I_2$, is spanned by

$$x^2 = \begin{bmatrix} 1 \\ -2 \end{bmatrix}$$

Therefore,

$$S = \begin{bmatrix} 1 & 1 \\ -1 & -2 \end{bmatrix}, \qquad S^{-1} = \begin{bmatrix} 2 & 1 \\ -1 & -1 \end{bmatrix}$$

and

$$T(t - t_0) = \begin{bmatrix} 1 & 1 \\ -1 & -2 \end{bmatrix} \begin{bmatrix} e^{-t} & 0 \\ 0 & e^{-2t} \end{bmatrix} \begin{bmatrix} 2 & 1 \\ -1 & -1 \end{bmatrix}$$

$$= \begin{bmatrix} 2e^{-(t-t_0)} - e^{-2(t-t_0)} & e^{-(t-t_0)} - e^{-2(t-t_0)} \\ -2e^{-(t-t_0)} + 2e^{-2(t-t_0)} & -e^{-(t-t_0)} + 2e^{-2(t-t_0)} \end{bmatrix}$$

EXAMPLE

We will show now how a very simple system with real entries can be diagonalized only into a system with complex entries. Consider a harmonic oscillator

$$\dot{x}_1 = x_2, \qquad \dot{x}_2 = -x_1,$$

$$A = \begin{bmatrix} 0 & 1 \\ -1 & 0 \end{bmatrix}, \qquad \det(\lambda I_2 - A) = \lambda^2 + 1, \qquad \lambda_1 = i, \quad \lambda_2 = -i$$

The basis in C^2 with respect to which A is transformed into the matrix

$$\begin{bmatrix} i & 0 \\ 0 & -i \end{bmatrix}$$

is described by the following matrix

$$S = \begin{bmatrix} 1 & 1 \\ i & -i \end{bmatrix}$$

Clearly, the state vector with respect to the new basis has complex entries, while the original state vector is real. The new state vector is not, in the terminology of quantum mechanics, an observable.

Exercises

*1. Let T_1 and T_2 be diagonalizable linear operators mapping a linear space X into itself. Show that if T_1 and T_2 commute, there is a basis for X each vector of which is an eigenvector of T_1 and also an eigenvector of T_2.

2. The matrix e^{tA} is actually a polynomial in A,

$$e^{tA} = \sum_{i=0}^{n-1} g_i(t) A^i, \qquad A^0 = I_n$$

Find differential equations for the functions $g_i(t)$, $i = 0, ..., n - 1$.

3. What can you say about e^A if A is triangular?

4.5. More on Reduction

There are operators which are not diagonalizable. Can we obtain matrix representations of these operators so that the matrices are reasonably simple? The answer to this question lies in the theory concerned with the *Jordan canonical form*.

We could develop this theory using the results obtained in the previous section, especially the decomposition of Theorem 4.6. It is possible to decompose the spaces $S_1, ..., S_q$ themselves even further, so that the operators induced in these new subspaces can be represented by particularly simple matrices relative to appropriate bases. We will not take this path, however, because it requires a rather long introduction. The alternative treatment given below is of a purely algebraic nature, and shows clearly how to reduce a matrix to Jordan canonical form. We start by looking at diagonalization from a rather different standpoint.

Let T be an operator mapping X, an n-dimensional linear space, into itself, and let T be its matrix relative to some basis. If $M(\lambda) = (\lambda I_n - T)^{-1}$, we know that

$$\det M(\lambda) = \frac{1}{\det(\lambda I_n - T)} \qquad (4.21)$$

that is,

$$\det M(\lambda) = \frac{1}{(\lambda - \lambda_1)(\lambda - \lambda_2) \cdots (\lambda - \lambda_n)} \qquad (4.22)$$

where λ_1, λ_2, ..., λ_n are the eigenvalues of T which we assume to be distinct. The entries of $M(\lambda)$ consist, then, of the ratio of a polynomial in λ (which can be proved to be of order less than n) and the polynomial $(\lambda - \lambda_1) \cdots (\lambda - \lambda_n)$. It follows that $M(\lambda)$ can be expanded in partial fractions

$$M(\lambda) = \sum_{i=1}^{n} \frac{M_i}{\lambda - \lambda_i} \qquad (4.23)$$

the matrices M_i, $i = 1, ..., n$, are constant. Since $(\lambda I_n - T)M(\lambda) = I_n$, it follows that

$$\sum_{i=1}^{n} \frac{(\lambda I_n - T)M_i}{\lambda - \lambda_i} = I_n \qquad (4.24)$$

If (4.24) is multiplied by $\lambda - \lambda_p$, with $1 \leqslant p \leqslant n$, and then the limit is taken as λ tends to λ_p in both sides, we obtain readily

$$(\lambda_p I_n - T)M_p = 0 \qquad (4.25)$$

which of course holds for $p = 1, ..., n$. Let us choose in the matrix M_p one nonzero column vector, x^p. Clearly,

$$(\lambda_p I_n - T)x^p = 0$$

it follows that x^p is simply an eigenvector of T (this should dispel any doubts as to whether we could find in each matrix M_p, $p = 1, ..., n$, one nonzero column vector). Therefore, by inverting the matrix $(\lambda I_n - T)$ and expanding it into partial fractions, we obtain eigenvectors of T, that is, a basis with respect to which the matrix of the operator T is diagonal.

EXAMPLE

Consider the matrix

$$T = \begin{bmatrix} 1 & 0 \\ 3 & 2 \end{bmatrix}$$

Since

$$\lambda I_2 - T = \begin{bmatrix} \lambda - 1 & 0 \\ -3 & \lambda - 2 \end{bmatrix}$$

$$\det(\lambda I_2 - T) = (\lambda - 1)(\lambda - 2)$$

$$(\lambda I_2 - T)^{-1} = \frac{1}{(\lambda - 1)(\lambda - 2)} \begin{bmatrix} \lambda - 2 & 0 \\ 3 & \lambda - 1 \end{bmatrix}$$

$$= \frac{M_1}{\lambda - 1} + \frac{M_2}{\lambda - 2}$$

with

$$M_1 = \begin{bmatrix} 1 & 0 \\ -3 & 0 \end{bmatrix}, \qquad M_2 = \begin{bmatrix} 0 & 0 \\ 3 & 1 \end{bmatrix}$$

Clearly, $\begin{bmatrix} 1 \\ -3 \end{bmatrix}$ is an eigenvector corresponding to $\lambda_1 = 1$ and $\begin{bmatrix} 0 \\ 1 \end{bmatrix}$ an eigenvector corresponding to $\lambda_2 = 2$.

A similar method can be employed to find simple matrices for operators which are not diagonalizable. We will present first the bare algebraic steps, while leaving for later a discussion of the connection between the procedure and the decomposition of Theorem 4.6.

Let T be a linear operator which maps X, an n-dimensional space, into itself, and let T be its matrix relative to a basis. A special case of a 6×6 matrix will be employed to present the method, mainly so that the principles of the argument do not become obscured by the notation. Let T then be a 6×6 constant matrix, with only three distinct eigenvalues, λ_1, λ_2, and λ_3, of multiplicity 1, 3, and 2 respectively. Therefore,

$$\det(\lambda I_6 - T) = (\lambda - \lambda_1)(\lambda - \lambda_2)^3(\lambda - \lambda_3)^2$$

The partial fraction expansion of $M(\lambda)$ can proceed as before

$$M(\lambda) = \frac{M_{11}}{\lambda - \lambda_1} + \sum_{i=1}^{3} \frac{M_{2i}}{(\lambda - \lambda_2)^i} + \sum_{j=1}^{2} \frac{M_{3j}}{(\lambda - \lambda_3)^j} \qquad (4.26)$$

Note that terms appear now which contain powers such as $(\lambda - \lambda_2)^2$ in their denominators. The matrices M_{11}, M_{2i}, and M_{3j} are constants. Since

$$(\lambda I_6 - T) M(\lambda) = I_6$$

$$(\lambda I_6 - T) \left\{ \frac{M_{11}}{\lambda - \lambda_1} + \sum_{i=1}^{3} \frac{M_{2i}}{(\lambda - \lambda_2)^i} + \sum_{j=1}^{2} \frac{M_{3j}}{(\lambda - \lambda_3)^j} \right\} = I_6 \qquad (4.27)$$

If we write

$$(\lambda I_6 - T) = (\lambda - \lambda_2) I_6 + (\lambda_2 I_6 - T)$$

we obtain

$$\frac{(\lambda_2 I_6 - T)M_{21} + M_{22}}{\lambda - \lambda_2} + \frac{(\lambda_2 I_6 - T)M_{22} + M_{23}}{(\lambda - \lambda_2)^2}$$

$$+ \frac{(\lambda_2 I_6 - T)M_{23}}{(\lambda - \lambda_2)^3} + M_{21} + (\lambda I_6 - T)M = I_6 \qquad (4.28)$$

where M represents the terms in (4.28) which do not include λ_2. Let λ tend to λ_2. The right-hand side of (4.28) is a constant matrix, I_6, while some terms in the left-hand side become unbounded as λ approaches λ_2. It follows that the coefficients of those terms *must* vanish

$$(\lambda_2 I_6 - T)M_{21} + M_{22} = 0$$

$$(\lambda_2 I_6 - T)M_{22} + M_{23} = 0 \qquad (4.29)$$

$$(\lambda_2 I_6 - T)M_{23} = 0$$

We make now the crucial assumption that it is possible to choose nonzero columns from each of the matrices M_{21}, M_{22}, and M_{23} in such a way that all these columns have the same position in the corresponding matrix: If the first column is chosen in M_{21}, the first column should be chosen in M_{22} and M_{23}. There is no reason to expect that this is possible in general; other cases will be treated later. We construct with the columns which have been chosen from M_{21}, M_{22}, and M_{23} the nonzero vectors in R^6, x^{21}, x^{22}, and x^{23} respectively; these vectors satisfy

$$(\lambda_2 I_6 - T)x^{23} = 0$$

$$(\lambda_2 I_6 - T)x^{22} = -x^{23} \qquad (4.30)$$

$$(\lambda_2 I_6 - T)x^{21} = -x^{22}$$

In the same way, if M_{31} and M_{32} are nonzero, nonzero column vectors x^{31} and x^{32} extracted from M_{31} and M_{32} satisfy the equations

$$(\lambda_3 I_6 - T)x^{32} = 0$$

$$(\lambda_3 I_6 - T)x^{31} = -x^{32} \qquad (4.31)$$

and a nonzero column vector extracted from M_{11}, x^{11}, satisfies

$$(\lambda_1 I_6 - T)x^{11} = 0 \tag{4.32}$$

We prove now that these 6 vectors, x^{11}, x^{21}, x^{22}, x^{23}, x^{31}, and x^{32} are independent, so that they form a basis for R^6. It so happens that the matrix T takes a very simple form when transformed to this basis; the reasons for this are of course the relationships (4.30), (4.31), and (4.32) which exist between the matrix and the members of the basis.

Suppose the six vectors form a dependent set, so that there are scalars c_1, ..., c_6, at least one of which is different from zero, so that

$$c_1 x^{11} + c_2 x^{21} + c_3 x^{22} + c_4 x^{23} + c_5 x^{31} + c_6 x^{32} = 0.$$

Suppose, for instance, that c_1 is not zero. Consider the matrix $Z = (\lambda_2 I_6 - T)^3 (\lambda_3 I_6 - T)^2$. Since the two factors commute, we have $Zx^{23} = Zx^{32} = 0$ and a bit of thought will convince us that $Zx^{22} = Zx^{23} = Zx^{31}$, so that

$$0 = c_1 Zx^{11} = c_1(\lambda_2 I_6 - T)^3(\lambda_3 I_6 - T)^2 x^{11}$$

$$= c_1(\lambda_2 I_6 - T)^3(\lambda_3 I_6 - T)(\lambda_3 I_6 - \lambda_1 I_6)x^{11}$$

$$= c_1(\lambda_2 - \lambda_1)^3(\lambda_3 - \lambda_1)^2 x^{11}$$

Since $\lambda_2 \neq \lambda_1$, $\lambda_3 \neq \lambda_1$, and x^{11} is not a zero vector, $c_1 = 0$, which is a contradiction. If another scalar c_i, $i = 2, ..., 6$, had been supposed nonzero, a different matrix would have been used instead of Z to show that this scalar is indeed zero. The vectors do form a basis for R^6.

Of course, to each of these vectors corresponds a vector in the original 6-dimensional space X in which T is defined, and this set of vectors in X is a basis for this space. The matrix of the operator T with respect to this basis can be computed easily. Since

$$Tx^{11} = \lambda_1 x^{11}, \qquad Tx^{21} = \lambda_2 x^{21} + x^{22}, \qquad Tx^{22} = \lambda_2 x^{22} + x^{23}$$

$$Tx^{23} = \lambda_2 x^{23}, \qquad Tx^{31} = \lambda_3 x^{31} + x^{32}, \qquad Tx^{32} = \lambda_3 x^{32}$$

the matrix of T relative to the new basis is

$$\begin{bmatrix} \lambda_1 & 0 & 0 & 0 & 0 & 0 \\ 0 & \lambda_2 & 0 & 0 & 0 & 0 \\ 0 & 1 & \lambda_2 & 0 & 0 & 0 \\ 0 & 0 & 1 & \lambda_2 & 0 & 0 \\ 0 & 0 & 0 & 0 & \lambda_3 & 0 \\ 0 & 0 & 0 & 0 & 1 & \lambda_3 \end{bmatrix} \tag{4.33}$$

This is, of course, also equal to $S^{-1}TS$, where S is the matrix with columns x^{11}, x^{21}, x^{22}, x^{23}, x^{31}, and x^{32}. It is known as the *Jordan canonical form* of the matrix T. Note that the previous development is quite general, and has proved that a matrix can be transformed into the Jordan canonical form provided that all the matrices which appear as coefficients in the partial fraction expansion of the matrix $M(\lambda)$ are nonzero.

EXAMPLE

Let

$$T = \begin{bmatrix} 1 & 1 \\ -1 & 3 \end{bmatrix}, \qquad \det(\lambda I_2 - T) = (\lambda - 2)^2$$

$$(\lambda I_2 - T)^{-1} = \frac{1}{(\lambda - 2)^2} \begin{bmatrix} \lambda - 3 & 1 \\ -1 & \lambda - 1 \end{bmatrix}$$

$$= \frac{1}{\lambda - 2} \begin{bmatrix} 1 & 0 \\ 0 & 1 \end{bmatrix} + \frac{1}{(\lambda - 2)^2} \begin{bmatrix} -1 & 1 \\ -1 & 1 \end{bmatrix}$$

We can choose x^{11} and x^{12} as

$$x^{11} = [0 \quad 1]', \qquad x^{12} = [1 \quad 1]'$$

so that

$$S = \begin{bmatrix} 0 & 1 \\ 1 & 1 \end{bmatrix}, \qquad S^{-1}TS = \begin{bmatrix} 2 & 0 \\ 1 & 2 \end{bmatrix}$$

What can we do if one or more of the coefficient matrices in the partial fraction expansion of $M(\lambda)$ is identically zero? Let us go back to the partial fraction expansion (4.26), and let us assume that all the coefficient matrices but the matrix M_{23} are nonzero. The vector x^{23} cannot then be chosen as before, so that the set of Equations (4.30) becomes

$$(\lambda_2 I_6 - T)x^{22} = 0$$
$$(\lambda_2 I_6 - T)x^{21} = -x^{22}$$

(4.34)

It should be clear that we do not have a basis anymore, but just five independent vectors. Where do we get another which is independent of the other five, and so that the resulting matrix is simple? We take another vector from M_{22}, $\hat{x}^{22}$, which of course satisfies the first equation in (4.34), so that the basis is $\{x^{11}, x^{21}, x^{22}, \hat{x}^{22}, x^{31}, x^{32}\}$; the equations which determine the new matrix are

$$Tx^{11} = \lambda_1 x^{11}, \qquad Tx^{21} = \lambda_2 x^{21} + x^{22}, \qquad Tx^{22} = \lambda_2 x^{22}$$

$$T\hat{x}^{22} = \lambda_2 \hat{x}^{22}, \qquad Tx^{31} = \lambda_3 x^{31} + x^{32}, \qquad Tx^{32} = \lambda_3 x^{32}$$

so that this matrix is

$$
\begin{bmatrix}
\lambda_1 & 0 & 0 & 0 & 0 & 0 \\
0 & \lambda_2 & 0 & 0 & 0 & 0 \\
0 & 1 & \lambda_2 & 0 & 0 & 0 \\
0 & 0 & 0 & \lambda_2 & 0 & 0 \\
0 & 0 & 0 & 0 & \lambda_3 & 0 \\
0 & 0 & 0 & 0 & 1 & \lambda_3
\end{bmatrix}
\tag{4.35}
$$

which again is said to be in Jordan canonical form. The vectors which form the new basis can be proved to be independent in much the same way as before.

We already have gained enough information so as to be able to generalize some of the knowledge we possess concerning the Jordan canonical form:

i. A basis can be associated with any operator which maps a finite-dimensional space into itself, so that the matrix of the operator relative to it is in Jordan canonical form. This follows simply from the fact that the constructions used above can be employed in the case of an arbitrary operator T.

ii. The eigenvalues of the operator occur in the diagonal of the Jordan canonical form. Each eigenvalue appears there as many times as the value of its multiplicity.

iii. The Jordan canonical form is the direct sum of several *elementary Jordan matrices* of the following type:

$$
\begin{bmatrix}
\lambda_i & & & & & \\
1 & \lambda_i & & & & \\
& 1 & \cdot & & & \\
& & \cdot & \cdot & \cdot & \\
& & & \cdot & \lambda_i & \\
& & & & 1 & \lambda_i
\end{bmatrix}
$$

The zero entries, as usual, have been omitted. For instance, the matrix (4.33) is the direct sum of the following elementary Jordan matrices:

$$
[\lambda_1], \qquad
\begin{bmatrix}
\lambda_2 & 0 & 0 \\
1 & \lambda_2 & 0 \\
0 & 1 & \lambda_2
\end{bmatrix}, \qquad
\begin{bmatrix}
\lambda_3 & 0 \\
1 & \lambda_3
\end{bmatrix}
$$

while the matrix (4.35) is the direct sum of

$$
[\lambda_1], \qquad
\begin{bmatrix}
\lambda_2 & 0 \\
1 & \lambda_2
\end{bmatrix}, \qquad
[\lambda_2], \qquad
\begin{bmatrix}
\lambda_3 & 0 \\
1 & \lambda_3
\end{bmatrix}
$$

Note that more than one elementary Jordan matrix can be associated with each eigenvalue.

iv. We can summarize the structure of a Jordan canonical form as follows: The space X is n-dimensional; we assume T to have q distinct eigenvalues, $q \leqslant n$. With the eigenvalue λ_i we associate a matrix T_i, $i = 1, ..., q$, of dimension $n_i \times n_i$; n_i is the multiplicity of λ_i , so that the Jordan canonical form T is the direct sum of the matrices T_i :

$$T = \begin{bmatrix} T_1 & & & & & \\ & T_2 & & & & \\ & & \cdot\cdot & & & \\ & & & T_i & & \\ & & & & \cdot\cdot & \\ & & & & & T_q \end{bmatrix}_{n \times n} \tag{4.36}$$

Each of these matrices T_i is also the direct sum of several $(s(i))$ elementary matrices

$$T_i = \begin{bmatrix} T_{i1} & & & & & \\ & T_{i2} & & & & \\ & & \cdot\cdot & & & \\ & & & T_{ij} & & \\ & & & & \cdot\cdot & \\ & & & & & T_{i\,s(i)} \end{bmatrix}_{n_i \times n_i} , \qquad i = 1, ..., q \tag{4.37}$$

Each of the elementary Jordan matrices is of the form

$$T_{ij} = \begin{bmatrix} \lambda_i & & & & \\ 1 & \lambda_i & & & \\ & 1 & \cdot\cdot & & \\ & & \cdot\cdot\cdot & \lambda_i & \\ & & & 1 & \lambda_i \end{bmatrix}_{n_{ij} \times n_{ij}} \tag{4.38}$$

The reader should be able to identify the matrices T_i and T_{ij} in (4.33) and (4.35).

EXAMPLE

$$T = \begin{bmatrix} 3 & 0 & 0 & 0 & 0 & 0 \\ 1 & 3 & 0 & 0 & 0 & 0 \\ 0 & 1 & 3 & 0 & 0 & 0 \\ 0 & 0 & 0 & 3 & 0 & 0 \\ 0 & 0 & 0 & 1 & 3 & 0 \\ 0 & 0 & 0 & 0 & 0 & 4 \end{bmatrix}$$

There are two distinct eigenvalues, $\lambda_1 = 3$ and $\lambda_2 = 4$. T_1 is the matrix obtained from T by deleting the 6th column and 6th row; T_2 is simply the 1×1 matrix [4]. Then,

$$T_{11} = \begin{bmatrix} 3 & 0 & 0 \\ 1 & 3 & 0 \\ 0 & 1 & 3 \end{bmatrix}$$

$$T_{12} = \begin{bmatrix} 3 & 0 \\ 1 & 3 \end{bmatrix}$$

$$T_{21} = [4]$$

We see that $s(1) = 2$, $s(2) = 1$.

A few words are needed to relate these results to our previous work on decomposition of spaces. The matrices T_i, $i = 1, ..., q$ are clearly matrices of the operators induced by T on the subspaces $S_1, S_2, ..., S_q$ of Theorem 4.6. These are particularly simple matrices because the bases for these subspaces were chosen in a suitable way. As mentioned at the beginning of this section, these subspaces can themselves be partitioned into other subspaces; it so happens that the induced operators in these elementary subspaces have as matrices relative to suitable bases elementary Jordan matrices. The theory of this partition is quite lengthy, and we will not cover it; the interested reader is referred to any of the texts on linear spaces mentioned in the previous chapters.

Consider now Equation (4.17), and assume that the matrix A is not diagonalizable. We can carry out exactly the same steps as when it was diagonalizable; S would now be the matrix which describes the change of basis from the standard basis in R^n to the basis relative to which the matrix takes its Jordan canonical form.

We must, then, derive an explicit form for the exponential of tT, if T is in Jordan canonical form. Clearly, by Equation (3.54), it is sufficient for us to compute the exponential of the time multiplying an elementary Jordan canonical form.

If this is as in (4.38), direct computation gives

$$
e^{t\,T_{ij}} = e^{t\lambda_i}
\begin{bmatrix}
1 & 0 & & & & & & 0 \\
t & 1 & & & & & & \\
\dfrac{t^2}{2!} & t & & & & & & \\
\vdots & \dfrac{t^2}{2!} & \cdot & & & \vdots & & \\
 & \vdots & \cdot & 1 & & & & \\
 & & \cdot & t & 1 & & & \\
\dfrac{t^{n_i-1}}{(n_i-1)!} & & & \dfrac{t^2}{2} & t & 1 & &
\end{bmatrix}
\tag{4.39}
$$

The matrix in the right-hand side of (4.39) is of course of dimension $n_i \times n_i$.

EXAMPLE

Let the matrix A in (4.17) be given by

$$
A = \begin{bmatrix} 1 & 1 \\ -1 & 3 \end{bmatrix}
$$

The corresponding Jordan canonical form is

$$
S^{-1}AS = \begin{bmatrix} 2 & 0 \\ 1 & 2 \end{bmatrix}
$$

with

$$
S^{-1} = \begin{bmatrix} -1 & 1 \\ 1 & 0 \end{bmatrix}
$$

Then,

$$
e^{(t-t_0)S^{-1}AS} = e^{2(t-t_0)} \begin{bmatrix} 1 & 0 \\ (t-t_0) & 1 \end{bmatrix}
$$

$$
= \begin{bmatrix} e^{2(t-t_0)} & 0 \\ (t-t_0)e^{2(t-t_0)} & e^{2(t-t_0)} \end{bmatrix}
$$

so that the solution of (4.17) is

$$
X(t) = S e^{(t-t_0)S^{-1}AS} S^{-1}
$$

$$
= e^{2(t-t_0)} \begin{bmatrix} (1-t+t_0) & t-t_0 \\ -(t-t_0) & (1-t+t_0) \end{bmatrix}
$$

Exercises

1. Show all possible Jordan canonical forms for matrices with each of the following characteristic polynomials

    i. $(\lambda - 2)^4 (\lambda - 1)$
    ii. $(\lambda - i)^3 (\lambda + i)^3$
    iii. $(\lambda - \lambda_1)(\lambda - \lambda_2)^2 (\lambda - \lambda_3)^3$

*2. Let the minimal polynomial of a matrix A be equal to its characteristic polynomial, and let A have q distinct eigenvalues. Prove that the Jordan canonical form of this matrix is the direct sum of q elementary Jordan matrices. Note that this implies that all the coefficient matrices M_i in an expansion such as (4.26) are nonzero.

4.6. Functions of Matrices

We want to end this chapter by generalizing slightly the procedure used in Section 4.4 for the computation of the exponential of a diagonalizable matrix. Needless to say, there are similar procedures for matrices which are not diagonalizable; they are, however, quite beyond the scope of out treatise, so the interested reader is referred to the literature for their study [1, 2].

Consider, as motivation, the matrix equation

$$\frac{d^2X}{dt^2} + BX = 0, \qquad X(t_0) = I_n \tag{4.40}$$

where X and B are $n \times n$ matrices. We could, of course, by defining appropriate new variables, put (4.40) in the form of first-order matrix equation. However, by analogy with the scalar case, we may wish to write the solution of (4.40) as

$$X(t) = \cos \sqrt{B}\, t \tag{4.41}$$

where $\sqrt{B}$ and $\cos \sqrt{B}$ have to be defined properly. We see that there is much to gain by devising a way of defining functions such as these. Of course, in some cases, it is indeed possible to make these definitions by means of series of matrices, as was done in the case of the exponential function.

Let T be an $n \times n$, diagonalizable matrix. It is possible to write it as $T = SVS^{-1}$, since $V = S^{-1}TS$; the matrix V is diagonal:

$$V = \begin{bmatrix} \lambda_1 & & & \\ & \lambda_2 & & \\ & & \ddots & \\ & & & \lambda_n \end{bmatrix}$$

the diagonal entries are not necessarily distinct. It follows easily that for any integer k,

$$T^k = SV^kS^{-1}$$

that is,

$$T^k = S \begin{bmatrix} \lambda_1{}^k & & & & \\ & \lambda_2{}^k & & & \\ & & \cdot & & \\ & & & \cdot & \\ & & & & \lambda_n{}^k \end{bmatrix} S^{-1} \qquad (4.42)$$

In the same way, if $g(\mu) = \sum_{i=0}^m a_i\mu^i$, then,

$$g(T) = S^{-1}g(V)S = S \begin{bmatrix} g(\lambda_1) & & & & \\ & g(\lambda_2) & & & \\ & & \cdot & & \\ & & & \cdot & \\ & & & & g(\lambda_n) \end{bmatrix} S^{-1} \qquad (4.43)$$

Suppose now that $f(\mu)$ is represented by a convergent series of the type $\sum_{i=0}^\infty a_i\mu^i$, and that the series converges in a disk which contains the set

$$\{\lambda : |\lambda| \leqslant \max_i |\lambda_i|\}$$

Then,

$$f(T) = S^{-1}f(V)S = S \begin{bmatrix} f(\lambda_1) & & & & \\ & f(\lambda_2) & & & \\ & & \cdot & & \\ & & & \cdot & \\ & & & & f(\lambda_n) \end{bmatrix} S^{-1}$$

For instance,

$$\sin T = S \begin{bmatrix} \sin \lambda_1 & & & & \\ & \sin \lambda_2 & & & \\ & & \cdot & & \\ & & & \cdot & \\ & & & & \sin \lambda_n \end{bmatrix} S^{-1}$$

The square root function $\sqrt{T}$ can be defined in a similar way; further background is needed in this case, though, to be acquired in Chapter 5.

Exercises

1. If A is nonsingular, find an expression for A^{-1} such as (4.43). Assume that the eigenvalues of A are distinct.

2. Define the logarithm of a matrix. What conditions has the matrix to satisfy so as to have a logarithm?

SUPPLEMENTARY NOTES AND REFERENCES

§4.2. Many good references exist on the subject of matrix algebra. The writer has found helpful in his own studies the books by Gantmacher [*1*] and the series of articles by Frame [*2*]. A rather more elementary, but very lucid, treatment is given in

R. Bellman, "Introduction to Matrix Analysis." McGraw-Hill, New York, 1970.

It should be clear by now that, in order to compute the eigenvalues of an $n \times n$ matrix, an nth-order algebraic equation has to be solved. If the entries of the matrix are complex, the coefficient of this polynomial equation may also be complex. It so happens that there are no general ways of solving such equations, while there are well-known and proven ways of solving algebraic equations with real coefficients. In most cases our systems will have, however, a set of real matrices associated with them, while the complex matrices will appear only after the original, real matrices have been reduced. In this case, of course, the eigenvalues of a complex matrix are known beforehand, as being equal to the eigenvalues of its (real) predecesor.

§4.5. A treatment of the Jordan canonical form which is similar in spirit to (but deeper than) ours, can be found in

V. I. Smirnov, "Linear Algebra and Group Theory." McGraw-Hill, New York, 1961.

REFERENCES

1. F. R. Gantmacher, "The Theory of Matrices." Chelsea, New York, 1959.
2. J. S. Frame, Matrix functions and applications, Part I—Matrix operations and generalized inverses, *IEEE Spectrum* 1 (March 1964), 209–220.
   J. S. Frame, Part II—Functions of a matrix, *IEEE Spectrum* 1 (April 1964), 102–108.
   J. S. Frame, Part IV—Matrix functions and constitutent matrices, *IEEE Spectrum* 1 (June 1964), 123–131.
   J. S. Frame, Part V—Similarity reductions by rational or orthogonal matrices, *IEEE Spectrum* 1 (July 1964), 103–109.
3. G. Birkhoff and S. MacLane, "A Survey of Modern Algebra." Macmillan, New York, 1965.

5

Controllability, Observability

5.1. Introduction

This chapter will show a change in the basic orientation of the text. In Chapters 2, 3, and 4 we have presented material which is rather divorced from the discipline of engineering; as such, it can be found in any text on differential equations. In this chapter we start introducing concepts of a basic engineering nature; by the adoption of this new viewpoint, we will see how the properties of systems such as the ones studied in the previous chapters multiply, so that the systems themselves acquire rich and complex structures.

Our objective in this and the following chapters is to find conditions for the properties of the systems in terms of the coefficients which appear in the differential equations (3.1), that is, in terms of the matrices A, B, C, and D in (3.4). It is our aim to characterize a system without "solving" the corresponding differential equations; we want to be able to say if a system is controllable or not, stable or not, just by operating on the matrices which define its state equations. Of course, to derive the rules by which we can achieve this aim, we will make use of the explicit expressions for the state or the output of a system which were obtained in Chapters 3 and 4. Again we mention that these expressions are not especially suitable to compute with; they are useful because they indicate the form which the state or the output take in terms of functions, (such as the transition matrix) which have known properties (such as the group property), so that by the use of these forms and properties we can characterize the systems themselves in terms of the matrices which define their state equations.

The main results of this chapter are contained in Sections 5.2–5.8; Sections 5.9–5.14 are concerned with specialized subjects which can be skipped without loss of continuity.

We start this chapter by defining a concept of great engineering importance, controllability.

5.2. Controllability

Consider a system with state equations such as (3.4). Let the state vector at an arbitrary but fixed value of time t_0 equal a vector x^0. Suppose we are interested in determining an input u from some admissible class of inputs such that the state vector takes the value x^1 at a time $t_1 > t_0$. Before this task is attempted, however, the following question should be answered: *Is it possible* to transfer the system from a state x^0 at t_0 to the state x^1 at some later time t_1 by the use of an input in the class of admissible inputs?

As a matter of fact, we will answer a more general question. Our efforts will be concentrated in determining conditions on the matrices A and B (which are the only ones which have any bearing on the behavior of the state vector x) such that *any* state can be reached at a later time when the initial state is specified at a time t_0. The class of admissible inputs will be defined as composed of r-vectors with components which are piecewise-continuous functions of the time defined on half-open intervals contained in the basic interval I on which the equations (3.4) are defined. We could make this class considerably larger by including all vectors with entries which are continuous almost everywhere or even by extending the concept of integral so as to accomodate vectors whose entries are merely integrable in the Lebesgue sense. Actually, nothing is lost by restricting our inputs to vectors with piecewise-continuous components; if a linear system with finite-dimensional state space is such that any state can be reached from any other state in finite time, then this can be accomplished by the use of an input with piecewise-continuous entries. On the other hand, if a region of the state space cannot be connected to another region by means of the trajectories of a system corresponding to inputs with piecewise-continuous entries, then they cannot be joined at all.

Let, then, $x(t_0) = x^0$; $t_0 \in R$ and $x^0 \in R^n$ are values of the time and state which are arbitrary but fixed; of course t_0 is in the basic interval of definition I of the system. Let also a fixed value of the time t_1, $t_0 < t_1$, $t_1 \in I$, be given, and consider the question as to whether the state x can be made to take any value whatsover at t_1 by the application of an admissible control given that it has taken the value x^0 at t_0 . We know that

$$x(t_1; t_0, x^0; u) = \Phi(t_1, t_0) x^0 + \int_{t_0}^{t_1} \Phi(t_1, \lambda) B(\lambda) u(\lambda) d\lambda \qquad (5.1)$$

We claim that the key to the answer to this question is contained in the properties of the integral term in Equation (5.1); more specifically, we prove

THEOREM 5.1.   Let a transformation V with domain equal to the set of admissible inputs and range in the state space R^n be defined by

$$V(u) = \int_{t_0}^{t_1} \Phi(t_1, \lambda) B(\lambda) u(\lambda) d\lambda \qquad (5.2)$$

Then $x(t_1; t_0, x^0; u)$ can be made equal to any vector in R^n by choosing u correspondingly if and only if the transformation V is onto.

Proof.   If V is onto, the integral in the right-hand side of (5.1) can be made to take any value whatsoever by choosing the control u correspondingly from the class of admissible inputs. To have the left-hand side of (5.1) take the value x^1, the integral simply is made to take $x^1 - \Phi(t_1, t_0) x^0$. If $x(t_1; t_0, x^0; u)$ can be made to take any value, say x^1, by choosing u correspondingly, then the integral takes the value $x^1 - \Phi(t_1, t_0) x^0$; clearly

$$R^n = \{x : x = x^1 - \Phi(t_1, t_0) x^0, \quad x^1 \in R^n\}$$

The theorem follows.

Note that the integral in (5.2) equals the value of the state vector at time t_1 if it was at the origin at time t_0 . Theorem 5.1 is therefore equivalent to the following.

THEOREM 5.2.   Let $x(t_0) = x^0$. If t_1 is a fixed but arbitrary value of the time in the basic interval of definition of the system (3.4), $t_1 > t_0$, there is admissible control u^1 defined on $[t_0, t_1)$ such that

$$x(t_1; t_0, x^0; u^1) = x^1$$

with x^1 a fixed but arbitrary vector in R^n, provided that there is an admissible control u^2 defined on $[t_0, t_1)$ and associated with each vector $x^2 \in R^n$ such that

$$x(t_1; t_0, 0; u^2) = x^2$$

This theorem suggests that we can classify systems as to whether their state vectors can, or cannot, take values over all of R^n at a given time t_1 by the use of an admissible control, given that they were at the origin of this space at a time $t_0 < t_1$ [1].

Definition. A system is said to be *controllable* (or *completely controllable*) *over* $[t_0, t_1]$, $t_0 < t_1$, if to every state $x^1 \in R^n$ can be associated an admissible control u such that $x(t_1; t_0, 0; u) = x^1$; that is, such that this control transfers the origin of the state space (at t_0) to the vector x^1 (at t_1). Of course, both t_0 and t_1 are in the interval of definition I of the system.

If the state of a system can be transferred from the origin at t_0 to any other state in R^n at some time later, not necessarily specified, we give the following definition.

Definition. A system is said to be *controllable* (or *completely controllable*) *at* t_0 if there is a number $t_1 > t_0$ such that the system is completely controllable over $[t_0, t_1]$.

At last, we make our most general definition.

Definition. A system is said to be *controllable* (or *completely controllable*) if it is completely controllable at every time t_0 in its interval of definition (with the exception of the right-hand limit of $I = [t_a, t_b]$, the value t_b, for which there cannot be $t_1 > t_b$, $t_1 \in I$).

Most of our efforts with regard to the subject of controllability will consist in obtaining conditions on the matrices A and B such that the system is completely controllable. This will prove to be a comparatively simple task if these matrices are constant.

We show now that the definitions given above are not trivial in the sense that there are systems for which some states cannot be reached at all from the origin in a finite time by the action of an admissible input. This will be done, of course, by means of some examples.

EXAMPLE

Let

$$A = \begin{bmatrix} 0 & 1 \\ 0 & 1 \end{bmatrix}, \quad B = \begin{bmatrix} 1 & -1 \\ 0 & 0 \end{bmatrix}$$

that is,

$$\dot{x}_1 = x_2 + u_1 - u_2, \qquad \dot{x}_2 = x_2$$

If $x_2(t_0) = 0$, $x_2(t) = 0$, $t \geq t_0$, so that many states in R^2 (all of those states for which x_2 is not zero) cannot be reached from the origin. This system is not controllable at any t_0.

EXAMPLE

The system with scalar input u,

$$\dot{x}_1 = u, \qquad \dot{x}_2 = u$$

is not controllable at any t_0. If $x(t_0) = 0$, then $x_1(t) = x_2(t)$ for any t, so that all states in R^2 for which $x_1 \neq x_2$ cannot be reached at any time.

Constant systems (that is, systems with constant coefficients) have the interesting property that they are controllable whenever they are controllable at any t_0, which of course simplifies the testing of these systems with respect to controllability. Indeed, consider a constant system characterized by matrices A and B. Remember that $\Phi(t, \lambda) = T(t - \lambda)$. Let the system be controllable at t_0, that is, given $x^1 \in R^n$, let $t_1 > t_0$ and an admissible control u^1 defined over $[t_0, t_1)$ exist such that

$$\int_{t_0}^{t_1} T(t_1 - \lambda) B(\lambda) u(\lambda) \, d\lambda = x^1 \tag{5.3}$$

Let τ be an arbitrary real number, and compute the response $x(t_1 - \tau; t_0 - \tau, 0; u^2)$, with u^2 defined by $u^2(t) = u^1(t + \tau)$, $t \in [t_0 - \tau, t_1 - \tau)$. This is

$$\int_{t_0 - \tau}^{t_1 - \tau} T(t_1 - \tau - \lambda) B u^1(\lambda + \tau) \, d\lambda$$

and can easily be shown, by putting $\mu = \lambda + \tau$, to equal the expression in the left-hand side of (5.3), that is, to equal x^1. It follows, of course, that if the system is controllable at t_0, it is also controllable at $t_0 - \tau$, with $\tau \in R$ and otherwise arbitrary, since if $x^1 \in R^n$ and $x(t_0) = 0$, there is a value of the time, $t_1 - \tau$ (which of course is larger than $t_0 - \tau$), and an admissible control, u^2, defined on $[t_0 - \tau, t_1 - \tau)$, such that the state $x(t_1 - \tau; t_0 - \tau, 0; u^2)$ equals x^1. We have proved:

THEOREM 5.3. A constant system is controllable if it is controllable at any time t_0 (for instance, at $t = 0$).

Our aim is now to obtain conditions on the matrices A and B so that a general (not necessarily constant) linear system be controllable. The first step will be to obtain an intermediate result, in the sense that the condition for controllability to be derived first involves the transition matrix $\Phi(t, \tau)$ and the matrix B; this is easily transformed into a condition for A and B when these matrices are constant. The general case is more complicated, and will be treated later on in the chapter. Before we embark in this program, we will give a short introduction to *positive definite* matrices.

5.3. A Necessary and Sufficient Condition for Controllability

Let T be an operator mapping R^n into itself, and T its matrix relative to a given basis. If $T = T'$, that is, if T is symmetric, we call the matrix T *positive definite* if

$$(x, Tx) > 0 \tag{5.4}$$

for all $x \in R^n$ such that $x \neq 0$.

If $(x, Tx) \geqslant 0$ for all $x \in R^n$, T is said to be *positive semidefinite*.

There are many criteria to investigate whether a given matrix is positive definite ([2], pp. 306–308; see also Section A.4). We will not need these, however; our main interest lies in the following theorem [3].

THEOREM 5.4. A positive definite (and then symmetric) matrix is nonsingular.

Proof. If $(x, Tx) > 0$ for all $x \neq 0$, it follows from the Cauchy–Schwarz inequality that if x is not zero,

$$0 < (x, Tx) \leqslant \| x \| \| Tx \| \tag{5.5}$$

that is, $\| Tx \| \neq 0$, from which it follows that $Tx \neq 0$ if $x \neq 0$, i.e., the null space of T consists of the zero vector alone. Then T has an inverse.

Positive definite or positive semidefinite matrices with complex entries can be defined in a similar way to the above.

A first theorem on controllability is as follows.

THEOREM 5.5. A system is controllable over $[t_0 , t_1]$ if and only if the symmetric matrix

$$W(t_0 , t_1) = \int_{t_0}^{t_1} \Phi(t_1 , t) B(t) B'(t) \Phi'(t_1 , t) \, dt \tag{5.6}$$

is positive definite.

Proof. If $W(t_0 , t_1)$ is positive definite, it is invertible. Let $x^1 \in R^n$, and put

$$u^1(t) = B'(t) \Phi'(t_1 , t) W^{-1}(t_0 , t_1) x^1 \tag{5.7}$$

$t \in [t_0 , t_1)$, so that

$$x(t_1; t_0 , 0; u^1) = \int_{t_0}^{t_1} \Phi(t_1 , \lambda) B(\lambda) B'(\lambda) \Phi'(t_1 , \lambda) W^{-1}(t_0 , t_1) x^1 d\lambda$$

$$= W(t_0 , t_1) W^{-1}(t_0 , t_1) x^1 = x^1 \tag{5.8}$$

so that the origin at t_0 can be transferred to any state x^1 at t_1 by the action of a control defined over $[t_0 , t_1)$, namely the control (5.7). The system is controllable over $[t_0 , t_1]$.

Let now the system be controllable over $[t_0 , t_1]$ and assume $W(t_0 , t_1)$ is not positive definite. Then there is a nonzero vector x^1, such that $(x^1, W(t_0 , t_1) x^1) = 0$ (as a matter of fact, there are many such vectors). It follows that

$$(x^1, W(t_0 , t_1) x^1) = \int_{t_0}^{t_1} (x^1, \Phi(t_1 , t) B(t) B'(t) \Phi'(t_1 , t) x^1) dt$$

$$= \int_{t_0}^{t_1} (B'(t) \Phi'(t_1 , t) x^1, B'(t) \Phi'(t_1 , t) x^1) dt$$

$$= \int_{t_0}^{t_1} \| B'(t) \Phi'(t_1 , t) x^1 \|^2 dt = 0 \tag{5.9}$$

where the norm (2.32), generated by the inner product operation in R^n, is used.

Since B and $\Phi(t_1 , t)$ are continuous functions of time, the norm inside the integral in (5.9) is also continuous; we have a situation in which the integral over the interval $[t_0 , t_1]$ of a nonnegative, continuous function is zero, which is known to imply that the integrand is identically equal to zero over $[t_0 , t_1]$; this in turn implies that

$$B'(t) \Phi'(t_1 , t) x^1 = 0, \qquad t \in [t_0 , t_1] \tag{5.10}$$

Since the system is controllable over $[t_0 , t_1]$, there is a control u such that

$$x(t_1; t_0 , 0; u) = \int_{t_0}^{t_1} \Phi(t_1 , \lambda) B(\lambda) u(\lambda) d\lambda = x^1 \tag{5.11}$$

Let us compute the norm of x^1; remember that, as a consequence of our assumption that $W(t_0 , t_1)$ is not positive definite, this norm is (supposed to be) nonzero.

However, (5.10) implies

$$(x^1, x^1) = \| x^1 \|^2 = \left(x^1, \int_{t_0}^{t_1} \Phi(t_1, \lambda) B(\lambda) u(\lambda) \, d\lambda \right)$$

$$= \int_{t_0}^{t_1} (B'(\lambda) \Phi'(t_1, \lambda) x^1, u(\lambda)) \, d\lambda$$

$$= 0, \tag{5.12}$$

which contradicts the fact that $x^1 \neq 0$. Hence, our assumption that $W(t_0, t_1)$ is not positive definite is false. The theorem follows.

A simple consequence of this theorem is

THEOREM 5.6.   A system is controllable over $[t_0, t_1]$ if and only if $W(t_0, t_1)$ is nonsingular.

Proof.   If the system is controllable over $[t_0, t_1]$, then $W(t_0, t_1)$ is, by Theorem 5.5, positive definite and thus invertible. If $W(t_0, t_1)$ is invertible, the system is controllable, as follows from the first part of the proof of Theorem 5.5.

We can characterize controllability at t_0 and controllability by a simple extension of these theorems. However, before doing this, we would like to investigate some consequences of the results just obtained. Our main intention is to characterize controllability over $[t_0, t_1]$ by means of properties of the matrices $\Phi(t, t_0)$ and B on this interval; this will enable us to say, for instance, whether controllability over $[t_0, t_1]$ implies controllability over a smaller interval $[t_0, t_2]$, $t_2 < t_1$.

Consider the $n \times n$ matrix function $\Phi(t_1, \cdot) B(\cdot)$. The row vectors will be called $g^1, ..., g^n$; remember that the row vectors of a matrix are the column vectors formed by transposing the rows. Then $W(t_0, t_1)$ is an $n \times n$ matrix with ijth entry

$$\int_{t_0}^{t_1} (g^i(t), g^j(t)) \, dt \tag{5.13}$$

The vectors $g^1, ..., g^n$ are in the space of r vectors whose entries are real-valued functions of the time defined over the interval $[t_0, t_1]$. Let us investigate the independence of these vectors. Since addition of vectors h and g in this space is defined as

$$(h + g)(t) = h(t) + g(t) \qquad \text{for all} \quad t \in [t_0, t_1]$$

the n vectors $g^1, ..., g^n$ are linearly dependent if scalars $c_1, ..., c_n$ can be found, not all of which are zero, such that

$$c_1 g^1(t) + \cdots + c_n g^n(t) = 0 \tag{5.14}$$

for all $t \in [t_0, t_1]$. We will say in this case that the vectors $g^1, ..., g^n$ are linearly dependent *over* $[t_0, t_1]$; it will happen sometimes, as below in Theorem 5.8 and elsewhere, that it is necessary to investigate the dependence of these vectors when considered as vector-valued functions defined on another interval such as $[t_2, t_3]$, with $t_0 < t_2 < t_3 < t_1$; then we use the concept of dependence over $[t_2, t_3]$.

We show a connection between these ideas and properties of the matrix $W(t_0, t_1)$. Multiply Equation (5.14) by $g^1(t)$, and integrate the resulting expression from t_0 to t_1 ; repeat this using $g^2(t), ..., g^n(t)$. The result is simply a set of n linear homogeneous equations for the scalars $c_1, ..., c_n$. It is well known that this set has nonzero solutions if and only if the determinant of the system is zero. Since the matrix of coefficients is $W(t_0, t_1)$, we can say that the vectors $g^1, ..., g^n$ are independent if and only if $W(t_0, t_1)$ is nonsingular. We deduce from this and Theorem 5.6 the following theorem.

THEOREM 5.7. A system is controllable over $[t_0, t_1]$ if and only if the rows of $\Phi(t_1, \cdot) B(\cdot)$ are independent over $[t_0, t_1]$, in the sense that no nonzero scalars $c_1, ..., c_n$ exist such that

$$c_1 g^1(t) + \cdots + c_n g^n(t) = 0$$

for all $t \in [t_0, t_1]$.

This theorem can be stated in another way. We can say that the contention of Theorem 5.7 is true if and only if for all nonzero column vectors a,

$$\Phi(t_1, t) B(t)a \neq 0$$

for all $t \in [t_0, t_1]$.

Note that if a system is controllable over $[t_0, t_1]$, it may happen that a set of scalars $c_1, ..., c_n$, not all of which are zero, may exist so that $c_1 g^1(t) + \cdots + c_n g^n(t) = 0$ for all t in a subinterval of $[t_0, t_1]$; we have, therefore,

THEOREM 5.8. If a system is controllable over $[t_0, t_1]$, it is not necessarily controllable over every subinterval in $[t_0, t_1]$.

For instance, if a system is controllable over $[t_0, t_1]$, it is not necessarily controllable over $[t_0, t_2]$, $t_2 < t_1$. It is, however, controllable over $[t_0, t_3]$, $t_3 > t_1$.

We give now some theorems which characterize controllability at t_0 and controllability.

THEOREM 5.9.   A system is controllable at t_0 if and only if there is a $t_1 > t_0$ such that $W(t_0, t_1)$ is nonsingular.

THEOREM 5.10.   A system is controllable at t_0 if and only if there is a $t_1 > t_0$ such that the rows of $\Phi(t_1, \cdot) B(\cdot)$ are linearly independent over $[t_0, t_1]$.

THEOREM 5.11.   A system is controllable if and only if for each t_0 in its basic interval of definition I there is a number $t_1 > t_0$, $t_1 \in I$, such that $W(t_0, t_1)$ is nonsingular. The number t_1 depends in general on t_0.

THEOREM 5.12.   A system is controllable if and only if for each t_0 in I there is a number t_1, $t_1 \in I$, such that the rows of $\Phi(t_1, \cdot) B(\cdot)$ are linearly independent over $[t_0, t_1]$.

Exercises

1.   Show that if A is positive definite, then A^{-1} is positive definite.

*2.   Show that if A is positive definite, there is a number k such that

$$(x, Ax) \geqslant k \| x \|^2$$

for all $x \in R^n$.

*3.   Suppose that A and B are symmetric, invertible matrices. Show that if the matrix $A - B$ is positive definite, then $B^{-1} - A^{-1}$ is also positive definite.

*4.   Prove that the rows of a matrix Z are independent over an interval $[t_0, t_1]$ if and only if the rows of TZ, T being a constant, nonsingular matrix, are independent over the same interval.

*5.   Use the result of Exercise 4 to show that a system is controllable over $[t_0, t_1]$ if and only if the symmetric matrix

$$\hat{W}(t_0, t_1) = \int_{t_0}^{t_1} \Phi(t_0, t) B(t) B'(t) \Phi'(t_0, t) \, dt$$

is nonsingular.

6.   Show that the matrix $\hat{W}$ in Exercise 5 satisfies the differential equation

$$\frac{d\hat{W}}{dt_0} = -B(t_0) B'(t_0) + A(t_0)W + WA'(t_0)$$

$$W(t_1, t_1) = 0$$

Derive a differential equation satisfied by the matrix W in Theorem 5.5.

7.   Prove the following result, to be used in Exercise 8: A necessary and sufficient condition that there exists an $n \times r$ matrix-valued function $V(t)$, with entries which are of integrable square over $[t_0, t_1]$, such that for some $t_1 > t_0$ the matrix

$$\int_{t_0}^{t_1} H(\lambda) V(\lambda) \, d\lambda$$

is nonsingular, is that for some $t_1 > t_0$ the matrix

$$\int_{t_0}^{t_1} H(\lambda) \, H'(\lambda) \, d\lambda$$

is nonsingular. H is a matrix-valued function of t whose entries are of integrable square over $[t_0 , t_1]$ for any given finite $t_1 > t_0$.

8. Consider the system

$$\dot{x}(t) = B(t) \, u(t)$$

Show by the use of the result in Exercise 7 that this system is controllable over $[t_0 , t_1]$ if and only if

$$M(t_0 , t_1) = \int_{t_0}^{t_1} B(\lambda) \, B'(\lambda) \, d\lambda$$

is nonsingular. The control u is assumed to have square-integrable components over $[t_0 , t_1]$.

9. By putting $y(t) = \Phi^{-1}(t, t_0) \, x(t)$, derive a controllability condition for

$$\dot{x}(t) = A(t) \, x(t) + B(t) \, u(t)$$

by using the result of Exercise 8.

10. Controllability has been defined in the text as a binary quality in the sense that a given system is or is not controllable over an interval $[t_0 , t_1]$. Suppose, however, that the matrix $W(t_0 , t_1)$ has a very small determinant. One would expect that the corresponding system would be difficult, somehow, to control; or, at least, it would be more difficult to control than a system for which the det $W(t_0 , t_1)$ is larger. Investigate these matters. See also

C. D. Johnson, Optimization of a certain quality of complete controllability and observability for linear dynamical systems, *Trans. ASME* 91, 228–238 (1969).

5.4. A Necessary and Sufficient Condition for the Controllability of Constant Systems

When the matrices A and B are constant, it is possible to transform the condition on $W(t_0 , t_1)$ proved in the previous section into a simple condition on the matrices A and B themselves. We follow [*1*] and [*4*].

THEOREM 5.13. A constant system is controllable if and only if the *controllability matrix*

$$Q = [B \ \ AB \ \ A^2B \ \cdots \ A^{n-1}B] \tag{5.15}$$

has rank n. Note that since the dimensions of A and B are $n \times n$ and $n \times r$ respectively, the matrix Q has n rows and rn columns.

Proof. Since the system is constant, it is controllable if it is controllable at $t = 0$. Hence, we need to consider the characteristics of the matrix $W(0, t_1)$.

Let the rank of Q be n, and assume that the system is not controllable at $t = 0$, that is, that $W(0, t_1)$ is singular for all $t_1 > 0$. This in turn implies that there is a nonzero vector x^1 associated with each $t_1 > 0$ such that $W(0, t_1) x^1 = 0$. Let us fix t_1, and derive, just as in the proof of Theorem 5.5, that

$$B'\Phi'(t_1, t)x^1 = 0, \qquad t \in [0, t_1] \tag{5.16}$$

Since $\Phi(t_1, t) = T(t_1 - t) = e^{(t_1-t)A}$, Equation (5.16) is equivalent to

$$B'e^{(t_1-t)A'}x^1 = 0 \tag{5.17}$$

Differentiating $(n - 1)$ times with respect to t and making $t = t_1$ in the resulting expressions we obtain that

$$B'(A')^i x^1 = 0, \qquad i = 0, 1, ..., n - 1 \tag{5.18}$$

which implies that x^1 is orthogonal to all the columns of the matrix Q, since the columns of this matrix are the rows of the matrices $B'(A')^i$, $i = 0, 1, ..., n - 1$.

Since the rank of Q is n, there are, however, n independent columns, $y^1, ..., y^n$, which are a basis for R^n; therefore,

$$x^1 = \sum_{j=1}^{n} (x^1, y^j) y^j = 0$$

since $(x^1, y^j) = 0$, $j = 1, ..., n$. This contradicts our assumption that x^1 is not zero, so that $W(0, t_1)$ is nonsingular for the particular t_1 chosen; the system is controllable over $[0, t_1]$, and then controllable at 0. Note, however, that since t_1 was arbitrarily chosen, the system is controllable over *any* interval $[0, t_1]$ (no matter how small).

We prove now the converse statement. Let the system be controllable, and assume the rank of Q to be less than n. Then there is a nonzero vector x^1 which satisfies (5.18). A consequence of the Hamilton–Cayley equation proved in Chapter 4 is that the nth and higher powers of a matrix are a linear combination of the powers of lesser order, so that the expression (5.18) is actually true for any i. Therefore

$$B'e^{(t_1-t)A'}x^1 = 0, \qquad t \in [0, t_1] \tag{5.19}$$

it follows that x^1, a nonzero vector, is in the null space of $W(0, t_1)$, which is therefore singular. Since t_1 has been arbitrarely chosen, this contradicts our assumption that the system is controllable (since then there would have have to be at least one value $t_1 > t_0$ for which $W(0, t_1)$ is non-singular). The rank of the matrix Q is therefore n.

EXAMPLE

Let A and B be as in the first example of this section. In this case, the matrix Q is

$$\begin{bmatrix} 1 & -1 & 0 & 0 \\ 0 & 0 & 0 & 0 \end{bmatrix}$$

which is of rank 1; we conclude, as we did in a more laborious way before, that the system is not controllable.

In the case of the system of the second example in this section, Q becomes $\begin{bmatrix} 1 & 0 \\ 1 & 0 \end{bmatrix}$, again of rank 1.

EXAMPLE

Consider the system with

$$A = \begin{bmatrix} 1 & 1 \\ 0 & 3 \end{bmatrix}, \qquad B = \begin{bmatrix} b_1 \\ b_2 \end{bmatrix}$$

and let us investigate the conditions on b_1 and b_2 so that the system be controllable. The matrix Q is in this case

$$\begin{bmatrix} b_1 & b_1 + b_2 \\ b_2 & 3b_2 \end{bmatrix}$$

which is of rank 2 if and only if

$$3b_1b_2 - b_1b_2 - b_2{}^2 = b_2(2b_1 - b_2) \neq 0$$

Therefore, the system is controllable if and only if $b_2 \neq 0$ and $2b_1 \neq b_2$.

A direct consequence of the proof of this theorem is:

THEOREM 5.14.   If a constant system is controllable, the origin can be transferred to any state in R^n in an arbitrarily short interval of time by means of an admissible control.

We examine now some consequences and extensions of Theorem 5.13. The determination of the rank of Q by numerical methods may be cumbersome when the dimension of the matrix is large; the problem is equivalent to the one of finding out which of the column vectors (or row vectors) are independent, and is treated in detail in the Appendix.

There are two results, however, which can be helpful in reducing the amount of work involved in the determination of the rank of Q; they are presented in the following two theorems.

THEOREM 5.15.   Let m be the degree of the minimum polynomial of A. Then the matrix

$$Q_1 = [B \; AB \; \cdots \; A^{m-1}B] \qquad (5.20)$$

has the same rank as the matrix Q defined by (5.15).

Proof. If $n = m$, then $Q_1 = Q$ and the theorem follows trivially. Assume then that $m < n$. It follows that the matrices A^m, A^{m+1}, ..., A^{n-1} are linear combinations of powers of A of lesser degree. The corresponding columns in Q are therefore linear combinations of the columns in Q_1. The theorem follows.

The other theorem is by Chen, Desoer, and Niederlinski [5].

THEOREM 5.16.   Let

$$Q_i = [B \; AB \; \cdots \; A^{i-1}B]$$

so that $Q_n = Q$. If k is the least integer such that rank of Q_k = rank of Q_{k+1}, then for all integers $s \geqslant k$, the rank of Q_s equals the rank of Q_k. Hence, to compute the rank of Q_n it is sufficient to determine Q_k and compute its rank.

Proof.   It is a well-known fact that if the dimensions of two subspaces of a vector space are equal, and one subspace is a subset of the other, then the two subspaces are equal; this was suggested in Chapter 2 as an exercise. Here the ranks of Q_k and Q_{k+1} are equal, and the range of Q_{k+1} contains the range of Q_k, since the vectors in the range of Q_k are the linear combinations of the columns of the matrices B, AB, ..., $A^{k-1}B$, while those in the range of Q_{k+1} are the linear combinations of the columns of these matrices and of A^kB as well. These subspaces are therefore identical; call them W. If $x \in W$, $Ax \in W$, since x is a linear combination of the columns of Q_k while Ax is a linear combination of the columns of Q_{k+1}. It follows that for any integer $s \geqslant k$ the columns of A^sB are in W, and since all the columns of Q_k are columns of Q_s the ranges of both matrices, and their ranks, are equal.

EXAMPLE

Let

$$A = \begin{bmatrix} 1 & -1 & 1 & -1 \\ 0 & 1 & -1 & 1 \\ 0 & 0 & 1 & 0 \\ 0 & 0 & 0 & 1 \end{bmatrix}, \qquad B = \begin{bmatrix} 0 \\ 0 \\ 1 \\ 1 \end{bmatrix}$$

Then

$$AB = \begin{bmatrix} 0 \\ 0 \\ 1 \\ 1 \end{bmatrix} = B$$

that is, $k = 1$, and the rank of

$$[B \ AB \ A^2B]$$

is one, since the rank of AB is one and $AB = B$.

Exercises

1. Consider a constant system with scalar input (that is, for which $r = 1$). Prove, by the use of Theorem 5.15, that the system is not controllable if the degree of the minimal polynomial is less than n.

*2. Show that the subspace spanned by the column vectors of Q is an invariant subspace of A.

3. Prove this generalization of Exercise 1: A linear system with scalar input fails to be controllable if and only if there exists a real polynomial $p(s)$ of degree less than n such that $p(A)B = 0$.

5.5. Transformation and Controllability

We show now that if the matrix A is in Jordan canonical form then it is quite simple to determine whether a constant system is controllable. We prove first that controllability is preserved under a change of basis.

THEOREM 5.17.   The constant system

$$\dot{x} = Ax + Bu \tag{5.21}$$

is controllable if and only if the system obtained from (5.21) by putting $x = Sz$, with S a nonsingular matrix, is controllable.

Proof.   Let the system (5.21) be controllable. Then given $x^1 \in R^n$ and $t_1 > 0$, an admissible control u^1 can be found defined on $[0, t_1)$ such that

$$x(t_1; 0, 0; u^1) = x^1 \tag{5.22}$$

This implies that given a vector $z^1 \in R^n$ and $t_1 > 0$, an admissible control u^2 can be found defined on $[0, t_2)$ such that

$$z(t_1; 0, 0; u^2) = z^1 \tag{5.23}$$

Indeed, put $x^1 = Sz^1$, and find u^1 such that (5.22) is satisfied. Then this control serves as u^2 in (5.23). If the transformed system (with z as state vector) is controllable, then (5.21) is proved to be controllable by the same argument, using the matrix S^{-1} instead of the matrix S.

Let the matrix A in (5.21) be in Jordan canonical form. We assume, as in (4.36), that A has q distinct eigenvalues. B is an $n \times r$ matrix. We will use the same notation for the Jordan canonical form as in Chapter 4, while we will partition the matrix B correspondingly:

$$
A = \begin{bmatrix} A_1 & & & & \\ & A_2 & & & \\ & & \cdot & & \\ & & & \cdot & \\ & & & & A_q \end{bmatrix}_{n \times n} , \qquad B = \begin{bmatrix} B_1 \\ B_2 \\ \vdots \\ B_q \end{bmatrix}_{n \times r}
$$

$$
A_i = \begin{bmatrix} A_{i1} & & & & \\ & A_{i2} & & & \\ & & \cdot & & \\ & & & \cdot & \\ & & & & A_{is(i)} \end{bmatrix}_{n_i \times n_i} , \qquad B_i = \begin{bmatrix} B_{i1} \\ B_{i2} \\ \vdots \\ B_{is(i)} \end{bmatrix}_{n_i \times r} \qquad (5.24)
$$

$$
A_{ij} = \begin{bmatrix} \lambda_i & & & & \\ 1 & \lambda_i & & & \\ & 1 & \cdot & & \\ & & & \lambda_i & \\ & & & 1 & \lambda_i \end{bmatrix}_{n_{ij} \times n_{ij}} , \qquad B_{ij} = \begin{bmatrix} b_{fij} \\ b_{2ij} \\ \vdots \\ b_{lij} \end{bmatrix}_{n_{ij} \times r}
$$

The matrices B_{ij} which correspond to each elementary Jordan matrix are composed of *row* vectors b_{kij} ; the first row of B_{ij} is b_{fij} while its last last is b_{lij} .

THEOREM 5.18 [6].   If the matrix A in (5.21) is in Jordan canonical form, the system is controllable if and only if for each $i = 1, ..., q$ the set of $s(i)$ r-dimensional row vectors

$$
b_{fi1} , b_{fi2} , ..., b_{fis(i)} \qquad (5.25)
$$

is a linearly independent set. Note that this set is composed of all the first rows of the matrices B_{ij} associated with a given submatrix A_i .

Proof.   The system (5.21) is controllable if and only if the rows of the matrix $\Phi(t_1 - t) B = e^{(t_1 - t)A}B$ are independent over an interval $[0, t_1]$;

these are independent if and only if the rows of $e^{tA}B$ are independent over the whole real line. Note that it follows from (5.24) that

$$
e^{tA}B = \begin{bmatrix} e^{tA_{11}}B_{11} \\ e^{tA_{12}}B_{12} \\ \vdots \\ e^{tA_{qs(q)}}B_{qs(q)} \end{bmatrix}
$$

the exponentials of the elementary Jordan matrices are given by (4.39).

Assume now that one of the sets (5.25) is linearly dependent for some i. It follows from (4.39) that the first row of $e^{tA_{ij}}B_{ij}$ is $b_{fij}e^{\lambda_i t}$; therefore, the first rows of $e^{tA_{ij}}B_{ij}$, $j = 1, ..., s(i)$, are linearly dependent, since they all share the same subindex i, and thus the same eigenvalue λ_i. The system (5.21) is not controllable. Hence, if the system is controllable, the sets (5.25) are independent for all i.

Suppose now that all these sets are linearly independent. All rows of $e^{tA}B$ in a submatrix associated with an eigenvalue λ_i are independent of any row in a submatrix associated with any other eigenvalue, since functions such as $p_1(t)\, e^{\lambda_j t}$ and $p_2(t)\, e^{\lambda_i t}$, with $p_1(t)$ and $p_2(t)$ nonzero polynomials and $\lambda_i \neq \lambda_j$, are independent over any interval. We have to prove, therefore, that the rows in the submatrix associated with the eigenvalue λ_i are independent for $i = 1, ..., q$. Indeed, the rows of $e^{tA_{ij}}B_{ij}$ are

$$
e^{\lambda_i t}(b_{fij})
$$

$$
e^{\lambda_i t}(tb_{fij} + b_{2ij})
$$

$$
e^{\lambda_i t}(\tfrac{1}{2}t^2 b_{fij} + tb_{2ij} + b_{3ij})
$$

$$
\vdots
$$

from which it follows (after a bit of algebra) that if the sets (5.25) are linearly independent, then all the rows of the matrices $e^{tA_{ij}}B_{ij}$, $j = 1, ..., s(i)$, are independent. The theorem follows.

If $r = 1$, that is, if u is an scalar input, the matrix B has dimensions $n \times 1$, that is, the vectors b_{fij} have only one component. It follows that the set $b_{fij}, ..., b_{fis(i)}$ have to consist of only one vector, that is, $s(i) = 1$ for all i. Hence, a system with a scalar input is controllable if and only if:

  i. there is only one Jordan block associated with each eigenvalue;

  ii. the components of B which correspond to the first row of each Jordan block are nonzero. Several proofs exist of this restricted theorem [4].

EXAMPLE

The system with matrices

$$
A = \begin{bmatrix} 2 & & & & \\ 1 & 2 & & & \\ & 0 & 1 & & \\ & & 1 & 1 & \\ & & & 0 & 4 \end{bmatrix}, \qquad B = \begin{bmatrix} a \\ 0 \\ b \\ 1 \\ d \end{bmatrix}
$$

is completely controllable if and only if a, b, and d are nonzero.

If A is diagonalizable, then each matrix A_{ij} is a 1×1 matrix, so that each matrix B_{ij} consists of one row only, which implies that all the rows associated with each eigenvalue λ_i have to be independent if the system is to be controllable. If all eigenvalues are distinct, there is only one row associated with each eigenvalue; the system is controllable in this case if and only if these rows are nonzero. We have proved that if the matrix A is diagonal with distinct eigenvalues, the system is controllable if and only if all the rows of B are nonzero.

5.6. Controllability and Transfer Functions

We can readily derive a condition for the controllability of a constant system in terms of transfer functions associated with the system. Consider again

$$
\dot{x} = Ax + Bu \tag{5.26}
$$

and take the Laplace transform of both sides; assume $x(0) = 0$. Of course, the transform of a vector is a vector with entries equal to the transforms of the corresponding entries of the original vector. Then,

$$
sX(s) = AX(s) + BU(s)
$$

so that

$$
X(s) = (sI_n - A)^{-1}BU(s) \tag{5.27}
$$

The transforms of the input and state vectors are related by means of the matrix $(sI_n - A)^{-1}B$, provided that $x(0) = 0$. As a matter of fact, this is the transform of the matrix $e^{tA}B$, as can easily seen by writing

$$
x(t) = \int_0^t e^{(t-\lambda)A}Bu(\lambda)\, d\lambda
$$

THEOREM 5.19.   The system (5.26) is controllable if and only if the matrix $(sI_n - A)^{-1}B$ has linearly independent rows, that is, there are no scalars c_1, ..., c_n, not all of which are zero, such that

$$c_1 V^1(s) + \cdots + c_n V^n(s) = 0$$

for all s for which $(sI_n - A)^{-1}B$ is defined. Here $V^1(s)$, ..., $V^n(s)$ are the row vectors of this matrix.

Proof.   Let the system (5.26) be controllable, and let v^1, ..., v^n be the (independent) row vectors of the matrix $e^{tA}B$. Let V^1, ..., V^n be their transforms. Assume that these are dependent, that is, assume there are constants c_1, ..., c_n, not all of which are zero, such that

$$c_1 V^1(s) + \cdots + c_n V^n(s) = 0$$

for all s for which $(sI_n - A)^{-1}B$ is defined. This implies that $c_1 v^1(t) + \cdots + c_n v^n(t) = 0$ for $t \in [0, \infty)$ because the Laplace transformation happens to be invertible, that is, its null space consists only of the zero vector. (Note that the domain of the Laplace transformation here is the set of row vectors with entries which are functions of the time on $[0, \infty)$). It follows, of course, that the row vectors v^1, ..., v^n are dependent, which contradicts our initial assumption, so that the rows of $(sI_n - A)^{-1}B$ are independent. The sufficiency of this condition for the controllability of (5.26) is proved similarly.

We assume now that the input to the system (5.26) is scalar, that is, that the dimension r of the input vector equals unity. The matrix $(sI_n - A)^{-1}B$ is of dimension $n \times 1$. Suppose that this matrix is calculated by means of Cramer's rule. Then, if no entry of this matrix has poles which coincides with one or more zeros of the same entry, we say that the matrix $(sI_n - A)^{-1}B$ has *no cancellation*. We will not prove the following theorem [7].

THEOREM 5.20.   The system (5.26) with scalar input is controllable if and only if the matrix $(sI_n - A)^{-1}B$ has no cancellation.

EXAMPLE

Consider the system

$$\dot{x}_1 = -2x_2 + 2u$$

$$\dot{x}_2 = -x_1 - 3x_2 + u$$

Let us check first if this system is controllable. Since

$$A = \begin{bmatrix} 0 & -2 \\ 1 & -3 \end{bmatrix}, \qquad B = \begin{bmatrix} 2 \\ 1 \end{bmatrix}$$

it follows that

$$[B\ AB] = \begin{bmatrix} 2 & -2 \\ 1 & -1 \end{bmatrix}$$

which is of rank one; the system is not controllable. On the other hand, the transfer function relating the transforms of x_1 and u can be easily computed as

$$\frac{s + 2}{(s + 1)(s + 2)}$$

in which a cancellation occurs.

5.7. A Condition for the Controllability of Time-Varying Systems

We have developed in the previous sections a rather complete collection of necessary and sufficient conditions for the controllability of constant systems. We give now a lone sufficient condition for the controllability of systems whose matrices A and B depend explicitly on the time. The following presentation is taken form Chang [8].

THEOREM 5.21.   Consider the system $\dot{x} = Ax + Bu$, with A and B defined on an interval I. Assume that A and B have entries which are $(n - 2)$ and $(n - 1)$ times continuously differentiable respectively. Let

$$B_1 = B$$

$$B_i = -AB_{i-1} + \dot{B}_{i-1}, \qquad i = 2, ..., n$$

Define

$$Q(t) = [B_1(t)\ B_2(t)\ \cdots\ B_n(t)], \qquad t \in I$$

Then the system is controllable at t_0 if the rank of $Q(t)$ equals n at some $t_1 > t_0$, t_0, $t_1 \in I$.

Proof.   Remember that a system is controllable over $[t_0, t_1]$ if and only if the rows of $\Phi(t_1, \cdot)\,B(\cdot)$ are independent over $[t_0, t_1]$. It will be convenient in this proof to make use of the fact that this condition is satisfied if and only if the rows of $\Phi(t_0, \cdot)\,B(\cdot)$ are independent over $[t_0, t_1]$, which follows by multiplying $\Phi(t_1, \cdot)\,B(\cdot)$ by the nonsingular matrix $\Phi(t_0, t_1)$.

Assume that the rank of $Q(t)$ is n for some $t_1 > t_0$ but that the system is not controllable over $[t_0, t_1]$. Then the rows of $\Phi(t_0, \cdot)\,B(\cdot)$ are not independent over $[t_0, t_1]$, that is, there is a constant, nonzero column vector y such that

$$y'\Phi(t_0, t)\,B(t) = 0, \qquad t \in [t_0, t_1]$$

Let $y'\Phi(t_0, t) \equiv \alpha(t_0, t)$, so that

$$\alpha(t_0, t) B(t) = 0, \qquad t \in [t_0, t_1] \tag{5.28}$$

It should be remembered from Chapter 3 that the matrix $\Phi(t_0, t) = \Phi^{-1}(t, t_0)$ satisfies the equation $\dot{\Phi} = -\Phi A$, from which it follows that $\alpha(t_0, t)$ satisfies

$$\dot{\alpha}(t_0, t) = -\alpha(t_0, t) A(t)$$

with initial condition $\alpha(t_0, t_0) = y \neq 0$. Hence, by the uniqueness theorem of homogeneous system presented in Chapter 3, $\alpha(t_0, t) \neq 0$ for all $t \in [t_0, t_1]$.

If we differentiate (5.28)

$$\dot{\alpha}(t_0, t) B(t) + \alpha(t_0, t) \dot{B}(t) = -\alpha(t_0, t) A(t) B(t) + \alpha(t_0, t) \dot{B}(t)$$
$$= \alpha(t_0, t) B_2(t) = 0$$

In a similar way, we can show that

$$\alpha(t_0, t) B_i(t) = 0, \qquad i = 3, ..., n$$

for $t \in [t_0, t_1]$. Since $\alpha(t_0, t_1)$ is not zero, it follows that all the columns of the matrix $Q(t_1)$ are orthogonal to a nonzero vector; this implies that the rank of $Q(t_1)$ cannot be n because if this were the case it would have n independent columns; it cannot have these because n independent columns cannot be orthogonal to a nonzero vector. Indeed, let $b^1,..., b^n$ be these columns. If $(x, b^i) = 0$, $i = 1, ..., n$, then $x = 0$. We have contradicted our assumption that the rank of $Q(t_1)$ is n, so that it follows that the system is controllable over $[t_0, t_1]$, that is, it is controllable at t_0.

It can be proved [8] that the condition of this theorem is also necessary if the components of A and B can be expanded in a Taylor series at each point of their domain of definition. Of course, if A and B are constant, the condition of this theorem coincides with the condition for controllability of constant systems.

A different proof of this theorem has been given by Silverman and Meadows [9].

EXAMPLE

Consider Hill's equation with a forcing term

$$\ddot{y}(t) + [a - f(t)] y(t) = b(t) u(t)$$

where f is a periodic function. Put $x_1 = y$, $x_2 = \dot{y}$; then

$$\dot{x}_1(t) = x_2(t)$$
$$\dot{x}_2(t) = -[a - f(t)] x_1(t) + b(t) u(t) \tag{*}$$

It may be that the periodicity of f influences in some interesting ways the controllability of (∗). However,

$$Q(t) = \begin{bmatrix} 0 & -b(t) \\ b(t) & \dot{b}(t) \end{bmatrix}$$

which implies (∗) is controllable at any t_0 such that $b(t) \neq 0$ for some $t > t_0$. The periodic system (∗) apparently has no controllability properties which are consequence of its periodicity.

5.8. Observability

It is necessary to study now a concept which is in many ways related to that of controllability, *observability*. Due to this relationship, most of the results already obtained will readily provide corresponding theorems and conditions for the observability of systems. Let

$$\dot{x} = Ax + Bu, \qquad y = Cx + Du \tag{5.29}$$

and assume that u and y are known over an interval $[t_0, t_1]$. A problem of engineering interest which has been encountered already in Chapter 1 is to determine the initial condition on the state vector, $x(t_0)$, from the values of the input and output vectors over $[t_0, t_1]$.

Consider now the *unforced system without direct transmission from input to output*

$$\dot{x} = Ax, \qquad y = Cx \tag{5.30}$$

We show that if it is possible to determine $x(t_0)$ from the values of y over $[t_0, t_1]$ for the system (5.30), it is possible to determine $x(t_0)$ from the values of y and u over $[t_0, t_1]$ for (5.29). In other words, the input does no play a significant role in the solution of this problem. Indeed, assume that $x(t_0)$ can be determined from the values of y over $[t_0, t_1]$ for (5.30). This means that Equation (5.31b) determines $x(t_0)$ uniquely

$$x(t) = \Phi(t, t_0) x(t_0) \tag{5.31a}$$

$$y(t) = C(t) \Phi(t, t_0) x(t_0), \qquad t \in [t_0, t_1] \tag{5.31b}$$

We can write for the output of (5.29)

$$y(t) - C(t) \int_{t_0}^{t} \Phi(t, \lambda) B(\lambda) u(\lambda) \, d\lambda - D(t) u(t)$$

$$= C(t) \Phi(t, t_0) x(t_0), \qquad t \in [t_0, t_1] \tag{5.32}$$

Clearly, if the output y and the input u are known over $[t_0, t_1]$, the left-hand side of (5.32) is known, and Equation (5.32) is of the same form as (5.31b); if this determines $x(t_0)$ uniquely, so does (5.32). Therefore, it is sufficient for us to consider the problem of determining $x(t_0)$ from the values of y over $[t_0, t_1]$ for the system (5.30).

Definition [*1*].   A system such as (5.30) is *observable* (or *completely observable*) on $[t_0, t_1]$ if the state at t_0 can be determined from the knowledge of the values of the output y on $[t_0, t_1]$; of course, t_0 and t_1 are to be in the interval of definition I of the system. Note that for (5.30) to be observable on $[t_0, t_1]$, *any* initial state in the state space should be able to be determined from knowledge of y on $[t_0, t_1]$.

Definition.   A system is said to be *observable* (or *completely observable*) at t_0, $t_0 \in I$, if there is $t_1 > t_0$, $t_1 \in I$, such that the system is observable on $[t_0, t_1]$.

Definition.   A system is *observable* (or *completely observable*) if it is observable at any $t_0 \in I$ (with the obvious exception of the right-hand limit of this interval).

Again, we show that these definitions are not trivial by means of an example.

EXAMPLE

Let

$$A = \begin{bmatrix} 2 & 0 \\ 0 & 2 \end{bmatrix}, \qquad C = \begin{bmatrix} 0 & 1 \\ 0 & 1 \end{bmatrix}$$

then

$$\Phi(t, t_0) = \begin{bmatrix} e^{2(t-t_0)} & 0 \\ 0 & e^{2(t-t_0)} \end{bmatrix}$$

and

$$y(t; x^0, 0; 0) = \begin{bmatrix} 0 & 0 \\ 0 & e^{2(t-t_0)} \end{bmatrix} x^0 \tag{*}$$

If the components of y are y_1, y_2 and those of x^0 are $x_1{}^0$, $x_2{}^0$, the expression (*) is equivalent to

$$y_1(t) = 0$$

$$y_2(t) = e^{2(t-t_0)} x_2{}^0$$

The knowledge of y_1 and y_2 over an interval $[t_0, t_1]$ (indeed, at a single point of this interval) determines $x_2{}^0$; however, there is no way to ascertain $x_1{}^0$ from the values of y_1 and y_2 over an interval.

THEOREM 5.22.   The system (5.30) is observable on $[t_0, t_1]$ if and only if the matrix

$$M(t_0, t_1) = \int_{t_0}^{t_1} \Phi'(t, t_0) C'(t) C(t) \Phi(t, t_0) dt$$

is nonsingular.

Proof.   Assume $M(t_0, t_1)$ is singular. Then, since

$$y(t) = C(t) \Phi(t, t_0) x^0, \qquad t \in [t_0, t_1]$$

it follows that,

$$\int_{t_0}^{t_1} \Phi'(t, t_0) C'(t) y(t) dt = M(t_0, t_1) x^0 \tag{5.33}$$

If $M(t_0, t_1)$ is singular, then for a given y defined on $[t_0, t_1]$, there will be in general several solutions x^0 which satisfy (5.33), so the system is not completely observable at t_0. If $M(t_0, t_1)$ is nonsingular, then x^0 can be determined uniquely from (5.33), that is, from a knowledge of the values of y over $[t_0, t_1]$. The theorem follows.

Of course, this theorem implies that the system is observable on $[t_0, t_1]$ if and only if $M(t_0, t_1)$ is positive definite and if and only if *the columns* of $C(\cdot) \Phi(\cdot, t_0)$ are independent over $[t_0, t_1]$. The system is observable at t_0 if and only if these conditions are satisfied at t_0 for some $t_1 > t_0$, and it is observable if and only if they are satisfied at very t_0 in I (with the exception of the right-hand limit) for some $t_1 > t_0$, $t_1 \in I$.

It should be apparent that there are similarities between the conditions for observability and controllability. As a matter of fact, we can write by inspection the theorems and conditions on observability from the corresponding ones on controllability. Before we embark on this program, we prove

THEOREM 5.23.   If a constant system is observable at $t_0 = 0$, then it is observable at any time t_0, and is therefore observable.

Proof.   If the system is completely observable at 0, then the matrix

$$M(0, t_1) = \int_0^{t_1} e^{tA'} C' C e^{tA} dt \tag{5.34}$$

is nonsingular for some $t_1 > t_0$. Put $\lambda = t + t_0$. Then,

$$M(0, t_1) = M(t_0, t_1 + t_0) = \int_{t_0}^{t_1 + t_0} e^{A'(\lambda - t_0)} C' C e^{A(\lambda - t_0)} d\lambda$$

is nonsingular, and the system is completely observable at t_0.

The following theorem provides the tools for the derivation of many results concerning observability from corresponding theorems on controllability.

THEOREM 5.24 [1].   Consider the systems

$$\dot{x} = Ax, \qquad y = Cx \tag{5.35}$$

and

$$\dot{z} = -A'z + C'u \tag{5.36}$$

The system (5.36) is known as the *adjoint* system of (5.35). Then (5.35) is observable on $[t_0, t_1]$ if and only if its adjoint system is controllable over $[t_0, t_1]$.

Proof.   We know from the results of Chapter 3 that the transition matrix of the adjoint system (5.36) is $\Phi(\tau, t)'$ if $\Phi(t, \tau)$ is the corresponding transition matrix of (5.35). The system (5.36) is, therefore, controllable over $[t_0, t_1]$ if and only if the rows of $\Phi(\cdot, t_1)' C'(\cdot)$ are independent over this interval, which is equivalent to the property that the columns of $C(\cdot) \Phi(\cdot, t_1)$ be independent over the the same interval. It should be clear that if we multiply this matrix by the nonsingular matrix $\Phi(t_1, t_0)$ as follows

$$C(t) \Phi(t, t_1) \Phi(t_1, t_0) = C(t) \Phi(t, t_0)$$

the columns of the resulting matrix are independent over $[t_0, t_1]$ if and only if the columns of the first matrix have this property. Of course, this is recognized as being the condition for the observability of (5.35) on $[t_0, t_1]$.

We derive readily from Theorem 5.14:

THEOREM 5.25.   If the initial state x^0 of a constant system can be ascertained from the values of the output for $t \in [0, t_1]$, it can be also be obtained from the values of the output on $[0, t^1]$, where t^1 is an arbitrarily small value of the time.

Theorem 5.13 applied to the system (5.36) implies:

THEOREM 5.26.   A constant system is observable if and only if the matrix

$$P = [C' \quad A'C' \quad A'^2C' \quad \cdots \quad A'^{(n-1)}C']$$

has rank n.

Of course, Theorems 5.15 and 5.16 apply in this case as well.

It is left to the reader to extend the results of Theorem 5.21 on the controllability of time-varying systems to corresponding conditions for the observability of such systems.

As a last topic in this section, we obtain some results on the observability of constant systems when the matrix A of such a system is in Jordan canonical form. We prove first:

THEOREM 5.27.   A constant system

$$\dot{x} = Ax, \qquad y = Cx \tag{5.37}$$

is observable if and only if the system obtained from (5.37) by making $x = Sz$, with S a nonsingular $n \times n$ matrix, is observable.

Proof.   If x^0 can be determined from the values of y for $t \in [0, t_1]$, z^0 can be determined from the same values by putting $z^0 = S^{-1}x^0$. If z^0 can be determined from the values $y(t)$, $t \in [0, t_1]$, $x^0 = Sz^0$.

We present now the counterpart of Theorem 5.18. Let A be in Jordan form in the system (5.37). We partition the matrix C as follows:

$$A = \begin{bmatrix} A_1 & & & \\ & A_2 & & \\ & & \ddots & \\ & & & A_q \end{bmatrix}_{n \times n}$$

$$C = [C_1 \ C_2 \ \cdots \ C_q]_{m \times n}$$

$$A_i = \begin{bmatrix} A_{i1} & & & \\ & A_{i2} & & \\ & & \ddots & \\ & & & A_{is(i)} \end{bmatrix}_{n_i \times n_i} \tag{5.38}$$

$$C_i = [C_{i1} \ C_{i2} \ \cdots \ C_{is(i)}]_{m \times n_i}$$

$$A_{ij} = \begin{bmatrix} \lambda_i & & & & \\ 1 & \lambda_i & & & \\ & 1 & \ddots & & \\ & & \ddots & \lambda_i & \\ & & & 1 & \lambda_i \end{bmatrix}_{n_{ij} \times n_{ij}}$$

$$C_{ij} = [c_{fij} \ \cdots \ c_{lij}]_{m \times n_{ij}}$$

The vectors c_{fij} , ..., c_{lij} are column vectors of dimension m.

THEOREM 5.28 [6].   The system (5.37) is observable if and only if for
each $i = 1, ..., q$, the set of $s(i)$ m-dimensional column vectors

$$c_{li1}, c_{li2}, ..., c_{lis(i)}$$

is a linearly independent set.

The proof is similar to the proof of Theorem 5.18.

5.9. The Controllability and Observability of Composite Systems

Suppose that a system is composed of several subsystems connected
in some way. If the subsystems are all controllable (or observable), is
the overall system controllable (or observable)? If one or more of the
subsystems are not controllable (or observable), can the overall system
still be controllable (or observable)? We will answer these questions in
this section for a configuration known as the *parallel connection* of two
systems. We follow Chen and Desoer [6].
Let the two systems S_a and S_b given by

$$\dot{x}^k = A_k x^k + B_k u^k$$
$$y^k = C_k x^k, \quad k = a, b \tag{5.39}$$

Without loss of generality, it can be assumed that the matrices A_k,
$k = a, b$, are in Jordan canonical form and that the matrices D_k are
identically zero. We say that a system S is the *parallel connection* of S_a
and S_b if its output y is

$$y = y^a + y^b \tag{5 40}$$

and if its input u is

$$u = u^a = u^b \tag{5.41}$$

A schematic representation of such a system S is indicated in Figure

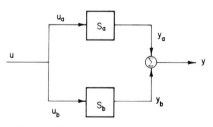

FIGURE 5.1.   The system S_a and S_b are connected in parallel to form the system S.

5.1. Clearly, the dimension of the input and of the output vectors should be the same for both systems if the parallel connection of them is to be well defined. If this is the case, we can define a state vector x for S as follows:

$$\dot{x} = \begin{bmatrix} \dot{x}^a \\ \dot{x}^b \end{bmatrix} = \begin{bmatrix} A_a & 0 \\ 0 & A_b \end{bmatrix} \begin{bmatrix} x^a \\ x^b \end{bmatrix} + \begin{bmatrix} B_a \\ B_b \end{bmatrix} u$$

$$y = [C_a \ C_b] \begin{bmatrix} x^a \\ x^b \end{bmatrix} = [C_a \ C_b] [x]$$

(5.42)

We assume that all the matrices in (5.42) are partitioned as indicated in Theorems 5.18 and 5.28. Let

$$\Lambda_a = \{\lambda_{ai} : i = 1, ..., q_a\} \quad \text{and} \quad \Lambda_b = \{\lambda_{bi} : i = 1, ..., q_b\}$$

be the sets of distinct eigenvalues of A_a and A_b respectively, and let the vectors components of the matrices B_{aij} and B_{bij} be designated by b_{afij}, ... and b_{bfij}, ...; similarly for the components of the matrices C_{aij} and C_{bij}.

If S is controllable (or observable), then S_a and S_b must be controllable (or observable). On the other hand, assume that S_a and S_b are controllable and observable. If the eigenvalues of A_a and A_b are not the same, that is, if Λ_a and Λ_b are disjoint, then (5.42) is already partitioned in the form of Theorems 5.18 and 5.28, and S is controllable and observable, since the conditions of these theorems are automatically satisfied for S if they are satisfied for S_a and S_b. We have

THEOREM 5.29 [6].   If the eigenvalues of A_a and A_b are distinct, the system S is controllable (observable) if and only if S_a and S_b are controllable (observable).

If Λ_a and Λ_b are not disjoint, Equations (5.42) are not in the form of Theorems 5.18 and 5.28, but can be put easily in this form by rearranging the Jordan blocks. Then,

THEOREM 5.30 [6].   Let $\Lambda_a \cap \Lambda_b \neq \emptyset$. If S_a and S_b are controllable (observable), then S is controllable (observable) if and only if each pair of common eigenvalues, say $\lambda_{a\alpha} = \lambda_{b\beta}$ is such that the set of $s_a(\alpha) + s_b(\beta)$ row vectors

$$b_{af\alpha 1}, b_{af\alpha 2}, ..., b_{af\alpha s_a(\alpha)}; \quad b_{bf\beta 1}, b_{bf\beta 2}, ..., b_{bf\beta s_b(\beta)}$$

is a linearly independent set. [The set of $s_a(\alpha) + s_b(\beta)$ column vectors

$$c_{al\alpha 1}, c_{al\alpha 2}, ..., c_{al\alpha s_a(\alpha)}; \quad c_{bl\beta 1}, c_{bl\beta 2}, ..., c_{bl\beta s_b(\beta)}$$

is a linearly independent set.]

If u is a scalar input, and if S_a and S_b are controllable (observable), S is controllable (observable) if and only if $\Lambda_1 \cap \Lambda_2 = \varnothing$. This follows directly from the fact that S is controllable (observable) if and only if each distinct eigenvalue has only one Jordan block associated with it, provided that S_a and S_b are controllable (observable).

The reader is referred to the literature for analogous results concerning the tandem [6, 10] and feedback [11] connections of subsystems.

5.10. Normal Systems

This concept was introduced by LaSalle [12]. A system

$$\dot{x} = Ax + Bu, \qquad y = Cx + Du \tag{5.43}$$

is said to be *normal* if for $k = 1, 2, ..., r$ it is controllable by all inputs of the form

$$u^k = \begin{bmatrix} 0 \\ 0 \\ \vdots \\ u_k \\ \vdots \end{bmatrix} \tag{5.44}$$

that is, if the origin can be transferred to any state in R^n in finite time by the use of a control in which all but one of the entries are zero. Note that we demand that this requirement be satisfied by controls with only the first entry different from zero, by controls with the second entry different from zero, etc. Clearly, normality is a property associated with the matrices A and B. If the columns of B are $b^1, ..., b^r$, then our definition is equivalent to saying that (5.43) is normal if each of the following systems is controllable:

$$\begin{aligned} \dot{x} &= Ax + b^1 u \\ &\vdots \\ \dot{x} &= Ax + b^r u \end{aligned} \tag{5.45}$$

One can, of course, write the necessary and sufficient condition for controllability over $[t_0, t_1]$ of each of the systems (5.45) in terms of the transition matrix of (5.43) and the vectors $b^1, ..., b^r$: the matrices

$$W_k(t_1, t_0) = \int_{t_0}^{t_1} \Phi(t_1, t) \, b^k(t) \, b'^k(t) \, \Phi'(t_1, t) \, dt, \qquad k = 1, ..., r \tag{5.46}$$

should be nonsingular. If A and B are constant matrices, (5.43) is normal if and only if the rank of each of the following r $n \times n$ matrices is n:

$$[b^1 \; Ab^1 \; \cdots \; A^{n-1}b^1]$$
$$\vdots$$
$$[b^r \; Ab^r \; \cdots \; A^{n-1} \; b^r]$$

Note that if a system is normal, it is completely controllable, since if the matrices (5.46) are nonsingular then the matrix $W(t_0, t_1)$ is nonsingular; also note that the inverse statement does not hold, that is, a completely controllable system is not necessarily normal.

5.11. Output Controllability

Kreindler and Sarachik [1] have introduced the concept of *output controllability*. In their terminology, what we have called controllability would be called *state controllability*. A system is *output controllable* over $[t_0, t_1]$ if every output y^1 can be attained at t_1 given that the state at t_0 equals zero. Clearly, if $D = 0$, the role of the matrix $\Phi(t_1, t) B(t)$ in the definition of $W(t_0, t_1)$ is taken by $C(t) \Phi(t_1, t) B(t)$, so that we have:

THEOREM 5.31. A system is output controllable if and only if the matrix

$$V(t_0, t_1) = \int_{t_0}^{t_1} C(t) \Phi(t_1, t) B(t) B'(t) \Phi'(t_1, t) C'(t) \, dt$$

is nonsingular, provided the matrix $D = 0$.

THEOREM 5.32. If the matrices A, B, and C are constant, and $D = 0$, the system is completely output controllable if and only if the matrix

$$[CB \; CAB \; CA^2B \; \cdots \; CA^{n-1}B]$$

has range n.

In [1], Kreindler and Sarachik also study the output controllability of systems with direct transmission between input and output, that is, systems for which $D \neq 0$.

5.12. Total Controllability. Total Observability

Theorem 5.14 and 5.25 indicate that, if a constant system is controllable (or observable) over an interval $[t_0, t_1]$, then it is controllable (or observable) over any other interval no matter how small. It was pointed out in Theorem 5.8 that this is not necessarily true for the controllability of systems which are not constant; this result, which is extended easily to cover observability, leads us to make the following definitions [*1*, *9*, *13 – 15*].

Definition.   Let a linear system be defined on a basic interval I. The system is *totally controllable on a subinterval* $J \subset I$ if it is controllable on every subinterval of J. The name *differentially controllable* is used synonymously with totally controllable.

Definition.   A linear system is *totally observable (differentially observable)* on J if it is observable on every subinterval of J.

The following theorem is fairly obvious:

THEOREM 5.33.   A system is totally controllable (observable) on J if and only if the rows of the matrix $\Phi(t_1, \cdot) B(\cdot)$ [the columns of the matrix $C(\cdot) \Phi(\cdot, t_0)$] are independent on every subinterval of J.

A more subtle condition is given in this last theorem. We prove only the sufficiency of its contention, remarking that its necessity can also be demonstrated [*15*].

THEOREM 5.34.   Let the matrices A and B have entries which are $(n-2)$ and $(n-1)$ times continuously differentiable respectively. Define the $n \times n$ matrix Q as in Theorem 5.21. Then the system is totally controllable on J if and only if Q has rank n almost everywhere on this set.

Proof of sufficiency.

Let $z^1, ..., z^n$ be row vectors of n entries which are functions of a real variable, and which have $(n-1)$ continuous derivatives. We define the Wronskian matrix for these vector functions as follows:

$$W = \begin{bmatrix} z^1 & \dot{z}^1 & \cdots & \overset{(n-1)}{z^1} \\ \vdots & \vdots & & \\ z^n & \dot{z}^n & \cdots & \overset{(n-1)}{z^n} \end{bmatrix} \tag{5.47}$$

Note that each entry of this matrix is a row vector of n components. We show first that the rows of W are linearly dependent over an interval J if the rows of the matrix

$$Z = \begin{bmatrix} z^1 \\ \vdots \\ z^n \end{bmatrix} \tag{5.48}$$

are linearly dependent over J. Clearly, if there are scalars $c_1, ..., c_n$, not all of which are zero, such that

$$c_1 z^1(t) + \cdots + c_n z^n(t) = 0, \qquad t \in J$$

then

$$c_1 \dot{z}^1(t) + \cdots + c_n \dot{z}^n(t) = 0, \qquad t \in J$$

$$\vdots$$

$$c_1 \overset{(n-1)}{z^1}(t) + \cdots + c_n \overset{(n-1)}{z^n}(t) = 0, \qquad t \in J$$

and the contention follows easily.

Let us compute now the Wronskian matrix of the rows of the matrix $\Phi(t, \tau) B(\tau)$. Let X be a fundamental matrix of the system. Then

$$\frac{d}{d\tau} \Phi(t, \tau) B(\tau) = X(t) \frac{d}{d\tau} (X^{-1}(\tau) B(\tau))$$

$$= X(t) \left(\frac{d}{d\tau} (X^{-1}(\tau)) B(\tau) + X^{-1}(\tau) \frac{d}{d\tau} B(\tau) \right)$$

$$= \Phi(t, \tau) \left(-A(t) B(t) + \frac{d}{d\tau} B(\tau) \right)$$

since $d/dt(X^{-1}) = -X^{-1}A$. In general,

$$\frac{d^k}{d\tau^k} (\Phi(t, \tau) B(\tau)) = \Phi(t, \tau) B_{k+1}(\tau) \tag{5.49}$$

where B_k, $k = 1, ..., n$, are the matrices defined in Theorem 5.21. Of course, (5.49) implies that the Wronskian matrix of the rows of $\Phi(t, \tau)B(\tau)$ is

$$W = \Phi(t, \tau) Q(\tau) \tag{5.50}$$

Let now the matrix Q have rank n almost everywhere on an interval J, and assume that the rows of $\Phi(t, \tau) B(\tau)$ are linearly dependent (as functions of τ, over J), so that the system is not totally controllable. As shown above, this implies that the rows of W are also dependent on J, that is, the rank of W is less than n everywhere on J. Since $\Phi(t, \tau)$ is everywhere nonsingular, the matrix Q must have rank less than n everywhere on J, which contradicts our original assumption. The theorem follows.

Total observability can be characterized in a similar way.

5.13. The Attainable Set

Let the state vector of a system be defined as usual by the equation

$$\dot{x} = Ax + Bu \qquad (5.51)$$

We have previously characterized the set of admissible inputs as those inputs whose entries are piecewise-continuous functions of the time. It should be recognized that we might, however, be interested in studying the set of responses of the system (5.51) when the set of admissible inputs is perhaps more restricted, such as the set of all vector-valued functions with r components each of which is a real-valued function of the time with range the interval $[-1, 1]$.

We say that a point $x^1 \in R^n$ is *attainable* from x^0 if there exists a control u from a class U of admissible controls such that $x(t_0) = x^0$ is transferred by u to $x(t_1) = x^1$ for some $t_1 > t_0$ [16]. The set of all points which are attainable from x^0 is called *the attainable set from x^0*. The set of points which are attainable at $t_1 > t_0$ is called *the attainable set at t_1 from x^0 at t_0*. For instance, if the class U is the set of vectors with piecewise-continuous components considered in the previous sections, and the system (5.51) is controllable, then the attainable set from x^0 is, for arbitrary x^0, the whole of the space R^n.

Most of the interesting properties of the attainable set show up when the class U is composed of control vectors whose components are measurable functions (besides satisfying other conditions, such as the magnitude of each component of the control vector being bounded by a number M), and are beyond the scope of this text. Here we will only prove a simple, but basic, theorem.

Definition.   A set $S \subseteq R^n$ is said to be *convex* if for every x and y in S, and for every real nonnegative λ such that $\lambda \leqslant 1$, the point $\lambda x + (1 - \lambda)y$ is also in S.

Convex sets appear in many areas of control theory. We prove:

THEOREM 5.35.   Consider a system such as (5.51). If the set U of admissible controls is convex, then the attainable set from x^0 at t is convex for all t and all x^0.

Proof.   This theorem follows trivially from the explicit expression for the state x of system (5.51) and the fact that if u^1 and u^2 are in U, then $\lambda u^1 + (1 - \lambda)\, u^2$ are also there, for $0 \leqslant \lambda \leqslant 1$.

A class U which satisfies the conditions of the theorem is the one characterized by $|\, u_j\, | \leqslant M, j = 1, ..., r$; clearly, if u^1 and u^2 are in U,

$\lambda u^1 + (1 - \lambda) u^2$ are also in U, since then $|\lambda u_j{}^1 + (1 - \lambda) u_j{}^2| \leqslant M$, $j = 1, ..., r$.

5.14. The Phase-Canonical Form

Consider once more a constant system

$$\dot{x} = Ax + bu \tag{5.52}$$

with u an scalar input, so that b is a $n \times 1$ matrix. We show that if the system (5.52) is controllable, then there is a nonsingular matrix S such that the system derived from (5.52) by making $x = Sz$ has the form

$$\dot{z} = \hat{A}z + \hat{b}u \tag{5.53}$$

with $\hat{A} = S^{-1}AS$, $\hat{b} = S^{-1}b$, and

$$\hat{A} = \begin{bmatrix} 0 & 1 & 0 & \cdots & 0 \\ 0 & 0 & 1 & \cdots & 0 \\ \cdot & \cdot & \cdot & & \vdots \\ \cdot & \cdot & \cdot & & 1 \\ \cdot & \cdot & \cdot & & 1 \\ -a_0 & -a_1 & -a_2 & \cdots & -a_{n-1} \end{bmatrix}, \quad \hat{b} = \begin{bmatrix} 0 \\ \cdot \\ \cdot \\ \cdot \\ 0 \\ 1 \end{bmatrix} \tag{5.54}$$

Here a_0, a_1,..., a_{n-1} are the coefficients of the characteristic polynomial of A,

$$\det(\lambda I_n - A) = \lambda^n + a_{n-1}\lambda^{n-1} + \cdots + a_1\lambda + a_0$$

The form (5.53), with matrices $\hat{A}$ and $\hat{b}$ as in (5.54), is said to be the *phase-canonical form* of the controllable system (5.52).

We show the existence of the matrix S which effects the transformation from (5.52) to (5.53). Consider the set of n vectors

$$\begin{aligned} e^n &= b \\ e^{n-1} &= (A + a_{n-1}I_n)b \\ e^{n-2} &= (A^2 + a_{n-1}A + a_{n-2}I_n)b \\ &\vdots \\ e^1 &= (A^{n-1} + a_{n-1}A^{n-2} + \cdots + a_1I_n)b \end{aligned} \tag{5.55}$$

We shall prove below in the lemma that if the system (5.52) is controllable, these n vectors are independent, and constitute then a basis for

R^n (or C^n). Relative to this new basis, the matrices A and B take on the forms (5.54); indeed,

$$Ae^n = Ab = e^{n-1} - a_{n-1}b = e^{n-1} - a_{n-1}e^n$$
$$Ae^{n-1} = A^2b + a_{n-1}Ab = e^{n-2} - a_{n-2}b = e^{n-2} - a_{n-2}e^n$$
$$\vdots$$
$$Ae^1 = -a_0e^n$$

The contention with respect to the form of $\hat{A}$ follows by inspection from these equalities; the contention concerning $\hat{b}$ follows from the first equality in (5.55). We only have to show that the vectors $e^1,..., e^n$ are independent.

LEMMA. If the constant system with scalar input (5.52) is controllable, the n vectors $e^1,..., e^n$ in (5.55) are linearly independent.

Proof. Assume that (5.52) is controllable (so that the vectors b, Ab,..., $A^{n-1}b$ are independent), and that the vectors $e^n,..., e^1$ in (5.55) are linearly dependent. Then there are scalars $c_1,..., c_n$, not all of which are zero, such that $\sum_{i=1}^n c_ie^i = 0$. However, we can write

$$\sum_{i=1}^n c_ie^i = \sum_{i=1}^n c_i(A^{n-i}b) + d \tag{5.56}$$

Suppose that the nonzero term with highest exponent for the matrix A in the sum in the right side of (5.56) is $c_k(A^{n-k}b)$. Then an examination of (5.55) should convince the reader that the vector d is in the subspace spanned by the vectors b, Ab,..., $A^{n-k-1}b$. Since the vectors b, Ab,...,$A^{n-1}b$ are independent, the left side of (5.56) cannot vanish, which contradicts our assumption that it is zero. Thus the vectors $e^1,..., e^n$ are independent, and the lemma follows.

SUPPLEMENTARY NOTES AND REFERENCES

§5.1. Some early references on controllability and observability are [*12*] and

R. E. Kalman, On the general theory of control systems, *Proceedings First IFAC Congress*, 481–492, 1960.

R. E. Kalman, Contributions to the theory of optimal control, *Bol. Soc. Mat. Mexicana* **5** (1960), 102–119.

§§5.2–5.3. A very interesting development of the theory of controllability can be found in

N. N. Krasovskii, On the theory of controllability and observability of linear dynamical systems, *Prikl. Mat. Meh.* **28** (1964), 3–14.

This approach, which is based on the theory connected with the so-called *L*-problem of moments, is in some ways of more interest than the one presented in the text, even though its study requires some background in elementary functional analysis. The problem of observability is related in this paper to some aspects of the theory of games. Other developments of this approach are given in

N. N. Krasovskii, Observation of a linear dynamical system and equations of retarded argument, *Differential Equations* **1** (1967), 1221–1225.

§5.4. Theorem 5.13 is proved in a different way in

P. L. Falb and M. Athans, A direct constructive proof of the criterion for complete controllability of time-invariant linear systems, *IEEE Trans. Automatic Control* **AC-9** (1964), 189–190.

Some important results are given in

W. G. Vogt and C. G. Cullen, The minimum number of inputs required for the complete controllability of a linear stationary dynamical system, *IEEE Trans. Automatic Control* **AC-12** (1967), 314.

We have not covered this material in the text because its development would entail going into the theory of the so-called invariant polynomials of a matrix. The main result of this paper is that given a constant $n \times n$ matrix A, an $n \times r$ constant matrix B can be associated with it so that the constant system $\dot{x} = Ax + Bu$ is controllable if and only if $r \geqslant p$, where p is the number of invariant polynomials of the matrix A. See [2] for information concerning invariant polynomials and related subjects. Other references to the subjects treated in this section are:

Hsin Chu, Some topological properties of complete controllability, *IEEE Trans. Automatic Control* **AC-11** (1966), 612–613.

Hsin Chu, A remark on complete controllability, *SIAM J. Control* **3** (1965), 439–442.

Some interesting results have been obtained by

D. A. Ford and C. D. Johnson, Invariant subspaces and the controllability and observability of linear dynamical systems, *SIAM J. Control* **6** (1968), 553–558.

§5.5.   Theorem 5.17 is proved in a different way by the same authors of [6] in

C. T. Chen and C. A. Desoer, A proof of controllability of Jordan form state equations, *IEEE Trans. Automatic Control* **AC-13** (1968), 195–196.

This new proof does not use the properties of the exponential matrix, and can therefore be applied to the controllability of discrete systems, as we will see in Chapter 7.

§5.7. Theorem 5.21 has been proved independently in

A. R. Stubberud, A controllability criterion for a class of linear systems, *IEEE Trans. Appl. Ind.* **68** (1964), 411–413.

L. M. Silverman and H. E. Meadows, Controllability and time-variable unilateral networks, *IEEE Trans. Circuit Theory* **CT-12** (1965), 308–313.

§5.8. Other references on observability are

R. D. Bonnel, An observability criterion for linear systems, *IEEE Trans. Automatic Control* **AC-11** (1966), 388–395.

§5.11. See also

C. T. Chen, Output controllability of composite systems, *IEEE Trans. Automatic Control* **AC-12** (1967), 201.

§5.13. A good reference on the attainable set is

E. Kreindler, Contributions to the theory of time-optimal control, *J. Franklin Inst.* **275** (1963), 314–344.

Most of the material presented in these references requires, however, knowledge of Lebesgue integration.

Admissible controls which are not simply all of those with piecewise continuous entries are considered also in the theory of *approximate controllability*. See

H. A. Antonsiewicz, Linear control systems, *Arch. Rational Mech. Anal.* **12** (1963), 313–324.

R. Conti, Contributions to linear control theory, *J. Differential Equations* **1** (1965), 427–445.

§5.14. Methods exist for the reduction to forms similar to (5.53) of systems with multidimensional control vectors as well as for time-varying systems. See, for instance,

L. M. Silverman, Transformation of time-variable systems to canonical (phase-variable) form, *IEEE Trans. Automatic Control* **AC-11** (1966), 300–303.

D. G. Luenberger, Canonical forms for linear multivariable systems, *IEEE Trans. Automatic Control* **AC-12** (1967), 290–293.

R. S. Bucy, Canonical forms for multivariable systems, *IEEE Trans. Automatic Control* **AC-13** (1968), 567–569.

C. E. Seal and A. R. Stubberud, Canonical forms for multiple-input time-variable systems, *IEEE Trans. Automatic Control* **AC-14** (1969), 704–707.

REFERENCES

1. E. Kreindler and P. E. Sarachik, On the concepts of controllability and observability of linear systems, *IEEE Trans. Automatic Control* **AC-9** (1964), 129–136.
2. F. R. Gantmacher, "The Theory of Matrices," Vol. 1, p. 306. Chelsea, New York, 1959.
3. P. R. Halmos, "Finite-Dimensional Vector Spaces." Van Nostrand, Princeton, New Jersey, 1958.
4. R. E. Kalman, Y. C. Ho, and K. S. Narendra, Controllability of linear dynamical systems. In "Contributions to Differential Equations," p. 182. Wiley, New York, 1961.
5. C. T. Chen, C. A. Desoer, and A. Niederlinski, Simplified conditions for controllability and observability of linear, time-invariant systems, *IEEE Trans. Automatic Control* **AC-11** (1966), 613–614.
6. C. T. Chen and C. A. Desoer, Controllability and observability of composite systems, *IEEE Trans. Automatic Control* **AC-12** (1967), 402–409.
7. S. Butman and R. Sivan, On cancellations, controllability and observability, *IEEE Trans. Automatic Control* **AC-9** (1964), 317–318.
8. A. Chang, An algebraic characterization of controllability, *IEEE Trans. Automatic Control* **AC-10** (1965), 112–113.
9. L. M. Silverman and H. E. Meadows, Controllability and observability in time-invariable linear systems, *SIAM J. Control* **5** (1967), 64–73.
10. E. G. Gilbert, Controllability and observability in multivariable control systems, *SIAM J. Control* **1** (1963), 128–151.
11. C. A. Desoer and C. Chen, Controllability and observability of feedback systems, *IEEE Trans. Automatic Control* **AC-12** (1967), 474–475.
12. J. P. LaSalle, The time-optimal control problem. In "Contributions to Differential Equations," p. 1. Princeton Univ. Press, Princeton, New Jersey, 1960.
13. L. Weiss, The concepts of differential controllability and differential observability, *J. Math. Anal. Appl.* **10** (1965), 442–449.
14. L. Weiss, Correction and addendum, The concepts of differential controllability and differential observability, *J. Math. Anal. Appl.* **13** (1966), 577–578.
15. L. M. Silverman and H. E. Meadows, Degrees of controllability in time-variable linear systems, *Proceedings of the National Electronics Conference* **21** (1965), 689–693.
16. E. Roxin, The existence of optimal controls, *Michigan Math. J.* **9** (1962), 109–119.

6

Synthesis

6.1. Introduction

We can treat at last the long-announced topic of synthesis, that is, the problem consisting in the determination of state descriptions of linear systems from their input–output characteristics. We will consider only systems without a direct connection between input and output, whose description is of the form

$$
\begin{aligned}
\dot{x} &= Ax + Bu \\
y &= Cx
\end{aligned}
\tag{6.1}
$$

The entries of the matrices A, B, and C will in this chapter be assumed continuous, real-valued functions of the time defined on the whole of the real line. The input u is r- and the output y is m-dimensional.

Let now $x(t_0)$ be zero, so that the output corresponding to an input u is determined by

$$
y(t; t_0, 0; u) = \int_{t_0}^{t} C(t)\, \Phi(t, \lambda)\, B(\lambda)\, u(\lambda)\, d\lambda
$$

for $t_0 \leqslant t$. This is, of course the zero-state response of the system (6.1). The $m \times r$ matrix H defined by $H(t, \lambda) = C(t)\, \Phi(t, \lambda)\, B(\lambda)$ for all t and λ will be said to be the *weighting pattern* of the system (6.1); note that the transition matrix Φ of this system is defined for all t, λ.

It was mentioned above that the object of this chapter is the development of methods for the derivation of state descriptions from input–

176

output characteristics. We ask now, can the weighting pattern be considered as such? Suppose that the ith entry of the input u is an impulse occurring at $\lambda = \tau$, $\tau \in R$. The corresponding output is the ith column of a matrix H_c defined by

$$H_c(t, \tau) = \begin{matrix} C(t)\,\Phi(t, \tau)\,B(\tau), & t \geqslant \tau \\ 0, & t < \tau \end{matrix} \qquad (6.2)$$

this is said to be the *impulse response matrix* of the system (6.1); it is clearly related to the weighting pattern in that they share common values in the set of R^2 defined by the inequality $t \geqslant \tau$. If H is a weighting pattern, we will say that the impulse response matrix H_c defined by

$$H_c(t, \lambda) = \begin{matrix} H(t, \lambda), & t \geqslant \lambda \\ 0, & t < \lambda \end{matrix}$$

is the impulse response matrix *derived from* the weighting pattern H.

Of course, the impulse response matrix is a measurable quantity, while the weighting pattern is not; remember that in Chapter 3 we have shown that the zero-state response of a system is determined, for any input, by the values which the weighting pattern takes on the region of R^2 defined by $t \geqslant \lambda$. There is no conceivable way of determining experimentally the values of the weighting pattern for $t < \lambda$, so that this matrix cannot be considered as an input–output characteristic.

The problem of synthesis, central in the present chapter, can therefore be described as follows: Given the impulse response matrix of a linear system, obtain a set of state equations such as (6.1) such that the impulse response matrix of this system equals the given impulse response matrix. It is said that such a system is a *realization* of the given matrix H_c; the system is said *to realize* the impulse response matrix.

Not all matrix-valued functions H_c are realizable by a system such as (6.1); it will be shown in Sections 6.2 and 6.6 that an impulse response matrix which can be realized has the form

$$H_c(t, \lambda) = \begin{matrix} \alpha(t)\,\beta(\lambda) & \text{for } t \geqslant \lambda \\ 0 & \text{for } t < \lambda \end{matrix}$$

that is, it equals for $t \geqslant \lambda$ the product of two matrices, $\alpha(t)$ and $\beta(\lambda)$, defined for all t and all λ respectively. This result suggests that we define the *naturally induced weighting pattern* associated with a realizable impulse response matrix H_c as the matrix $H(t, \lambda) = \alpha(t)\,\beta(\lambda)$, defined for all t and λ. Of course, $H_c(t, \lambda)$ is derived from $H(t, \lambda)$. It happens that for a class of systems, the analytic systems (to be defined below), the naturally induced weighting patterns have importance in the development of a theory of synthesis.

Definition.   A system such as (6.1) is said to be *analytic* if the entries of the matrices A, B, and C are analytic functions of the time, that is, if they can be expanded in a Taylor series in a neighborhood of every point of the time axis.

There are three remarks we would like to make concerning analytic systems:

i. The transition matrix, and then the weighting pattern, of an analytic system is analytic in the (t, λ) plane, that is, the entries of the transition matrix and of the weighting pattern of such a system can be expanded in a Taylor series in a neighborhood of each point of the (t, λ) plane.

ii. The values of the weighting pattern of an analytic system for all (t, λ) are determined by its values on any neighborhood, no matter how small, of *any point* in the (t, λ) plane. Note that the values of the weighting pattern are determined for all (t, λ) if the values of its entries and of all of their derivatives are known at a single point of the (t, λ) plane.

iii. Constant systems are analytic.

We leave the proofs of the contentions i and ii above to the reader. Suppose now that we are given an impulse response matrix H_c to be realized by a system such as (6.1), and that, somehow, this can be effected by means of an analytic system. It follows that the weighting pattern of the system which realizes H_c equals the naturally induced weighting pattern associated with this impulse response matrix, since both weighting patterns are completely determined by their values for $t \geqslant \lambda$, which are identical.

This situation should be contrasted with the one prevailing in more general cases, when the system realizing the given impulse response matrix is not analytic. Then the weighting pattern of the system is not necessarily identical in the domain of R^2 defined by $t > \lambda$, with the naturally induced weighting pattern associated with the impulse response; in fact, the weighting pattern of the system is quite arbitrary there.

An important conclusion that can be drawn from these considerations is that in the synthesis problem associated with analytic systems one can either specify the desired impulse response matrix or the weighting pattern naturally induced by it; the resulting system will realize both. It happens that it is simpler to formulate a theory of realizability when the weighting pattern is specified, and we proceed to do this in the next section; the resulting theory will be applied to the synthesis of weighting patterns and impulse response matrices by means of constant systems as well as by more general analytic systems.

It should be noted that one of the basic problems of synthesis is to find realizations of an impulse response matrix such that the dimensions of the state space of the system which realizes the matrix is as small as possible; it is emphasized that this requirement is due partly to a striving for economy in the simulation of the system (each extra dimension of the state space adds one integrator to an analog-computer simulation, as shown in the examples of Chapter 1), and partly to an effort to obtain simple controls (which depend on the state variables, and are of simpler structure if there are less of these) whenever the system is to be controlled. Consider now the class of all systems which realize a given impulse response matrix; likewise, consider the class of all systems whose weighting patterns equal the naturally induced weighting pattern associated with the given impulse response matrix. Of course, the members of this second class do belong to the first, while those in the first do not necessarily belong to the second; a system which realizes the impulse response matrix has essentially an arbitrary weighting pattern for $t < \lambda$, not necessarily equal to the naturally induced weighting pattern associated with the impulse response.

Assume now that in the first class there is a realization whose state space has the lowest dimension of all realizations in this first class, and that in the second class there is also a realization whose state space has the lowest dimension of all realizations in this second class. It follows from the previous considerations that, since the first class includes the second, the minimum dimension in the first class is not larger, and may be smaller, than the minimum dimension in the second class; it is important therefore that a theory of synthesis by means of general (not necessarily analytic) systems be developed by considering the impulse response matrices, not the weighting patterns, as the data for the synthesis problems; the resulting systems will generally be smaller than if the weighting patterns were specified, due to the freedom allowed by the essentially arbitrary nature of their weighting patterns for $t < \lambda$.

Of course, the two classes described above are identical when the given impulse response matrix can be realized by an analytic system; in this case nothing is sacrificed by considering the weighting patterns as the data of the synthesis problems. We proceed, therefore, to develop in the next section a theory of the synthesis of weighting patterns; we shall not make the specific assumption that these are analytic; we remark however that the theory has interest only under this assumption.

Exercise

*1. Prove contentions i and ii made with reference to analytic systems.

6.2. Some Results on the Synthesis of Weighting Patterns

Let an $m \times r$ matrix function H be defined for all t, λ. We say that this matrix is *w-realizable* if there is a triple of matrices (A, B, C) such that the corresponding system (6.1) has a weighting pattern equal to the given matrix for all t, λ. The system (6.1) is said to be a *w-realization* of H. The dimension n of its state vector is said to be the *dimension* of the *w-realization*. In this section we will derive an interesting result concerning the dimensions of different *w-realizations* of a matrix H.

The following theorem ([1], [2]) gives conditions as to whether a given matrix is or is not *w-realizable*.

THEOREM 6.1.   The $m \times r$ matrix H is *w-realizable* if and only if for all t and λ

$$H(t, \lambda) = \alpha(t)\,\beta(\lambda) \tag{6.3}$$

where

  i. $\alpha(t)$ and $\beta(\lambda)$ are $m \times n$ and $n \times r$ matrices, for some integer n.

  ii. The entries of α and β are continuous functions of their arguments.

Proof.   If H is as in (6.3), it is realized by the system defined by

$$\dot{x} = \beta u, \qquad y = \alpha x \tag{6.4}$$

since then, if $x_0 = 0$,

$$y(t; t_0, 0; u) = \int_{t_0}^{t} \alpha(t)\,\beta(\lambda)\,u(\lambda)\,d\lambda$$

Note that the state vector in this realization of H has dimension n. If H is *w-realizable* by a system such as (6.1) with matrices A, B, and C, then, if $\Phi(t, \lambda)$ is the corresponding transition matrix and μ is an arbitrary number,

$$H(t, \lambda) = C(t)\,\Phi(t, \lambda)\,B(\lambda)$$
$$= C(t)\,\Phi(t, \mu)\,\Phi(\mu, \lambda)\,B(\lambda) \tag{6.5}$$

which is a product of type (6.3). Clearly, since C and B are continuous functions of their arguments, and $\Phi(t, \lambda)$ is absolutely continuous, the functions α and β defined by $\alpha(t) = C(t)\,\Phi(t,\mu)$ and $\beta(\lambda) = \Phi(\mu,\lambda)\,B(\lambda)$ are continuous functions of their arguments.

We have been able, therefore, to characterize nicely the class of matrices H which are w-realizable. Later on in the chapter we will obtain conditions as to whether a given matrix is w-realizable by means of a *constant* system. Following Youla [2], we will refer to (6.3) as a *decomposition* of the matrix H.

If a matrix H is w-realizable, that is, if it is of the form (6.3), it should not be very difficult to obtain a decomposition of it. Our aim, however, is to realize such a matrix as the weighting pattern of a system with a state vector of as small a dimension as possible. We now go deeply into this matter.

It is necessary to recall some definitions of independence used extensively in Chapter 5.

The column vectors of α are vector-valued functions mapping R into R^m; likewise, the row vectors of β map R into R^r. The set of all possible column vectors of α is a linear space over the field R, with the operations of addition and scalar multiplication defined, as in previous occasions, as

$$(\alpha^1 + \alpha^2)(t) = \alpha^1(t) + \alpha^2(t)$$

$$(c\alpha^1)(t) = c\alpha^1(t)$$

with α^1, α^2 two members of the space. We say that $\alpha^1, \alpha^2, ..., \alpha^q$ are independent if there are scalars in R, $c_1, ..., c_q$, not all of which are zero, such that

$$c_1\alpha^1 + c_2\alpha^2 + \cdots + c_q\alpha^q = 0$$

which is equivalent to requiring

$$c_1\alpha^1(t) + c_2\alpha^2(t) + \cdots + c_q\alpha^q(t) = 0$$

for all $t \in R$. Of course, similar definitions and results are obtained for the space of row vectors of the matrix β. We shall say, as in Chapter 5, that the columns of α and the rows of β are independent when the column vectors of α and the row vectors of β, respectively, are independent. We can define now:

Definition. The decomposition (6.3) of the weighting pattern H is said to be *globally reduced* if the columns of α and the rows of β are linearly independent in the sense defined above.

We prove:

THEOREM 6.2 [2]. Every w-realizable matrix H possesses a globally reduced decomposition.

Proof.   Since H is w-realizable, it possesses a decomposition such as (6.3). Assume this is not globally reduced. Then either the columns of α or the rows of β, or both, are linearly dependent over R. Assume, for instance, that the n columns of α are dependent, while the n rows of β are independent. It follows that there is a nonzero vector c such that

$$\alpha(t)c = 0, \qquad t \in R \tag{6.6}$$

Suppose there are k such vectors, c^1, ..., c^k, which are linearly independent (clearly, $1 \leqslant k \leqslant n$). Form a nonsingular $n \times n$ matrix P with these vectors as the last k columns

$$P = [p^1 \; \cdots \; p^{n-k} \; c^1 \; c^2 \; \cdots \; c^k] \tag{6.7}$$

p^1, ..., p^{n-k} are arbitrary n-vectors chosen such that P be nonsingular; this is always possible since the vectors c^1, ..., c^k are independent. Then, if we write

$$\alpha(t)\,\beta(\lambda) = \alpha(t)\,PP^{-1}\beta(\lambda) \tag{6.8}$$

we see that, from (6.6),

$$\alpha(t)P = [\alpha(t)\,p^1 \; \cdots \; \alpha(t)\,p^{n-k}0 \; \cdots \; 0] \tag{6.9}$$

for all $t \in R$. Hence, if we define P_0 as $P_0 = [p^1 \; p^2 \; \cdots \; p^{n-k}]$,

$$\alpha(t)\,\beta(\lambda) = \alpha(t)\,P_0\hat{\beta}(\lambda) \tag{6.10}$$

where $\hat{\beta}(\lambda)$ is the $(n-k) \times r$ matrix formed with the first $(n-k)$ rows of $P^{-1}\beta(\lambda)$. The vectors αp^1, ..., αp^{n-k}, which are in the column space of the matrix α, are independent; this follows from:

   i. The vectors p^1, ..., p^{n-k} are independent.

   ii. The null space of $\alpha(t)$ is spanned by the vectors c^1, ..., c^k for all $t \in R$.

   iii. The vectors p^1, ..., p^{n-k}, c^1, ..., c^k are independent, so that no vector p^1, ..., p^{n-k}, or nonzero linear combinations of them, is in the null space of $\alpha(t)$ for all $t \in R$.

   iv. The equality

$$a_1\alpha(t)\,p^1 + \cdots + a_{n-k}\alpha(t)\,p^{n-k} = 0$$

is therefore satisfied for all $t \in R$ only for $a_1 = \cdots = a_{n-k} = 0$.

If we put $\hat{\alpha}(t) = \alpha(t)P$, we obtain a globally reduced decomposition of H

$$H(t, \lambda) = \hat{\alpha}(t)\,\hat{\beta}(\lambda) \tag{6.11}$$

since the rows of β are linearly independent because the rows of β are linearly independent and P^{-1} is nonsingular. We see that if we start with a decomposition in which the columns of α are not independent, and the rows of β are, we can obtain a globally reduced decomposition. Clearly, if both the columns of α and the rows of β were not independent, the resulting matrix β would not have linearly independent rows, and the entire process would have to be repeated on β; the end result would be, as before, a globally reduced decomposition. The theorem follows.

Let now H have two decompositions, one being

$$H(t, \lambda) = \alpha(t)\,\beta(\lambda) \tag{6.12}$$

where α has n columns and β n rows, while the other is a globally reduced one

$$H(t, \lambda) = \alpha_0(t)\,\beta_0(\lambda) \tag{6.13}$$

here α_0 has n_0 columns and β_0 has n_0 rows. We will proceed to obtain a relationship between n and n_0. From (6.12) and (6.13) it follows that

$$\alpha(t)\,\beta(\lambda) = \alpha_0(t)\,\beta_0(\lambda) \tag{6.14}$$

Let $\varDelta$ be an interval in the real line to be defined below. If we multiply (6.14) by $\alpha_0'(t)$ and integrate with respect to t over $\varDelta$, we obtain

$$\left\{ \int_\varDelta \alpha_0'(t)\,\alpha(t)\,dt \right\} \beta(\lambda) = \left\{ \int_\varDelta \alpha_0'(t)\,\alpha_0(t)\,dt \right\} \beta_0(\lambda) \tag{6.15}$$

It can be proved [2] that the matrix $M_0 = \int_\varDelta \alpha_0'(t)\,\alpha_0(t)\,dt$ is nonsingular for some choice of the interval $\varDelta$. Therefore, we can write (6.15) as

$$\beta_0(\lambda) = U\beta(\lambda) \tag{6.16}$$

with

$$U = M_0^{-1} \int_\varDelta \alpha_0'(t)\,\alpha(t)\,dt \tag{6.17}$$

In the same way, we can prove that

$$\alpha_0(t) = \alpha(t)V \tag{6.18}$$

where

$$V = \left\{ \int_{\varDelta'} \beta(\lambda)\,\beta_0'(\lambda)\,d\lambda \right\} N_0^{-1}$$

$$N_0 = \int_{\varDelta'} \beta_0(\lambda)\,\beta_0'(\lambda)\,d\lambda \tag{6.19}$$

Remember that α_0 has n_0 independent columns. Since V is an $n \times n_0$ matrix, its rank cannot be higher than n. From (6.18) we conclude that the number of independent columns in α_0 cannot be higher than the rank of V. Then,

$$n \geqslant n_0 \tag{6.20}$$

We see that a globally reduced decomposition is characterized by a number n_0 not larger than the number n which characterizes any other decomposition. If (6.12) were a reduced decomposition, that is, if α had *n independent* columns, the argument employed to prove (6.20) could be reversed to show that $n \leqslant n_0$, from which it follows that any globally reduced decomposition is characterized by *the same* integer n_0, dependent only on the matrix H, to be called the *order* of this matrix.

It should be clear from the proof of Theorem 6.1 that a decomposition such as (6.12) gives rise to a w-realization of dimension n, while any globally reduced decomposition gives rise to a w-realization of dimension n_0. It follows that the dimension of the w-realization associated with a globally reduced decomposition is the least among all systems which w-realize a given matrix function H, that is, among all systems whose weighting pattern is identical with the given one. Such a w-realization is said to be w-*minimal*.

Note, however, that this by no means implies that a system associated with a globally reduced decomposition has a state vector of minimum dimension among all systems which are zero-state equivalent to the one with the given weighting pattern H; the weighting pattern of these systems does not need to be identical with H for all t and λ, but only for $t \geqslant \lambda$, so that in general systems which are zero-state equivalent to the one with weighting pattern H do not need to be w-realizations of H. As mentioned in Section 6.1, the class of these systems is larger than the class of systems which w-realize H (there are more systems whose weighting pattern coincides with H for $t \geqslant \lambda$ only than systems whose weighting patterns coincide with H for all t and λ), so that it is possible that one may find systems whose weighting patterns coincide with H only for $t \geqslant \lambda$ and whose state vector is of dimension less that the number n_0. Indeed, examples of this situation will be shown in Section 6.6. On the other hand, it was shown in Section 6.1 that if a weighting pattern is w-realizable by an analytic system, then all systems which are zero-state equivalent to the one with weighting pattern H are w-realizations of this pattern; that is, if the weighting pattern of an analytic system equals H for $t \geqslant \lambda$, then it equals H also for $t < \lambda$. It follows that n_0 is indeed the minimum dimension of an analytic system which is zero-state equivalent to a system with weighting pattern H. Of course, we must require H to be w-realizable by an analytic system.

As we shall see in the following sections, it is not difficult to construct

w-realizations of a w-realizable analytic weighting pattern; this task is very simple if the pattern is w-realizable by means of a constant system. However, there are in general no means for determining if the resulting system is or is not w-minimal, other than by ascertaining if the corresponding decomposition is or is not globally reduced. The following theorem, though, provides a sufficient condition for w-minimality, and ties this concepts with some fundamental ideas of controllability and observability.

THEOREM 6.3.   A controllable and observable w-realization of a weighting pattern H is w-minimal.

Proof.   We show that the decomposition associated with a controllable and observable w-realization is globally reduced. Let the w-realization be defined by matrices A, B, and C. Many decompositions can be generated for the weighting pattern H; all are characterized by matrices α and β of the form

$$\alpha(t) = C(t)\, \Phi(t, \mu)$$
$$\beta(\lambda) = \Phi(\mu, \lambda)\, B(\lambda) \tag{6.21}$$

for different values of the scalar $\mu \in R$. Of course, if a decomposition defined by a fixed value of μ, $\bar{\mu}$, is globally reduced, then all decompositions, for any value of μ, are globally reduced; this is a consequence of the nonsingularity of the transition matrix of the system and of the group property of this matrix.

We show first the independency of the rows of β, for all values of μ:

   i. It should be remembered from our results of Chapter 5 that if a system is controllable at t_0, $t_0 \in R$, then there is a number $t_1 > t_0$ such that the rows of the matrix $\Phi(t_1, \cdot)\, B(\cdot)$ are independent over $[t_0, t_1]$.

   ii. Since the controllability of a system over $[t_0, t_1]$ implies the controllability over $[t_0, t_2]$, $t_2 > t_0$, the controllability of the system at t_0 implies that the rows of $\Phi(t_2, \cdot)\, B(\cdot)$ are independent over $[t_0, t_2]$.

   iii. Since t_2 is any number higher than t_1, the rows of $\Phi(\mu, \cdot)\, B(\cdot)$ are independent for any $\mu > t_1$ over $[t_0, \infty)$ if the system is controllable at t_0.

   iv. Because of the nonsingularity of the transition matrix and the group property of this matrix, the controllability of the system at t_0 implies the independency of the rows of $\Phi(\mu, \cdot)\, B(\cdot)$ over $[t_0, \infty)$ for all $\mu \in R$.

   v. We recall now that if the system is controllable then by definition it is controllable at every t_0, so that in this case the rows of $\Phi(\mu, \cdot)\, B(\cdot)$

are independent over the whole of R, that is, independent in the sense of the theorem.

The proof of the independency of the columns of α follows from the the observability of the system by considering the adjoint system and showing, exactly as above, that the controllability of this system implies the independency of the rows of the matrix α' for all values of μ. The theorem follows.

Note that this theorem does not preclude the existence of w-minimal realizations which are not controllable and observable. Of course, w-realizations which are not w-minimal are not controllable and observable. An example of this situation is shown below.

EXAMPLE

Let a system with one input and one output have the following state equations; a_{11}, a_{22}, a_{33}, b_2, b_3, c_{11}, and c_{13} are real constants:

$$\dot{x}_1 = a_{11}x_1, \qquad \dot{x}_2 = a_{22}x_2 + b_2u, \qquad \dot{x}_3 = a_{33}x_3 + b_3u$$

$$y = c_{11}x_1 + c_{13}x_3$$

with b_3, c_{13} different from zero. The transfer function of this system is

$$[c_{11}\ 0\ c_{13}] \begin{bmatrix} \dfrac{1}{s - a_{11}} & & \\ & \dfrac{1}{s - a_{22}} & \\ & & \dfrac{1}{s - a_{33}} \end{bmatrix} \begin{bmatrix} 0 \\ b_2 \\ b_3 \end{bmatrix} = \dfrac{c_{13}b_3}{s - a_{33}}$$

This system is not controllable, since the behavior of the state variable x_1 cannot be influenced by the input, and is not observable, since the state variable x_2 does not influence the output; this is shown clearly in the signal flow graph of Figure 6.1.

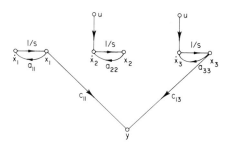

FIGURE 6.1. Flow graph of a system which is not controllable and observable.

These two state variables x_1 and x_2 do not influence the transfer function or the weighting pattern associated with it. A w-minimal w-realization of this weighting pattern is

$$\dot{x} = a_{33}x + b_3 u$$

$$y = c_{13}x_3$$

$c_{13}, b_3 \neq 0$, as shown in Figure 6.2.

Clearly, then, the first w-realization, which is not controllable and observable, is not w-minimal.

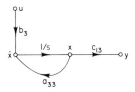

FIGURE 6.2.   Flow graph of a system with the same transfer function as the system of Figure 6.1.

We study next the synthesis problem for constant systems, that is, the determination of state-space realizations of weighting patterns (and then of impulse responses) which consist of systems with constant matrices A, B, and C. If a matrix is w-realizable in this form, our aim will be to obtain a w-minimal w-realization for it, that is, one in which the dimension of the state-vector equals the order n_0 of the given matrix. We will see in the next section that non-w-minimal w-realizations are not difficult to construct. Our objective will be, therefore, to transform such a w-realization into a w-minimal one.

6.3. Introduction to Synthesis

If an impulse response matrix is realized by means of a constant system, the naturally induced weighting pattern associated with the impulse response matrix is also w-realized by the system, so that in this case the two concepts, realizability and w-realizability, as well as the theories which can be built around them, are equivalent; in this and the next section we shall simply say realizable [realization] instead of w-realizable [w-realization].

A realization of a weighting pattern H consisting of a system with constant matrices A, B, and C will be said to be a *constant-coefficient realization*. We investigate first the conditions which a matrix H should satisfy so as to have a constant-coefficient realization. If this is the case, it is w-realizable and is clearly of the form $H(t, \lambda) = H(t - \lambda)$, since the transition matrix of the realization is of the form $\Phi(t - \lambda)$ and the matrices B and C are constant. However, the condition that H can be a function of $t - \lambda$ is not sufficient for realizability; it has to be strengthened somewhat, as shown by the following theorem.

THEOREM 6.4.   Let a weighting pattern H be dependent only on $t - \lambda$, $H(t, \lambda) = H(t - \lambda)$. Then, if a decomposition of H is defined by matrices α and β,

$$H(t - \lambda) = \alpha(t)\,\beta(\lambda), \qquad t, \lambda \in R$$

then H is realizable by a constant system if the matrices α and β are continuously differentiable functions.

Proof.   Assume this condition to be valid, and that, without loss of generality, the decomposition $\alpha(t)\,\beta(\lambda)$ is globally reduced. It follows that, for some intervals Δ and Δ', the following matrices M and N are nonsingular

$$M = \int_{\Delta} \alpha'(t)\,\alpha(t)\,dt \tag{6.22}$$

$$N = \int_{\Delta'} \beta'(\lambda)\,\beta(\lambda)\,d\lambda \tag{6.23}$$

Consider now

$$\frac{\partial}{\partial t} H(t - \lambda) = \frac{d}{dt} H(\tau)\Big|_{\tau = t - \lambda} \frac{d\tau}{dt} = \dot{\alpha}(t)\,\beta(\lambda)$$

$$\frac{\partial}{\partial \lambda} H(t - \lambda) = \frac{d}{d\tau} H(\tau)\Big|_{\tau = t - \lambda} \frac{d\tau}{d\lambda} = \alpha(t)\,\dot{\beta}(\lambda)$$

so that

$$\dot{\alpha}(t)\,\beta(\lambda) = -\alpha(t)\,\dot{\beta}(\lambda) \tag{6.24}$$

since $d\tau/dt = -d\tau/d\lambda = 1$.

We multiply now (6.24) by $\alpha'(t)$ and integrate over Δ; then,

$$\dot{\beta}(\lambda) = -U\beta(\lambda) \tag{6.25}$$

with

$$U = M^{-1} \int_{\Delta} \alpha'(t)\,\dot{\alpha}(t)\,dt \tag{6.26}$$

In a similar way, we multiply (6.24) by $\beta'(\lambda)$ and integrate over Δ'; then

$$\dot{\alpha}(t) = -\alpha(t)V \tag{6.27}$$

with

$$V = \left(\int_{\varDelta'} \beta(\lambda)\, \beta'(\lambda)\, d\lambda \right) N^{-1} \tag{6.28}$$

It follows simply from (6.24), by multiplying by $\beta'(\lambda)$ and integrating over $\varDelta'$, that $V = -U$. We write now explicit expressions for the solutions of (6.25) and (6.27)

$$\beta(\lambda) = e^{-U\lambda}K_1, \qquad \alpha(t) = K_2 e^{-Vt} \tag{6.29}$$

from which it follows that

$$\alpha(t)\, \beta(\lambda) = K_2 e^{-Vt} e^{-U\lambda} K_1 = K_2 e^{U(t-\lambda)} K_1$$

which shows that $H(t - \lambda)$ possesses the constant-coefficient realization $A = U$, $B = K_1$, $C = K_2$. The theorem follows.

We will assume in this and the following section that the input–output data are in the form of a *transfer function matrix* $Z(s)$, which is of course the Laplace transform of the impulse response matrix which it is desired to realize. Note that we are assuming that our realizations have no direct connection between input and output, that is, $D = 0$, so that, as indicated in Exercise 1 of this section, the degree of the numerators of the entries of $Z(s)$ are strictly *less* than the degrees of the corresponding denominators.

We will show in the rest of this section a simple method to obtain *nonminimal* realizations of a transfer function matrix, while in the next section we present an approach to the (more difficult) problem of transforming a nonminimal realization into a minimal one, that is, of extracting the controllable and observable "part" of a nonminimal realization. We follow Mayne [3].

Let the $m \times r$ matrix $Z(s)$ have entries $z_{ij}(s)$, $i = 1, ..., m$, $j = 1, ..., r$, and let the least common denominator of the ith row of $Z(s)$ be $b^i(s)$, a polynomial of degree n_i.[†] Then $z_{ij}(s)$ can be written as $z_{ij}(s) = a^{ij}(s)/b^i(s)$, where $a^{ij}(s)$ is a polynomial of degree (at most) $n_i - 1$, that is,

$$a^{ij}(s) = a^{ij}_{n_i} s^{n_i-1} + \cdots + a^{ij}_2 s + a^{ij}_1 \tag{6.30}$$

$$b^i(s) = s^{n_i} + \cdots + b^i_2 s + b^i_1 \tag{6.31}$$

[†] Without loss of generality, the polynomials $b^i(s)$, $i = 1, ..., m$, will be considered to be *monic*, that is, the coefficient of the largest power of s in each polynomial will be assumed to be equal to one.

A realization of the *transfer function* $z_{ij}(s)$, by means of a system with one input and one output is defined by the matrices A_i, B_{ij}, and C_i:

$$A_i = \begin{bmatrix} 0 & \cdots & 0 & -b_1{}^i \\ 1 & \cdots & 0 & -b_2{}^i \\ \vdots & & \vdots & \vdots \\ 0 & \cdots & 1 & -b_{n_i}^i \end{bmatrix}_{n_i \times n_i} \tag{6.32}$$

$$B_{ij} = [a_1^{ij} \cdots a_{n_i}^{ij}]'_{n_i \times 1} \tag{6.33}$$

$$C_i = [0 \cdots 0\ 1]_{1 \times n_i} \tag{6.34}$$

as can easily be seen by inspection (see also Chapter 1). Therefore, a realization of $Z(s)$ is

$$A = \begin{bmatrix} A_1 & & & & \\ & A_2 & & & \\ & & \cdot & & \\ & & & \cdot & \\ & & & & A_m \end{bmatrix}_{n \times n} \qquad \text{where} \quad n = \sum_{i=1}^{m} n_i \tag{6.35}$$

$$B = \begin{bmatrix} B_{11} & \cdots & B_{1r} \\ \vdots & & \vdots \\ B_{m1} & \cdots & B_{mr} \end{bmatrix}_{n \times r} \tag{6.36}$$

$$C = \begin{bmatrix} C_1 & & & & \\ & C_2 & & & \\ & & \cdot & & \\ & & & \cdot & \\ & & & & C_m \end{bmatrix}_{m \times n} \tag{6.37}$$

This is indeed a realization of $Z(s)$, since

$$Ce^{At}B = (C_i e^{A_i t} B_{ij})$$

and the Laplace transform of this general entry is $z_{ij}(s)$.

EXAMPLE

Let

$$Z(s) = \begin{bmatrix} \dfrac{1}{s(s+1)} & \dfrac{3}{s+2} & \dfrac{2}{s} \\[3mm] \dfrac{1}{s} & 0 & \dfrac{3}{s+4} \end{bmatrix}$$

$$= \begin{bmatrix} \dfrac{s+2}{s(s+1)(s+2)} & \dfrac{3s(s+1)}{s(s+1)(s+2)} & \dfrac{2(s+1)(s+2)}{s(s+1)(s+2)} \\[3mm] \dfrac{s+4}{s(s+4)} & 0 & \dfrac{3s}{s(s+4)} \end{bmatrix}$$

Hence,

$$A_1 = \begin{bmatrix} 0 & 0 & 0 \\ 1 & 0 & -2 \\ 0 & 1 & -3 \end{bmatrix}_{3\times3}$$

$$A_2 = \begin{bmatrix} 0 & 0 \\ 1 & -4 \end{bmatrix}_{2\times2}$$

$$B_{11} = \begin{bmatrix} 2 \\ 1 \\ 0 \end{bmatrix}_{3\times1}, \qquad B_{12} = \begin{bmatrix} 0 \\ 3 \\ 3 \end{bmatrix}_{3\times1}, \qquad B_{13} = \begin{bmatrix} 4 \\ 6 \\ 2 \end{bmatrix}_{3\times1}$$

$$B_{21} = \begin{bmatrix} 4 \\ 1 \end{bmatrix}_{2\times1}, \qquad B_{22} = \begin{bmatrix} 0 \\ 0 \end{bmatrix}_{2\times1}, \qquad B_{23} = \begin{bmatrix} 0 \\ 3 \end{bmatrix}_{2\times1}$$

$$C_1 = [0 \ \ 0 \ \ 1]_{1\times3}, \qquad C_2 = [0 \ \ 1]_{1\times2}$$

The realization is

$$A = \begin{bmatrix} 0 & 0 & 0 & 0 & 0 \\ 1 & 0 & -2 & 0 & 0 \\ 0 & 1 & -3 & 0 & 0 \\ 0 & 0 & 0 & 0 & 0 \\ 0 & 0 & 0 & 1 & -4 \end{bmatrix}_{5\times5}$$

$$B = \begin{bmatrix} 2 & 0 & 2 \\ 1 & 3 & 6 \\ 0 & 3 & 2 \\ 4 & 0 & 0 \\ 1 & 0 & 3 \end{bmatrix}_{5\times3}$$

$$C = \begin{bmatrix} 0 & 0 & 1 & 0 & 0 \\ 0 & 0 & 0 & 0 & 1 \end{bmatrix}_{2\times5}$$

We prove now that this type of realization is always observable (even though it may not be controllable). It is sufficient to prove the observability of a system with matrices A_i and C_i (remember that the matrix B plays no role with respect to observability considerations), since, if A and C are as in (6.35) and (6.37), the observability matrix of the system is

$$P = [C \ A'C' \ \cdots \ A'^{(n-1)} \ C'] =$$

$$
\begin{bmatrix}
C_1' & A_1'C_1' & A_1'^{(n-1)}C_1' \\
 C_2' & A_2'C_2' & A_2'^{(n-1)}C_2' \\
 & & \\
 & & \\
C_m' & A_m'C_m' & A_m'^{(n-1)}C_m'
\end{bmatrix}
$$

$$(6.38)$$

Due to its particular form, this matrix has rank n if each matrix

$$[C_i' \ A_i'C_i' \ \cdots \ A_i'^{(n-1)}C_i'] \tag{6.39}$$

has rank n_i, $i = 1, ..., m$. These are of course the observability matrices of the systems with matrices A_i, C_i, $i = 1, ..., m_i$; we write (6.39) explicitly as

$$
\begin{bmatrix}
0 & 0 & 0 & \cdots & 1 \\
 & & & \vdots & -b^i_{n_i} \\
 & & & \vdots & -b_{n_i-1} + b^{i2}_{n_i} \\
\vdots & \vdots & \vdots & 0 & \\
 & & & 1 & \\
 & 0 & & \vdots & \\
0 & 1 & -b^i_{n_i} & & \\
0 & 1 & -b^i_{n_i} & -b^i_{n_i-1} + b^{i2}_{n_i} & \\
1 & -b_{n_i} & -b^i_{n_i-1} + b^{i2}_{n_i} & -b^i_{n_i-2} + b^{i2}_{n_i-1} & -b^{i3}_{n_i} \cdots -b_2^i + \cdots + (-b^i_{n_i})^{n_i-1}
\end{bmatrix}
$$

$$(6.40)$$

The rank of this matrix is n (since its determinant equals 1 or -1). The component systems are observable, and the overall system is also observable.

There is another realization for the same transfer function matrix $Z(s)$, which is controllable. Let the least common denominator of the elements of the jth *column* of $Z(s)$ be $b^j(s)$, a polynomial of degree n_j. Then $z_{ij}(s)$ can be written as $z_{ij}(s) = a^{ij}(s)/b^j(s)$, where $a^{ij}(s)$ is a polynomial of degree (at most) $n_j - 1$, that is,

$$a^{ij}(s) = a^{ij}_{n_j}s^{n_j-1} + \cdots + a^{ij}_1 \tag{6.41}$$

$$b^j(s) = s^{n_j} + b^j_{n_j}s^{n_j-1} + \cdots + b^j_1 \tag{6.42}$$

A realization of the transfer function $z_{ij}(s)$, by means of a system with one input and one output is defined by the matrices A_j, B_j, C_{ij}:

$$A_j = \begin{bmatrix} 0 & 1 & \cdots & 0 \\ 0 & 0 & 1 & \cdots & 0 \\ & & \vdots & \\ -b_1^{\,j} & & \cdots & -b_{n}^{\,j} \end{bmatrix}_{n_j \times n_j} \tag{6.43}$$

$$B_j = [0 \cdots 1]'_{n_j \times 1} \tag{6.44}$$

$$C_{ij} = [a^{ij}_1 \cdots a^{ij}_{n_j}]_{1 \times n_j} \tag{6.45}$$

as shown in Chapter 1. A realization of $Z(s)$ is:

$$A = \begin{bmatrix} A_1 & & & \\ & A_2 & & \\ & & \ddots & \\ & & & A_m \end{bmatrix}_{n \times n} , \qquad n = \sum_{j=1}^{r} n_j \tag{6.46}$$

$$B = \begin{bmatrix} B_1 & & \\ & \ddots & \\ & & B_m \end{bmatrix}_{n \times r} \tag{6.47}$$

$$C = \begin{bmatrix} C_{11} & \cdots & C_{1r} \\ \vdots & & \vdots \\ C_{m1} & \cdots & C_{mr} \end{bmatrix}_{m \times n} \tag{6.48}$$

It is left to the reader to prove that this realization is controllable; the proof takes the same form as the proof of the observability of the previous realization of $Z(s)$.

Nonminimal realizations, therefore, can be obtained from a transfer function matrix practically by inspection. The problem of reducing the dimension of the state vector of such a realization so as to obtain a minimal one will be dealt with in the next section.

Exercises

*1. Prove that the transfer function matrix of a constant system such as (6.1) has a smaller number of finite zeros than of finite poles.

*2. Show that the transfer function matrix of the system

$$\dot{x} = Ax + Bu, \qquad y = Cx + Du$$

may have entries with as many poles as zeros. Give the necessary and sufficient conditions for this to happen (Hint: Consider

$$H(s) = (sI - A)^{-1} B + D$$

and expand it in powers of $1/s$.)

6.4. A Method of Synthesis

We present a modification by Mayne [3] of a method originally due to Kalman to obtain a minimal realization of a transfer function matrix from a nonminimal one. The starting point is either the controllable or the observable realizations obtained in the previous section. A description of the method will be followed by some examples, after which the proof of the validity of the procedure will be given.

Let $Z(s)$ be an $m \times r$ transfer function matrix, and consider an observable realization of it characterized by the matrices A, B, and C; A is $n \times n$. Let the columns of B be b^1, ..., b^r. The first step leading to the transformation of this realization into a minimal one consists in the construction of a matrix S of dimensions $n \times n_0$:

$$S = [\sigma^1 \sigma^2 \cdots \sigma^{n_0}]$$

We shall show below that the number of columns of the matrix S equals the order n_0 of the impulse response matrix associated with the transfer function matrix Z. The construction of S proceeds as follows. Without loss of generality, the first column of B is assumed to be nonzero; if it were zero, a relabeling of the input could make it so. Then:

  i. Put $\sigma^1 = b^1$

 ii. If $A\sigma^i$ is linearly independent of σ^1, ..., σ^i, then $\sigma^{i+1} = A\sigma^i$

 iii. The first columns of S are, then, $\sigma^1, A\sigma^1, A^2\sigma^1, ..., A^{p_1}\sigma^1$, for some integer $p_1 \geqslant 1$.

Suppose that $A^{p_1+1}\sigma^1$ is linearly dependent on the previous vectors. Then select another column of B, say b^k, such that it is independent of the columns of S already selected; put then $\sigma^{p_1+2} = b^k$.

iv. Apply the same procedure as in ii and iii with b^k as starting vector, that is construct the vectors Ab^k, A^2b^k, ..., $A^{p_2}b^k$, until $A^{p_2+1}b^k$ is found to be dependent on the previous vectors b^1, Ab^1, ..., Ab^k, ..., $A^{p_2}b^k$. Select then another vector from the remaining columns of B until no further independent vectors can be found. The matrix S will have the form

$$S = [b^1\ Ab^1\ \cdots\ A^{p_1}b^1\ b^k\ \cdots\ A^{p_2}b^k\ \cdots\ b^l\ \cdots\ A^{p_q}b^l] \qquad (6.49)$$

Note that, by construction, if A^pb^k is a column of S, for some p, then all the vectors b^k, Ab^k, ..., $A^{p-1}b^k$, are also columns of S.

The next step in the reduction procedure consists in finding a constant $n_0 \times n$ matrix V such that

$$VS = I_{n_0} \qquad (6.50)$$

There are many ways of computing such a matrix, since (6.50) represents nothing but a system of linear algebraic equations in the entries of V. A simple procedure to obtain this matrix numerically is shown in the Appendix.

We can now construct the minimal realization; its matrices A_0, B_0, and C_0 are:

$$A_0 = VAS \qquad (6.51)$$

$$B_0 = VB \qquad (6.52)$$

$$C_0 = CS \qquad (6.53)$$

Remember that the number of columns of S equals n_0, so that A_0 has dimensions $n_0 \times n_0$. We present several examples.

EXAMPLE

Let

$$Z(s) = \begin{bmatrix} \dfrac{1}{s+1} & \dfrac{1}{s+1} \\[2ex] \dfrac{1}{s+1} & \dfrac{1}{s+1} \end{bmatrix}$$

A completely observable realization is

$$A = \begin{bmatrix} -1 & 0 \\ 0 & -1 \end{bmatrix}, \qquad B = \begin{bmatrix} 1 & 1 \\ 1 & 1 \end{bmatrix}$$

$$C = \begin{bmatrix} 1 & 0 \\ 0 & 1 \end{bmatrix}$$

We start constructing S. We make $\sigma^1 = b^1 = \begin{bmatrix} 1 \\ 1 \end{bmatrix}$. Then $A\sigma^1 = \begin{bmatrix} -1 \\ -1 \end{bmatrix} = -\sigma^1$ so our first set of the form $\{b^1, ..., A^p{}_1 b^1\}$ consists of only one vector b^1. Since the only remaining vector in B is not independent of b^1 (it is equal to b^1), we conclude that

$$S = [b^1] = \begin{bmatrix} 1 \\ 1 \end{bmatrix}$$

and therefore $n_0 = 1$. We must find a 1×2 matrix V which satisfies (6.50). If $V = [a\ b]$, then

$$[a\ b]\begin{bmatrix} 1 \\ 1 \end{bmatrix} = a + b = 1$$

indicates that there will be an arbitrary parameter in V; put $V = [a\ 1 - a]$, with a arbitrary. Therefore, a minimal realization for $Z(s)$ is

$$A_0 = [a\ 1 - a]\begin{bmatrix} -1 & 0 \\ 0 & -1 \end{bmatrix}\begin{bmatrix} 1 \\ 1 \end{bmatrix} = [-1]$$

$$B_0 = [a\ 1 - a]\begin{bmatrix} 1 & 1 \\ 1 & 1 \end{bmatrix} = [1\ 1]$$

$$C_0 = \begin{bmatrix} 1 & 0 \\ 0 & 1 \end{bmatrix}\begin{bmatrix} 1 \\ 1 \end{bmatrix} = \begin{bmatrix} 1 \\ 1 \end{bmatrix}$$

The arbitrary parameter a does not appear in the realization; however, the arbitrary parameters which appear in V will, in general, also appear in the minimal realization.

EXAMPLE

Let

$$Z(s) = \begin{bmatrix} \dfrac{1}{s} & \dfrac{1}{s^2} \\ \dfrac{1}{s^2} & \dfrac{1}{s^3} \end{bmatrix}$$

A realization of this matrix is:

$$A = \begin{bmatrix} 0 & 0 & 0 & 0 & 0 \\ 1 & 0 & 0 & 0 & 0 \\ 0 & 0 & 0 & 0 & 0 \\ 0 & 0 & 1 & 0 & 0 \\ 0 & 0 & 0 & 1 & 0 \end{bmatrix}, \qquad B = \begin{bmatrix} 0 & 1 \\ 1 & 0 \\ 0 & 1 \\ 1 & 0 \\ 0 & 0 \end{bmatrix}$$

$$C = \begin{bmatrix} 0 & 1 & 0 & 0 & 0 \\ 0 & 0 & 0 & 0 & 1 \end{bmatrix}$$

Therefore,

$$
b^1 = \begin{bmatrix} 0 \\ 1 \\ 0 \\ 1 \\ 0 \end{bmatrix}, \quad
Ab^1 = \begin{bmatrix} 0 \\ 0 \\ 0 \\ 0 \\ 1 \end{bmatrix}, \quad
A^2b^1 = \begin{bmatrix} 0 \\ 0 \\ 0 \\ 0 \\ 0 \end{bmatrix}, \quad
Ab^2 = \begin{bmatrix} 0 \\ 1 \\ 0 \\ 1 \\ 0 \end{bmatrix} = b^1
$$

Since b^2 is independent of b^1 and Ab^1, we have

$$
S = \begin{bmatrix} 0 & 0 & 1 \\ 1 & 0 & 0 \\ 0 & 0 & 1 \\ 1 & 0 & 0 \\ 0 & 1 & 0 \end{bmatrix}
$$

and $n_0 = 3$. The following matrix V satisfies (6.50)

$$
V = \begin{bmatrix} 1 & 0 & -1 & 1 & 0 \\ 1 & 0 & -1 & 0 & 1 \\ 1 & 1 & 0 & -1 & 0 \end{bmatrix}
$$

Then,

$$
A_0 = VAS = \begin{bmatrix} 0 & 0 & 1 \\ 1 & 0 & 0 \\ 0 & 0 & 0 \end{bmatrix}
$$

$$
B_0 = VB = \begin{bmatrix} 1 & 0 \\ 0 & 0 \\ 0 & 1 \end{bmatrix}
$$

$$
C_0 = CS = \begin{bmatrix} 1 & 0 & 0 \\ 0 & 1 & 0 \end{bmatrix}
$$

is a minimal realization for $Z(s)$.

Note that the construction of the matrix S entails the determination of whether several sets of n-vectors are, or are not, linearly independent. As mentioned in Chapter 2, a numerical procedure to accomplish this is given in the Appendix.

If the matrix Z has been realized in the completely controllable form indicated in the previous section, a similar procedure exists, consisting of the following:

i. If $c^{1'}, ..., c^{m'}$ are the rows of C, a matrix T is constructed from the columns vectors $c^1, ..., c^m$ in the same way as before, with the c^i replacing the b^i and A' replacing A. The matrix T has n_0 columns.

ii. An $n \times n_0$ matrix U is found such that

$$T'U = I_{n_0} \tag{6.54}$$

iii. A minimal realization is given by

$$A_0 = T'AU \tag{6.55}$$

$$B_0 = T'B \tag{6.56}$$

$$C_0 = CU \tag{6.57}$$

EXAMPLE

A completely controllable realization of

$$Z(s) = \begin{bmatrix} \dfrac{1}{s+1} & \dfrac{1}{s+1} \\ \dfrac{1}{s+1} & \dfrac{1}{s+1} \end{bmatrix}$$

is

$$A = \begin{bmatrix} -1 & 0 \\ 0 & -1 \end{bmatrix}, \qquad B = \begin{bmatrix} 1 & 0 \\ 0 & 1 \end{bmatrix}, \qquad C = \begin{bmatrix} 1 & 1 \\ 1 & 1 \end{bmatrix}$$

Since

$$\sigma^1 = c^1 = \begin{bmatrix} 1 \\ 1 \end{bmatrix}, \qquad A'c^1 = \begin{bmatrix} -1 \\ -1 \end{bmatrix}$$

$A'c^1$ is $-c^1$ and $c^2 = c^1$; therefore, $n_0 = 1$ and

$$T = \begin{bmatrix} 1 \\ 1 \end{bmatrix}, \qquad U = \begin{bmatrix} 1 \\ 0 \end{bmatrix}$$

$$A_0 = [-1], \qquad B_0 = [1 \ 1], \qquad C_0 = \begin{bmatrix} 1 \\ 1 \end{bmatrix}$$

We show now that the procedures presented above do give a minimal realization when nonminimal, observable (or controllable) realizations are available.

THEOREM 6.5.   Let the constant system defined by the matrices A, B, C be observable. Then, the system defined by the matrices A_0, B_0, C_0 in (6.51), (6.52) and (6.53) is a minimal realization of this system. Similarly, if the system defined by A, B, C is controllable, then the system defined by the matrices A_0, B_0, C_0 in (6.55), (6.56), and (6.57) is a minimal realization of this system.

Proof. We shall prove only the part of this theorem which concerns the reduction of an observable realization into an observable and controllable (and then minimal) realization of the same transfer function matrix; the second part of the theorem can be proved in a similar way. Call n_1 the number of columns in S, since we have not yet proved that this number equals n_0; indeed this will be one of the consequences of the present theorem. Consider the system of dimension n_1

$$\dot{x} = VASx + VBu, \qquad y = CSx \tag{6.58}$$

It will be shown that (i) this system is controllable and observable; (ii) it has the same impulse response matrix as the one defined by the matrices A, B, C. It will follow, of course, that (6.58) is a minimal realization of the original system, that is, that $n_1 = n_0$.

Let us construct a nonsingular, $n \times n$ matrix M^{-1} as follows

$$M^{-1} = \underset{n_1 \quad n-n_1}{[S \quad W]} \tag{6.59}$$

where S is the matrix in (6.58) and W is any $n \times (n - n_1)$ matrix such that M^{-1} is nonsingular. The matrix M, the inverse of M^{-1}, will be partitioned:

$$M = \overset{n}{\begin{bmatrix} \tilde{V} \\ \hat{V} \end{bmatrix}} \begin{array}{l} \} \, n_1 \\ \} \, n-n_1 \end{array} \tag{6.60}$$

V is, of course, the same matrix which appears in (6.58), since, from (6.59) and (6.60), $VS = I_{n_1}$. We show next that

$$\hat{V}B = \hat{V}AB = \hat{V}A^2B = \cdots = \hat{V}A^kB = \cdots = 0 \tag{6.61}$$

that is, $\hat{V}A^kB = 0$ for all k. Indeed, since

$$MM^{-1} = \begin{bmatrix} VS & VW \\ \hat{V}S & \hat{V}W \end{bmatrix} = I_n \tag{6.62}$$

then $\hat{V}S$ equals the zero matrix of dimensions $(n_1 - n) \times n_1$. It follows that the columns of S are in the null space of $\hat{V}$. Since the columns of S are all the independent vectors in the set of all columns of the matrices B, AB, A^2B, ..., A^kB, ..., these are in the null space of $\hat{V}$, which implies (6.61).

Consider now

$$M^{-1}M = I_n = [S \ W] \begin{bmatrix} V \\ \hat{V} \end{bmatrix} = SV + W\hat{V}$$

so that

$$SV = I_n - W\tilde{V} \tag{6.63}$$

clearly,

$$SVB = (I_n - W\tilde{V})B = B \tag{6.64}$$

since $\tilde{V}B$ is a zero matrix. In the same way we show that

$$SVAB = AB$$

$$SVA^2B = A^2B$$

$$\vdots \tag{6.65}$$

$$SVA^kB = A^kB$$

$$\vdots$$

so that the controllability matrix of (6.58) is

$$Q = [VB \;\; VASVB \;\; (VAS)^2\, VB \;\cdots\; (VAS)^{n_1-1}\, VB]$$

$$= [VB \;\; VAB \;\; VA^2B \;\cdots\; VA^{n_1-1}B]$$

$$= V[B \;\; AB \;\cdots\; A^{n_1-1}B] \tag{6.66}$$

for instance, $(VAS)^2\, VB = VAS(VASVB) = VASVAB = VA^2B$.

We can prove very simply that this matrix has rank n_1, so that (6.58) is controllable. Indeed, the matrix in the square bracket in the right-hand side of (6.66) has a total of $n_1 r$ columns; it happens that n_1 of these are the columns of S. Clearly the highest exponent p_q in (6.49) cannot be higher than $n_1 - 1$ because this would imply that the total number of columns in S would be higher than n_1; if $p_q = n_1$, for instance, all of the $(n_1 + 1)$ vectors

$$b^l, \; Ab^l, \; ..., \; A^{n_1}b^l$$

would be columns of S. This is certainly impossible, so that p_q is at most equal to $n_1 - 1$, and all the columns of S are indeed columns of the matrix in square brackets in the right-hand side of (6.66). Of course, because of (6.50), when these columns are multiplied by V they are transformed into unit vectors so that a submatrix of Q is the unit matrix I_{n_1}. Of course, this implies that Q has rank n_1.

We shall not prove here the observability of the system (6.58), but give the proof below, where we consider these matters in a slightly more general setting. We mention only that the observability of (6.58) is a consequence of the observability of the realization with which we started, the one of dimension n and matrices A, B, C.

We show at last that this system (6.58) has the same impulse response matrix as the one characterized by the matrices A, B, C. The impulse response matrix of (6.58) is

$$CSe^{tVAS}VB = CS\left(I_n + \frac{t}{1!}VAS + \frac{t^2}{2!}(VAS)^2 + \cdots\right)VB$$

$$= CSVB + \frac{t}{1!}CSVASVB + \frac{t^2}{2!}CS(VAS)^2VB + \cdots$$

$$= CB + \frac{t}{1!}CAB + \frac{t^2}{2!}CA^2B + \cdots$$

$$= Ce^{tA}B \tag{6.67}$$

for $t \geqslant 0$; this follows from (6.65):

$$CSVA = CB$$

$$CSVASVB = CSVAB = CAB$$

$$CSVASVASVB = CSVASVAB = CSVA^2B = CA^2B, \ldots$$

The two impulse response matrices are indeed equal. The theorem follows.

This reduction method can be applied, suitably modified, to a realization which is neither controllable nor observable, so as to eventually separate it after several applications of the method into four subsystems with special properties.

We start with a system which is observable, and characterized by matrices A, B, C. Let us construct the matrix S as before, and also the matrix M^{-1} as in (6.59); we write n_0 instead of n_1, since we have shown that both numbers are equal. If we put $z = Mx$, with z the new and x the old state vector, we find that the system with z as state vector is characterized by the matrices $\hat{A}$, $\hat{B}$, and $\hat{C}$:

$$\hat{A} = MAM^{-1} \tag{6.68}$$

$$\hat{B} = MB \tag{6.69}$$

$$\hat{C} = CM^{-1} \tag{6.70}$$

that is, using (6.59) and (6.60)

$$\hat{A} = \begin{bmatrix} VAS & VAW \\ \tilde{V}AS & \tilde{V}AW \end{bmatrix} \tag{6.71}$$

$$\hat{B} = \begin{bmatrix} VB \\ \tilde{V}B \end{bmatrix} \tag{6.72}$$

$$\hat{C} = [CS \ \ CW] \tag{6.73}$$

We show that $\tilde{V}AS$ is a zero $n_0 \times n_0$ matrix 0_{n_0}. Indeed, by (6.62), $\tilde{V}S$ is a zero matrix, so that the null space of $\tilde{V}$ contains the range of S, which in turn contains the range of AS, so that the columns of AS are in the null space of $\tilde{V}$, which implies $\tilde{V}AS = 0_{n_0}$. Remember that $\tilde{V}B$ is also a zero matrix.

Define now two state vectors z^1 and z^2 of dimensions n_0 and $(n - n_0)$ respectively, and, again, define two systems, the first being the same as (6.58)

$$\begin{aligned} \dot{z}^1 &= (VAS)\, z^1 + VBu \\ y^1 &= CSz^1 \end{aligned} \tag{6.74}$$

while the second is

$$\begin{aligned} \dot{z}^2 &= (\tilde{V}AW)\, z^2 \\ y^2 &= CWz^2 \end{aligned} \tag{6.75}$$

The relationship between these systems and the bigger system defined by the matrices $\hat{A}$, $\hat{B}$, and $\hat{C}$ is depicted in Figure 6.3. The matrices of this complete system with state vector $z = [z^1 \ z^2]'$ are clearly $\hat{A}$, $\hat{B}$, and $\hat{C}$. Note that an extra input, $VAWz^2$, has to be added to (6.74) so that, unless W is chosen so as to make VAW a zero matrix, the two component systems (6.74) and (6.75) are interacting.

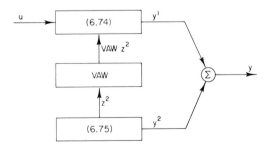

FIGURE 6.3. The subsystems (6.74) and (6.75), interacting through the matrix VAW, give rise to the system described by the matrices $\hat{A}$, $\hat{B}$, and $\hat{C}$.

The system (6.74) is, as shown in Theorem 6.5, controllable; the system (6.75) is certainly not controllable. Both systems are observable, as shown by the following argument. Clearly the system described by the matrices $\hat{A}$, $\hat{B}$, $\hat{C}$ is observable, since it has been obtained from the original system, which was assumed to be observable, by means of a nonsingular transformation. Suppose now that W is chosen so that VAW is a zero matrix; then the systems (6.74) and (6.75) are in parallel to form the bigger system described by the matrices $\hat{A}$, $\hat{B}$, $\hat{C}$; since this is observable, the components systems are also observable, as remarked in Chapter 5 when presenting Theorem 5.30.

It is said that the transformation M separates the original observable system into an observable, controllable subsystem, (6.74), and an observable, noncontrollable subsystem, (6.75). Note that if the original system were not observable, these subsystems would still be controllable, and noncontrollable respectively, even if in general they would not be observable.

A transformation can be constructed which, when applied to a system, permits the identification of observable and nonobservable subsystems, which are generally coupled, as before; this transformation, when applied to (6.74) will give rise to a controllable, observable part and a controllable, nonobservable part; when applied to (6.75) will give rise to an observable, noncontrollable part and a nonobservable, noncontrollable part. This transformation is characterized by a matrix N

$$N = \begin{bmatrix} T' \\ Q \end{bmatrix} \begin{matrix} \} \, n_0 \\ \} \, n-n_0 \end{matrix} \tag{6.76}$$

$$N^{-1} = [\underbrace{U}_{n_0} \quad \underbrace{\tilde{U}}_{n-n_0}] \tag{6.77}$$

where T and U are as in (6.54), Q is an $(n - n_0) \times n$ matrix such that N is nonsingular, and U and $\tilde{U}$ are appropriate partitions of N^{-1}. The matrices which characterize the transformed system with state vector $z = Nx$ are

$$\hat{A} = \begin{bmatrix} T'AU & 0 \\ QAU & QA\tilde{U} \end{bmatrix} \tag{6.78}$$

$$\hat{B} = \begin{bmatrix} T'B \\ QB \end{bmatrix} \tag{6.79}$$

$$\hat{C} = [CU \; 0] \tag{6.80}$$

The following two subsystems can be identified; the first is

$$\dot{z}^1 = (T'AU) z^1 + T'Bu, \qquad y^1 = CUz^1 \tag{6.81}$$

which is of course the same as the one defined by (6.55), (6.56), and (6.57). This system is observable. The second system is

$$\dot{z}^2 = (QAU)\, z^2 + QBu, \qquad y^2 = 0 \qquad (6.82)$$

which is not observable. We show the applications of these ideas in an example.

EXAMPLE

Consider the system

$$A = \begin{bmatrix} -1 & 0 & 0 & 0 \\ 0 & -1 & 0 & 0 \\ 0 & 0 & -1 & 0 \\ 0 & 0 & 0 & -1 \end{bmatrix}, \qquad B = \begin{bmatrix} 1 & 0 \\ 0 & 1 \\ 1 & 0 \\ 0 & 1 \end{bmatrix}$$

$$C = \begin{bmatrix} 1 & 1 & 0 & 0 \\ 0 & 0 & 1 & 1 \end{bmatrix}$$

which is neither controllable nor observable. We construct M^{-1} and M as described

$$M^{-1} = \begin{bmatrix} 1 & 0 & 0 & 0 \\ 0 & 1 & 0 & 0 \\ 1 & 0 & 1 & 0 \\ 0 & 1 & 0 & 1 \end{bmatrix}$$
$$\underbrace{\qquad\qquad}_{W}$$

$$M = \begin{bmatrix} 1 & 0 & 0 & 0 \\ 0 & 1 & 0 & 0 \\ -1 & 0 & 1 & 0 \\ 0 & -1 & 0 & 1 \end{bmatrix}\left.\begin{matrix} \\ \\ \\ \\ \end{matrix}\right\} V$$

so that

$$\hat{A} = MAM^{-1} = A = \begin{bmatrix} -1 & 0 & 0 & 0 \\ 0 & -1 & 0 & 0 \\ 0 & 0 & -1 & 0 \\ 0 & 0 & 0 & -1 \end{bmatrix}$$

$$\hat{B} = MB = \begin{bmatrix} 1 & 0 \\ 0 & 1 \\ 0 & 0 \\ 0 & 0 \end{bmatrix}$$

$$\hat{C} = CM^{-1} = \begin{bmatrix} 1 & 1 & 0 & 0 \\ 1 & 1 & 1 & 1 \end{bmatrix}$$

The controllable subsystem is characterized by

$$A_0 = VAS = \begin{bmatrix} -1 & 0 \\ 0 & -1 \end{bmatrix}, \qquad B_0 = VB = \begin{bmatrix} 1 & 0 \\ 0 & 1 \end{bmatrix}$$

$$C_0 = CS = \begin{bmatrix} 1 & 1 \\ 1 & 1 \end{bmatrix}$$

This system can be transformed into one from which a controllable and observable subsystem can be extracted. Let us construct the matrix T from A_0 and C_0

$$c^1 = \begin{bmatrix} 1 \\ 1 \end{bmatrix}, \qquad Ac^1 = -c^1, \qquad c^1 = c^2$$

that is, $n_0 = 1$, and

$$T = \begin{bmatrix} 1 \\ 1 \end{bmatrix}$$

We can construct N

$$N = \begin{bmatrix} 1 & 1 \\ 0 & 1 \end{bmatrix} \begin{matrix} \} \, T' \\ \} \, Q \end{matrix}$$

$$N^{-1} = \begin{bmatrix} 1 & -1 \\ 0 & 1 \end{bmatrix}$$

and the system obtained from A_0, B_0, and C_0 by putting $z = Nx$ is characterized by the matrices

$$\hat{A}_0 = N A_0 N^{-1} = A_0 = \begin{bmatrix} -1 & 0 \\ 0 & -1 \end{bmatrix}$$

$$\hat{B}_0 = N B_0 = \begin{bmatrix} 1 & 1 \\ 0 & 1 \end{bmatrix}$$

$$\hat{C}_0 = C_0 N^{-1} = \begin{bmatrix} 1 & 0 \\ 1 & 0 \end{bmatrix}$$

The observable (and controllable) subsystem is characterized by the matrices

$$A_{co} = [-1], \qquad B_{co} = \begin{bmatrix} 1 \\ 0 \end{bmatrix}, \qquad C_{co} = \begin{bmatrix} 1 \\ 1 \end{bmatrix}$$

while the controllable, but not observable, subsystem is characterized by

$$A_{c\phi} = [-1], \qquad B_{c\phi} = \begin{bmatrix} 1 \\ 1 \end{bmatrix}, \qquad A_{c\phi} = \begin{bmatrix} 0 \\ 0 \end{bmatrix}$$

The subindex c (or o) indicates controllability (or observability); the subindex ¢ (or ø) indicates lack thereof. In the same way, from A, B, and C above we extract a subsystem which is not controllable

$$A_1 = \begin{bmatrix} -1 & 0 \\ 0 & -1 \end{bmatrix}, \qquad B_1 = \begin{bmatrix} 0 & 0 \\ 0 & 0 \end{bmatrix}, \qquad C_1 = \begin{bmatrix} 0 & 0 \\ 1 & 1 \end{bmatrix}$$

We can change the basis so that the resulting system can be separated into a noncontrollable, observable subsystem and a noncontrollable, nonobservable subsystem. By inspection,

$$T = \begin{bmatrix} 1 \\ 1 \end{bmatrix}, \qquad N = \begin{bmatrix} 1 & 1 \\ 1 & 0 \end{bmatrix} \begin{matrix} \} \, T' \\ \} \, Q \end{matrix}$$

$$N^{-1} = \begin{bmatrix} 0 & 1 \\ 1 & -1 \end{bmatrix}$$

$$\hat{A}_1 = \begin{bmatrix} -1 & 0 \\ 0 & -1 \end{bmatrix}, \qquad \hat{B}_1 = \begin{bmatrix} 0 & 0 \\ 0 & 0 \end{bmatrix}$$

$$\hat{C}_1 = \begin{bmatrix} 0 & 0 \\ 1 & 1 \end{bmatrix} \begin{bmatrix} 0 & 1 \\ 1 & -1 \end{bmatrix} = \begin{bmatrix} 0 & 0 \\ 1 & 0 \end{bmatrix}$$

this can be separated into a noncontrollable, observable subsystem of dimension one described by the matrices

$$A_{\varrho o} = [-1], \qquad B_{\varrho o} = \begin{bmatrix} 0 \\ 0 \end{bmatrix}, \qquad C_{\varrho o} = \begin{bmatrix} 0 \\ 1 \end{bmatrix}$$

and a noncontrollable, nonobservable subsystem also of dimension one described by

$$A_{\varrho\emptyset} = [-1], \qquad B_{\varrho\emptyset} = \begin{bmatrix} 0 \\ 0 \end{bmatrix}, \qquad C_{\varrho\emptyset} = \begin{bmatrix} 0 \\ 0 \end{bmatrix}$$

6.5. Analytic Systems

It was mentioned in Section 6.1 that if an impulse response matrix derived from a weighting pattern is realized by means of an *analytic* system, then this system w-realizes the weighting pattern as well, so that, as in the case of constant systems, the two concepts, realization and w-realization, are equivalent. The material of Section 6.2 is seen to be quite relevant to the study of synthesis by means of analytic systems; for instance, a controllable and observable realization is minimal.

It is not difficult to prove that a weighting pattern is realizable (or w-realizable) by an analytic system [that is, a system such as (6.1) with analytic matrices] if and only if it is realizable in the sense of Theorem 6.1 and if its constituent matrices α and β are themselves analytic functions of their arguments.

Let H be, then, an analytic weighting pattern, and define [4]:

$$\Gamma_{ij}(t, \lambda) = \left(H_{kl}(t, \lambda) \right) \tag{6.83}$$

with

$$H_{kl}(t, \lambda) = \frac{\partial^k}{\partial t^k} \left(\frac{\partial^l}{\partial t^l} H(t, \lambda) \right)$$

$0 \leqslant k \leqslant i-1$, $0 \leqslant l \leqslant j-1$; Γ_{ij} is an $im \times jr$ matrix. Suppose there is an integer n such that $\Gamma_{nn}(t, \lambda)$ has rank n. Let us select an $n \times n$, nonsingular matrix $F(t, \lambda)$ from $\Gamma_{nn}(t, \lambda)$; note that t and λ are, for the moment, considered fixed. Define further

$$F^*(t, \lambda) = \frac{\partial}{\partial t} F(t, \lambda) \qquad (6.84)$$

and

$$G(t, \lambda) = F^*(t, \lambda) F^{-1}(t, \lambda) \qquad (6.85)$$

We show that $G(t, \lambda)$ does not depend on λ. Indeed, from (6.85),

$$F^*(t, \lambda) = G(t, \lambda) F(t, \lambda)$$

which implies that multiplying by $G(t, \lambda)$ is equivalent to differentiating with respect to t. Then,

$$\frac{\partial}{\partial \lambda} F^*(t, \lambda) = \frac{\partial}{\partial \lambda} \left[\frac{\partial}{\partial t} F(t, \lambda) \right] = \frac{\partial}{\partial t} \left[\frac{\partial}{\partial \lambda} F(t, \lambda) \right]$$

$$= G(t, \lambda) \frac{\partial}{\partial \lambda} F(t, \lambda) \qquad (6.86)$$

Differentiate now (6.85) with respect to λ

$$\frac{\partial}{\partial \lambda} G(t, \lambda) = \frac{\partial}{\partial \lambda} F^*(t, \lambda) F^{-1}(t, \lambda) + F^*(t, \lambda) \frac{\partial}{\partial \lambda} F^{-1}(t, \lambda)$$

$$= G(t, \lambda) \frac{\partial}{\partial \lambda} F(t, \lambda) F^{-1}(t, \lambda) + G(t, \lambda) F(t, \lambda) \frac{\partial}{\partial \lambda} F^{-1}(t, \lambda)$$

$$= G(t, \lambda) \frac{\partial}{\partial \lambda} [F(t, \lambda) F^{-1}(t, \lambda)] = 0 \qquad (6.87)$$

the matrix $G(t, \lambda)$ does not, after all, depend on λ. We show that $F(t, \lambda)$ is separable in the form

$$F(t, \lambda) = F(t, \tau) F^{-1}(\xi, \tau) F(\xi, \lambda) \qquad (6.88)$$

where ξ and τ are such that $F(\xi, \tau)$ is nonsingular. We can derive easily that

$$\frac{\partial}{\partial t} [F^{-1}(t, \tau) F(t, \lambda)] = -F^{-1}(t, \tau)(G(t, \tau) - G(t, \lambda)) F(t, \lambda) = 0 \quad (6.89)$$

since $G(t, \tau) = G(t, \lambda)$, so that $F^{-1}(t, \tau) F(t, \lambda)$ is not a function of t. Therefore, if $F(\xi, \tau)$ is nonsingular,

$$F^{-1}(t, \tau) F(t, \lambda) = F^{-1}(\xi, \tau) F(\xi, \lambda)$$

and (6.88) follows. The matrix H is also separable. Let $F_1(t, \lambda)(F_2(t, \lambda))$ be the matrix composed of those columns (rows) of the first m rows (r columns) of $\Gamma_{nn}(t, \lambda)$ which correspond to the columns (rows) of $F(t, \lambda)$. It may be verified that

$$H(t, \lambda) = F_1(t, \lambda)\, F^{-1}(t, \lambda)\, F(t, \lambda)\, F^{-1}(t, \lambda)\, F_2(t, \lambda)$$

It can be proved that $F_1(t, \lambda)\, F^{-1}(t, \lambda)$ does not depend on λ, and that $F^{-1}(t, \lambda)\, F_2(t, \lambda)$ does not depend on t, in a similar way as when we proved that $G(t, \tau)$ does not depend on τ. It follows that

$$H(t, \lambda) = F_1(t, \tau)\, F^{-1}(t, \tau)\, F(t, \lambda)\, F^{-1}(\xi, \lambda)\, F_2(\xi, \lambda)$$

(6.88) implies then that

$$H(t, \lambda) = F_1(t, \tau)\, F^{-1}(\xi, \tau)\, F_2(\xi, \lambda) \tag{6.90}$$

We can write by inspection a realization of the weighting pattern H: The role of the matrix A in (6.1) is played by the zero matrix 0_n, the role of $B(t)$ is played by $F^{-1}(\xi, \tau)\, F_2(\xi, t)$ while $C(t)$ is $F_1(t, \tau)$. This is, of course, a realization of dimension n; it also happens to be a w-minimal (and in this case, minimal) realization, which is controllable and observable as we show below.

We defined in Section 5.7 a controllability matrix $Q(t)$, such that if its rank was equal to the order of the system at some $t > t_0$, then the system was controllable at t_0. In the same way one could define an observability matrix $P(t)$ with similar properties. It so happens that if $H(t, \lambda)$ is realized by a system with controllability matrix $Q(t)$ and observability matrix $P(t)$, then $H(t, t) = P'(t)\, Q(t)$ [4], so that the realization is controllable and observable if $H(t, t)$ has rank equal to the dimension of the realization. In the case above, $H(t, t)$ does have, by assumption, dimension n, so that the realization derived from (6.90) is indeed controllable and observable and then minimal.

We can progress even further. The previous realization has a matrix A equal to 0_n, which may not be very desirable; also, if H is realizable by a constant system, the previous realization may not be constant. The basic procedure presented above can, however, be modified so as to avoid these drawbacks.

Suppose that $F(t, \lambda)$ has rank n for all t, λ (or at least for all $t \geqslant \lambda$). Then, if we write $F(t)$ for $F(t, t)$, etc, the system characterized by the matrices

$$A(t) = F^*(t)\, F^{-1}(t), \qquad B(t) = F_2(t), \qquad C(t) = F_1(t)\, F^{-1}(t) \tag{6.91}$$

also realizes H. Indeed, write

$$H(t, \lambda) = F_1(t, \lambda) F^{-1}(t, \lambda) F(t, \tau) F^{-1}(\xi, \tau) F(\xi, \lambda) F^{-1}(\xi, \lambda) F_2(\xi, \lambda)$$

and make $\tau = t$, $\xi = \lambda$. Then,

$$H(t, \lambda) = F_1(t, t) F^{-1}(t, t) F(t, \tau) F^{-1}(\lambda, \tau) F(\lambda, \lambda) F^{-1}(\lambda, \lambda) F_2(\lambda, \lambda)$$

$$= (F_1(t) F^{-1}(t))(F(t, \tau) F^{-1}(\lambda, \tau))(F_2(\lambda))$$

We show that $F(t, \tau) F^{-1}(\lambda, \tau) = \phi(t, \lambda)$ is the transition matrix of the matrix A in (6.91), from which the assertion follows obviously. Certainly $\phi(t, t) = I_n$. Moreover, $F(t, \tau)$ is a fundamental matrix of the homogeneous system defined by the matrix A, since, by (6.85)

$$G(t, \lambda) = G(t, t) = A(t) = F^*(t, t) F^{-1}(t, t)$$

and then

$$F^*(t, \tau) = \frac{\partial}{\partial t} F(t, \tau) = A(t) F(t, \tau)$$

The matrix $\phi(t, \lambda)$ is the transition matrix of A, and the system defined by (6.91) realizes H. This system is constant if H is realizable by a constant system.

6.6. The Reduction to Minimal Order in the General Case

It should be remembered that an impulse response matrix is said to be realizable if a linear system can be found such that its impulse response matrix equals the given one. The following theorem closely parallels Theorem 6.1.

THEOREM 6.6.   An impulse response matrix H_c is realizable if and only if

$$H_c(t, \lambda) = \alpha(t) \beta(\lambda) \tag{6.92}$$

for all t, λ such that $t \geqslant \lambda$. Here α and β are continuous functions. In this book we do not consider systems whose impulse response matrices are not continuous functions of their arguments.

Suppose that H_c is derived from a weighting pattern H. Then, unless the weighting pattern is w-realizable by means of a constant system, or by means of an analytic system, a system which has minimum order among all those which w-realize H does not in general have minimum order among all systems which realize H_c. We show this by means of an example.

EXAMPLE [5]

Let $r = m = 1$, and $H(t, \lambda) = \sum_{i=1}^{3} \alpha_i(t)\,\beta_i(\lambda)$, where the functions α_i, β_i, $i = 1, 2, 3$, are continuous and such that

$$\alpha_1(t) \begin{matrix} \neq 0, & t \in [-2, -1] \\ = 0, & \text{elsewhere} \end{matrix} \qquad \beta_1(\lambda) \begin{matrix} \neq 0, & \lambda \in [-4, -3] \\ = 0, & \text{elsewhere} \end{matrix}$$

$$\alpha_2(t) \begin{matrix} \neq 0, & t \in [3, 4] \\ = 0, & \text{elsewhere} \end{matrix} \qquad \beta_2(\lambda) \begin{matrix} \neq 0, & \lambda \in [1, 2] \\ = 0, & \text{elsewhere} \end{matrix}$$

$$\alpha_3(t) \begin{matrix} \neq 0, & t \in [5, 6] \\ = 0, & \text{elsewhere} \end{matrix} \qquad \beta_3(\lambda) \begin{matrix} \neq 0, & \lambda \in [7, 8] \\ = 0, & \text{elsewhere} \end{matrix}$$

The functions α_1, α_2, α_3 are independent over the real line; the functions β_1, β_2, β_3 are also independent over this set. It follows that a globally reduced w-realization of this weighting pattern has dimension *three*. However, the product $\alpha_3(t)\,\beta_3(\lambda)$ is zero for $t \geqslant \lambda$, so that there are realizations of the impulse response matrix derived from H of dimension *two*; indeed, this impulse response matrix can be considered to be derived from a weighting pattern $H_1(t, \lambda) = \sum_{i=1}^{2} \alpha_i(t)\,\beta_i(\lambda)$; a globally reduced realization of H_1 has dimension *two*.

The rest of this section will be concerned with the presentation of a method of reduction of a factorization of an impulse response matrix; that is, if $H_c(t, \lambda)$ is given in a factorized form, $H_c(t, \lambda) = \alpha(t)\,\beta(\lambda)$, we will learn how to operate on this factorization so as to make $H_c(t, \lambda)$ equal to $\bar{\alpha}(t)\,\bar{\beta}(\lambda)$; the number of columns of $\bar{\alpha}$ (which equals the number of rows of $\bar{\beta}$) is the least among all such factorizations.

Let, then, $H_c(t, \lambda) = \alpha(t)\,\beta(\lambda)$, with $\alpha(t)$ an $m \times n$ and $\beta(\lambda)$ an $n \times r$ matrix. Further, let $U(t)$ be the set of vectors x in R^n for which a square integrable function u exists such that

$$x = \int_{-\infty}^{t} \beta(\lambda)\,u(\lambda)\,d\lambda \tag{6.93}$$

while $V(t)$ is to denote the set of vectors x for which

$$\alpha(\tau)x = 0 \tag{6.94}$$

for all $\tau \geqslant t$.

Obviously, if $t_1 \leqslant t_2$, $U(t_1) \subseteq U(t_2)$ and $V(t_1) \subseteq V(t_2)$. We show:

LEMMA.   The set-valued functions U and V change at only finitely many times.

Proof.   It follows from the definitions that the sets $U(t)$ and $V(t)$ are subspaces of R^n. It was mentioned above that, if $t_2 \geqslant t_1$, $U(t_1) \subseteq U(t_2)$. This implies that, if the value of U changes at t', say, then the new value has to have a higher dimension than the value of U prior to the change. After at most n such changes the value of U has to coincide with the whole space R^n; afterward there can be no changes. The same argument can be applied to the set-valued function V. The lemma follows.

Let t_1, t_2, ..., t_m, with $t_1 < t_2 < \cdots < t_m$ be the values of time at which *either* U or V change. Then, the lemma just proved states that $U(t) = U(t_1)$, $-\infty < t \leqslant t_1$, $U(t) = U(t_i)$, $t_{i-1} < t \leqslant t_i$, $i = 2, ..., m$, $U(t) = U(t_{m+1})$, $t > t_m$, with t_{m+1} any number larger than t_m. A similar situation prevails for the function V.

Our intention is to make a direct sum decomposition of the space R^n. With this purpose, we define several subspaces by means of the following equalities

$$U(t_1) = U(t_1) \cap V(t_1) \oplus W(t_1) \tag{6.95}$$

$$U(t_i) = U(t_{i-1}) \oplus R(t_i) \oplus W(t_i), \qquad i = 2, 3, ..., m+1 \tag{6.96}$$

here $W(t_1)$ is any subspace satisfying (6.95) and $R(t_i)$ is any subspace of $V(t_i)$ of largest possible dimension which satisfies (6.96) for some $W(t_i)$. Let $R(t_1) = U(t_1) \cap V(t_1)$, and put

$$W = W(t_1) \oplus \cdots \oplus W(t_{m+1})$$

$$R = R(t_1) \oplus \cdots \oplus R(t_{m+1})$$

It follows from the previous definitions that $U(t_1) = R(t_1) \oplus W(t_1)$; $U(t_2) = R(t_1) \oplus W(t_1) \oplus R(t_2) \oplus W(t_2)$, ..., so that, at last,

$$U(t_{m+1}) = W \oplus R \tag{6.97}$$

Since $W \oplus R$ is a subspace of R^n, we can decompose R^n itself into the direct sum of $U(t_{m+1})$ and "$U(t_{m+1})$ perp," $U(t_{m+1})^\perp$

$$R^n = U(t_{m+1}) \oplus U(t_{m+1})^\perp \tag{6.98}$$

this concept was introduced in Chapter 2; $U(t_{m+1})^\perp$, the *orthogonal complement* of $U(t_{m+1})$, is the set of all vectors in R^n which are orthogonal to every vector in $U(t_{m+1})$. It follows that, if $x \in X$,

$$x = y + z \tag{6.99}$$

$y \in U(t_{m+1})$, $z \in U(t_{m+1})^\perp$. Clearly we can associate with every $x \in X$ a vector y in $U(t_{m+1})$, so actually we have defined an operator P such that $P(x) = y$; this *projection* operator has a range of, say, dimension $\bar{n}$.

We show that this operator provides the clue to the reduction of the factorization $H_c(t, \lambda) = \alpha(t)\,\beta(t)$ to one of lesser order.

THEOREM 6.7 [5].

i. For all t, λ such that $t \geqslant \lambda$,

$$\alpha(t)\,\beta(\lambda) = \alpha(t)\,P\beta(\lambda)$$

where P is the matrix of the projection operator defined above relative to the standard basis in R^n.

ii. If $\bar{\alpha}(t) = \alpha(t)P$, $\bar{\beta}(\lambda) = P\beta(\lambda)$, then $H_c(t, \lambda) = \bar{\alpha}(t)\,\bar{\beta}(\lambda)$, $t \geqslant \lambda$; the dimensions of $\bar{\alpha}$ and $\bar{\beta}$ are $m \times \bar{n}$, and $\bar{n} \times r$ respectively. The number $\bar{n}$ was defined above.

iii. If $H_c(t, \lambda) = \hat{\alpha}(t)\,\hat{\beta}(\lambda)$, $t \geqslant \lambda$, and $\hat{\alpha}$ and $\hat{\beta}$ are of dimensions $m \times \hat{n}$, $\hat{n} \times r$ respectively, $\bar{n} \leqslant \hat{n}$.

Proof. We only prove parts i and ii; the reader is referred to [5] for the proof of the minimality of the factorization characterized by $\bar{\alpha}$, $\bar{\beta}$.

To show i, it is sufficient to prove that

$$\int_{-\infty}^{t} \alpha(t)\,\beta(\lambda)\,u(\lambda)\,d\lambda = \int_{-\infty}^{t} \alpha(t)\,P\beta(\lambda)\,u(\lambda)\,d\lambda \qquad (6.100)$$

for all square integrable functions u. Let $z(t) = \int_{-\infty}^{t} \beta(\lambda)\,u(\lambda)\,d\lambda$. By definition, $z(t) \in U(t) = U(t_i)$ for some i. Then, since

$$U(t_i) = W(t_1) \oplus W(t_2) \oplus \cdots \oplus W(t_i) \oplus R(t_1) \oplus R(t_2) \oplus \cdots \oplus R(t_i) \qquad (6.101)$$

then $z(t)$ can be written as $z(t) = x + y$, where $x \in W(t_1) \oplus \cdots \oplus W(t_i)$, $y \in R(t_1) \oplus \cdots \oplus R(t_i)$. It follows that

$$Pz(t) = P(x + y) = Px = x$$

so that

$$\int_{-\infty}^{t} \alpha(t)\,\beta(\lambda)\,u(\lambda)\,d\lambda = \alpha(t)\,z(t) = \alpha(t)x + \alpha(t)\,y$$

If we were to show that $\alpha(t)\,y = 0$, then

$$\alpha(t)x = \alpha(t)\,Px = \alpha(t)\,Pz(t) = \int_{-\infty}^{t} \alpha(t)\,P\beta(\lambda)\,u(\lambda)\,d\lambda$$

for all square integrable u. Indeed, since $y \in R(t_1) \oplus \cdots \oplus R(t_i)$, $y \in V(t)$, since $R(t_1) \oplus \cdots \oplus R(t_i) \subset V(t)$, for $t > t_i$, by definition of the subspaces $R(t_i)$. It follows that $\alpha(t)\,y = 0$, and then the assertion i.

Assertion ii is a consequence of the fact that $P^2x = Px$ for all x in its domain, so that $P^2 = P$; then

$$\alpha(t) \, P\beta(\lambda) = \alpha(t) \, P^2\beta(\lambda) = \alpha(t) \, PP\beta(\lambda) \tag{6.102}$$

from which contention ii follows by taking $\bar{\alpha}(t) = \alpha(t)P$, $\bar{\beta}(\lambda) = P\beta(\lambda)$.

6.7. The Observer

We end this chapter with the treatment of a topic which is not directly related to structural problems, but which is very much connected with the practical implementation of the realizations of impulse response matrices or transfer functions.

If the only available description of a linear system is its impulse response or transfer function matrix, and the state variables cannot be measured directly, it is often desirable to generate a set of state variables for this system; this is done by first obtaining a realization of the system by means of the methods studied in the previous sections of this chapter and, second, by simulating this realization by means of analog computer components, as indicated in Figure 6.4.

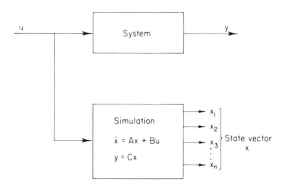

FIGURE 6.4.   Generation by means of analog simulation of a set of state variables for a system known by means of its transfer function or impulse response matrix.

It may happen, though, that there are better, more economical ways of generating this state vector; the theory of the *observer* is concerned with a way of accomplishing this goal for constant systems. We follow [6].

Suppose that we try to estimate the state vector x by means of the system shown in Figure 6.5; the systems which are labeled "observer" are analog simulations of the processes indicated in the boxes. The system in Figure 6.5 generates an estimate $\hat{x}$ of the state vector x, which

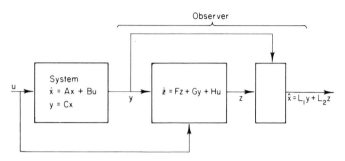

FIGURE 6.5. Generation of a set of state variables by means of an observer.

we will show later on to be, under some conditions, equal to x. We assume that the first system in Figure 6.5 has a realization characterized by constant matrices A, B, and C, and that C has rank m. The (constant) matrices in the observer of Figure 6.5 have the following dimensions:

$$F: p \times p, \quad p = n - m$$

$$G: p \times m, \qquad H: p \times r \tag{6.103}$$

$$L_1 : n \times m, \qquad L_2 : n \times p$$

Since $p < n$, the simulation in Figure 6.5 needs less integrators than that of Figure 6.4.

Suppose that we make H in Figure 6.5 equal to

$$H = TB \tag{6.104}$$

where T is a $p \times n$ matrix which is a solution of the equation

$$TA - FT = GC \tag{6.105}$$

here G is an unspecified $p \times m$ matrix. We assume for the moment that there is a solution T of (6.105), leaving for later the actual determination of such a solution. Put

$$e = z - Tx, \quad \text{or} \quad z = Tx + e \tag{6.106}$$

then

$$\dot{z} = T\dot{x} + \dot{e} = T(Ax + Bu) + \dot{e}$$

$$= Fz + Gy + Hu = FTx + Fe + Gy + Hu \tag{6.107}$$

so that we must have

$$\underbrace{(TA - FT)x}_{GC} + \underbrace{(TB - H)u}_{0} - \underbrace{Gy}_{GCx} + \dot{e} - Fe = 0 \qquad (6.108)$$

we find that the p-vector e is determined by

$$\dot{e} = Fe \qquad (6.109)$$

The estimate $\hat{x}$ of the state vector x is

$$\hat{x} = L_1 Cx + L_2 Tx + L_2 e = L(Wx + \hat{e}), \qquad (6.110)$$

where

$$L = [L_1 \ L_2]$$
$$W = \begin{bmatrix} C \\ T \end{bmatrix}, \qquad \hat{e} = \begin{bmatrix} 0 \\ e \end{bmatrix} \begin{matrix} \} \ n-p \\ \} \ p \end{matrix} \qquad (6.111)$$

Suppose, at last, that W is nonsingular; we will derive conditions which ensure precisely this later on. Then, we choose $L = W^{-1}$, so that

$$\hat{x} = x + W^{-1}e \qquad (6.112)$$

Since e, by (6.109), is simply

$$e(t) = \exp(F(t - t_0)) \, e(t_0) \qquad (6.113)$$

we see that $\hat{x}$ and x will be equal in the steady state if and only if the matrix F of the observer has eigenvalues which have only negative real parts. We can summarize the design procedure leading to the design of the observer as follows:

i. Choose an $(n - m) \times p$ matrix F with eigenvalues with negative real parts, and such that no eigenvalue of F equals an eigenvalue of A; as we will see below, this ensures that the equation (6.105) has a solution.

ii. Solve equation (6.105) after choosing G arbitrarily. (We will see below that G has to be chosen so that the observer is controllable with respect to the plant input.)

iii. Make $H = TB$.

iv. Form W and compute W^{-1}; partition this matrix into the matrices L_1 and L_2 .

Let us consider the equation (6.105); we will obtain at the same time some insight into the properties of the matrix W. From (6.105) we obtain easily

$$TA^0 - F^0 T = 0, \qquad TA - FT = GC$$

$$TA^2 - F^2 T = GCA + FGC \tag{6.114}$$

$$TA^n - F^n T = GCA^{n-1} + FGCA^{n-2} + \cdots + F^{n-2}GCA + F^{n-1}GC$$

Let now $f(\lambda) = \sum_{i=0}^{n} \alpha_i \lambda^i$, $\alpha_n = 1$, be the characteristic polynomial of A. Multiply the first equation in (6.114) by α_0, the second by α_1, and so on. Then

$$Tf(A) - f(F)T = GCf_1(A) + FGf_2(A) + \cdots + F^{n-1}GCf_n(A) \tag{6.115}$$

where

$$f_i(\lambda) = 1/\lambda(f_{i-1}(\lambda) - \alpha_{i-1}), \qquad i = 2, \ldots, n \tag{6.116}$$

which implies that, since $f(A) = 0$, and since $f(F)$ is nonsingular (A and F are not supposed to have common eigenvalues),

$$T = -f(F)^{-1} Q_0 \Omega P_p \tag{6.117}$$

where

$$Q_0 = [G \; FG \; \cdots \; F^{n-1}G] \tag{6.118}$$

$$P_p = [C' \; A'C' \; \cdots \; A'^{(n-1)}C'] \tag{6.119}$$

$$\Omega = \begin{bmatrix} \alpha_1 I_p & \alpha_2 I_p & \cdots & \alpha_{n-1} I_p & \alpha_n I_p \\ \alpha_2 I_p & \alpha_3 I_p & \cdots & \alpha_n I_p & 0_p \\ \vdots & \vdots & & \vdots & \vdots \\ \alpha_{n-1} I_p & \alpha_n I_p & & 0_p & 0_p \\ \alpha_n I_p & 0_p & & 0_p & 0_p \end{bmatrix} \tag{6.120}$$

The matrix Ω is nonsingular. We can write

$$W = \begin{bmatrix} C \\ -f^{-1}(F)Q_0 \Omega P_p \end{bmatrix} = \begin{bmatrix} I & 0 \\ 0 & -f^{-1}(F) \end{bmatrix} \begin{bmatrix} C \\ Q_0 \Omega P_p \end{bmatrix} \tag{6.121}$$

so that W is nonsingular if and only if the matrix

$$\hat{W} = \begin{bmatrix} C \\ Q_0 \Omega P_p \end{bmatrix}$$

is nonsingular. Note that Q_0 is an augmented controllability matrix for the observer, while P_p is the observability matrix for the original system. We prove at last:

THEOREM 6.8. If the matrix W is nonsingular, then the observer is controllable with respect to the system output y and the system is observable. If $m = 1$ (that is, if the output of the system is a scalar), these conditions are also sufficient for the invertibility of W.

Proof. The rank of $\tilde{Q}_0 = [G\,FG \cdots F^{p-1}G]$ is the same as that of Q_0. Suppose now that the observer is not controllable with respect to y; then the rank of $\tilde{Q}_0$, and thus of Q_0, is less than p, which implies that the rank of W, which is not higher than the rank of C plus the rank of $Q_0 \Omega P_p$, is less than $m + p = n$, that is, $\hat{W}$ and W are singular. The necessity of the observability of the system follows in a similar way. We will not prove the sufficiency of the condition for $m = 1$.

We have not really succeeded, then, for $m > 1$, in solving our problem completely, since we do not really know of ways of ensuring that the matrix W will be invertible. If $m = 1$, however, it is sufficient to select the realization of the system to be observable and the realization of the observer to be controllable (by choosing the matrix G in step ii of the design procedure in an appropriate way). If $m > 1$, the realizations of the system and of the observer have to be chosen in this manner (otherwise we are sure that W *will not* be invertible), even though it may happen that W turns out to be singular.

SUPPLEMENTARY NOTES AND REFERENCES

§6.1. The student with a background in passive network synthesis should be able to recognize the similitude between this and our synthesis problem. The interested reader should consult

H. H. Rosenbrock, Connection between network theory and the theory of linear dynamical systems, *Electronic Letters* 3 (1967), 296.

H. H. Rosenbrock, System matrices giving positive-real transfer-function matrices, *Proc. IEE* 115 (1968), 328–329.

R. E. Kalman, Irreducible realizations and the degree of a rational matrix, *SIAM J. Appl. Math.* 13 (1965), 520–544.

§6.2. The order of a matrix $H(t - \lambda)$ which is realizable by means of a time-invariant system equals the so-called *McMillan degree* of the transform $Z(s)$ of $H(t - \lambda)$. See

B. McMillan, Introduction to formal realizability theory, *Bell System Tech. J.* 31 (1952), 217–279, 541–600.

D. Hazony, "Elements of Network Synthesis." Van Nostrand, Princeton, New Jersey, 1963.

§6.4. An apparently very economical method of synthesis is presented in

H. H. Rosenbrock, Computation of minimal representations of a rational transfer-function matrix, *Proc. IEE* **115** (1968), 325–327.

H. H. Rosenbrock, Transformation of linear constant system equations, *Proc. IEE* **114** (1967), 541–544.

H. H. Rosenbrock, On linear system theory, *Proc. IEE* **114** (1967), 1353–1359.

As a matter of fact, the material presented in these references goes beyond the derivation of an effective synthesis method, but represents an original attempt at constructing a theory of linear, constant systems. The reader is urged to investigate Rosenbrock's ideas, which are both novel and useful.

Another synthesis method is presented in

B. L. Ho and R. E. Kalman, Effective construction of linear state-variable models from input–output functions, *Regelungstechnik* **14** (1966), 545–548.

There is a large body of knowledge concerned with the transformation of linear systems and the subsequent separation of the state space in regions representing the state spaces of controllable and observable, controllable but not observable, etc, subsystems. We present a very elementary treatment of this topic in Section 6.4; the interested reader should consult:

R. E. Kalman, Y. C. Ho, and K. S. Narendra, Controllability of linear dynamical systems, *Contrib. Differential Equations* **1** (1962), 189–213.

R. E. Kalman, Canonical structure of linear dynamical systems, *Proc. Nat. Acad. Sci. U.S.A.* **48** (1962), 596–600.

L. Weiss, Weighting patterns and the controllability and observability of linear systems, *Proc. Nat. Acad. Sc. U.S.A.* **51** (1964), 1122–1127.

L. Weiss and R. E. Kalman, Contributions to linear system theory, *Internat. J. Engng. Sci.* **3** (1965), 141–171.

L. Weiss, On the structure theory of linear differential systems, *SIAM J. Control* **6** (1968), 659–680.

L. Weiss and P. L. Falb, Doležal's theorem, linear algebra with continuously parametrized elements, and time-varying systems, *Math. Systems Theory* **3** (1969), 67–75.

§6.5. Further extensions of the ideas presented here appear in

L. M. Silverman, Stable realization of impulse response matrices, *IEEE Internat. Conv. Rec.* **15** (1967), 32–36.

L. M. Silverman, Synthesis of impulse response matrices by internally stable and passive realizations, *IEEE Trans. Circuit Theory* **CT-15** (1968), 238–245.

L. M. Silverman and H. E. Meadows, Equivalent realization of linear systems, *SIAM J. Appl. Math.* **17** (1969), 393–408.

The last two of these references deal with equivalence and synthesis for systems which are not analytic, and are an interesting extension and development of the ideas in [4]. One of the topics treated is the connection between synthesis, stability, controllability, and observability, which we will study, in a rather different form, in Chapter 8.

§6.6. See also

A. V. Balakrishnan, On the problem of deducing states and state-relations from input–output relations for linear time-varying systems, *SIAM J. Control* **5** (1967), 309–325.

§6.7. The concept of the observer is due to D. G. Luenberger. See

D. G. Luenberger, Observing the state of a linear system, *IEEE Trans. Military Electronics* **MIL-8** (1964), 74–80.

D. G. Luenberger, Observers for multivariable systems, *IEEE Trans. Automatic Control* **AC-11** (1966), 190–197.

REFERENCES

1. R. E. Kalman, Mathematical description of linear dynamical systems, *SIAM J. Control* **1** (1963), 152–192.
2. D. C. Youla, The synthesis of linear dynamical systems from prescribed weighting patterns, *SIAM J. Appl. Math.* **14** (1966), 527–549.
3. D. Q. Mayne, Computational procedure for the minimal realization of transfer-function matrices, *Proc. IEE* **115** (1968), 1383–1368.
4. L. M. Silverman and H. E. Meadows, Equivalence and synthesis of time-variable linear systems, *Proc. 1966 Fourth Allerton Conference on Circuit and Systems Theory* (1966), 776–784.
5. C. A. Desoer and P. Varaiya, The minimal realizations of a nonanticipative impulse response matrix, *SIAM J. Appl. Math.* **15** (1967), 754–764.
6. J. J. Bongiorno, Jr. and D. C. Youla, On observers in multivariable control systems, *Internat. J. Control* **8** (1968), 221–243.

7

Difference Systems

7.1. Introduction

The subject matter of this chapter is the study, much in the same way linear differential systems were treated in previous chapters, of another kind of linear systems, whose basic domain of definition is not a closed interval of the reals, but a subset of the set of integers. Instead of asking, for instance, what is the state of a differential system at $t = \sqrt{2}$ given that at $t = 0.5$ it was 0, we may ask, what is the state of a difference system at $k = 7$ given that at $k = 2$ it was 0?

We show now several examples of these systems, so as to lead the way into a process of abstraction and then into a general formulation of this class of system, similar to the description of linear differential systems by means of a set of state equations such as (3.1).

Consider, first, a system such as the one shown in Figure 7.1. Here the

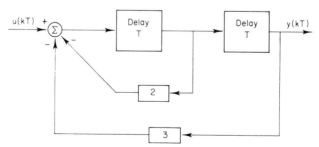

FIGURE 7.1. A difference system, showing delay elements and constant feedbacks.

220

input (and then the output) are defined only at discrete instants of *the time*, instants which we designate by kT, k being an integer and T an arbitrary, positive real number. The *delay*, a new component for us, simply delays the sequence of numbers at its input by T seconds, so that if its input is $w((k+1)T)$, then its output is $w(kT)$. It follows that the input and output sequences here are related by

$$y((k+2)T) = u(kT) - 2y((k+1)T) - 3y(kT) \qquad (7.1)$$

or

$$y((k+2)T) + 2y((k+1)T) + 3y(kT) = u(kT) \qquad (7.2)$$

We can transform this equation, much as we transformed high-order differential equations in Chapter 1, into a system of equations; put $x_1(kT) = y(kT)$, $x_2(kT) = y((k+1)T)$, so that

$$x_1((k+1)T) = x_2(kT)$$
$$x_2((k+1)T) = -3x_1(kT) - 2x_2(kT) + u(kT) \qquad (7.3)$$
$$y(kT) = x_1(kT)$$

Note that each of the first two equations gives the value of a (state) variable at $(k+1)T$ in terms of the state variables at the previous time kT and the values of the input at kT. We can, of course, write (7.3) in the form

$$x((k+1)T) = Ax(kT) + Bu(kT)$$
$$y(kT) = Cx(kT) \qquad (7.4)$$

where x is a two-dimensional state vector, u is a scalar input, y is a scalar output, and

$$A = \begin{bmatrix} 0 & 1 \\ -3 & -2 \end{bmatrix}, \qquad B = \begin{bmatrix} 0 \\ 1 \end{bmatrix}, \qquad C = [1 \ 0]$$

The system in Figure 7.2 makes use of other new components, the *sampler* and the *hold*. This difference system is composed of a continuous-time, *constant* differential system described by $\dot{x} = Ax + Bu$; u^1 is a continuous-time output which is sampled and transformed into $\hat{u}$; this

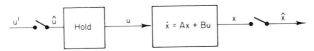

FIGURE 7.2. The transformation of a differential system into a difference system by means of the operations of sampling and holding.

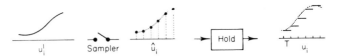

FIGURE 7.3. The operations of the sampler and hold on one of the components of the input u^1 to the system in Figure 7.2.

operation is shown, for the ith component of u^1, in Figure 7.3, where we can also see the effect of the *hold* operation. We have, therefore, transformed u^1, a continuous-time input, into u, another continuous-time input. At the output of the differential system, we sample the state vector x, and produce a sampled output, defined only at $t = kT$. It so happens that we can find a set of equations which relate the *sampled* input, $\hat{u}$ to the sampled output $\hat{x}$ in terms of the matrices A and B which define the continuous system. Indeed,

$$x(t) = e^{A(t-t_0)}x(t) + \int_{t_0}^{t} e^{A(t-\lambda)}Bu(\lambda)\, d\lambda$$

put $t = (k + 1)T$ and $t_0 = kT$. Then,

$$x((k + 1)T) = e^{AT}x(kT) + \int_{kT}^{(k+1)T} e^{A[(k+1)T-\lambda]}Bu(\lambda)\, d\lambda \qquad (7.5)$$

which, since $u(\lambda) = u(kT)$, $kT < \lambda < (k + 1)T$, becomes

$$e^{AT}x(kT) + \left\{ \int_{kT}^{(k+1)T} e^{A[(k+1)T-\lambda]}B\, d\lambda \right\} u(kT)$$

so that the sampled output is

$$\hat{x}((k + 1)T) = (e^{AT})\, \hat{x}(kT) + \hat{B}\hat{u}(kT) \qquad (7.6)$$

where

$$\hat{B} = \int_{kT}^{(k+1)T} e^{A[(k+1)T-\lambda]}B\, d\lambda = \int_{0}^{T} e^{A\mu}B\, d\mu \qquad (7.7)$$

Of course, an output could be defined as $y(kT) = Cx(kT)$, with C an appropriate matrix; the resulting combination of (7.6) and this equation is similar to (7.4); indeed, this general form will be used in all of our next developments. Before discussing this and other matters more fully, we present an example of the derivation of the matrices e^{AT} and $\hat{B}$ in (7.6).

EXAMPLE

If

$$A = \begin{bmatrix} 0 & 1 \\ 0 & 0 \end{bmatrix}, \qquad B = \begin{bmatrix} 0 \\ 1 \end{bmatrix}$$

$$e^{tA} = \begin{bmatrix} 1 & t \\ 0 & 1 \end{bmatrix} \qquad e^{AT} = \begin{bmatrix} 1 & T \\ 0 & 1 \end{bmatrix}$$

$$\hat{B} = \int_0^T \begin{bmatrix} 1 & \mu \\ 0 & 1 \end{bmatrix} \begin{bmatrix} 0 \\ 1 \end{bmatrix} d\mu = \begin{bmatrix} \frac{1}{2} T^2 \\ T \end{bmatrix}$$

It should be recognized that, once T is fixed, the systems (7.4) and (7.6) can be considered as relationships between vectors defined on the set of all integers, rather than on the values of kT. Indeed, in the case of (7.4), all systems with the same matrices A and B but different value of T, are essentially identical, in the sense that they share exactly the same properties; the interest lies only in the dynamical progression of the systems as k goes from 1 to 2, from 2 to 3, etc. If T is known, the same can be said about systems such as (7.6); we are led to define our general *difference* linear system by the equations

$$x(k + 1) = A(k) x(k) + B(k) u(k)$$
$$y(k) = C(k) x(k) + D(k) u(k), \qquad k \in I \tag{7.8}$$

Note that we have added an extra term to the output representing direct transmission between input and output. Here x is an n-vector, the state vector, $A(k)$ is an $n \times n$ matrix, $B(k)$ an $n \times r$ matrix, $C(k)$ an $m \times n$ matrix, $D(k)$ an $m \times r$ matrix. The input $u(k)$ is an r-vector, the output $y(k)$ an m-vector. All of these matrices and vectors are defined on a subset I of the set of integers of the type $I = \{p, p + 1, p + 2, ..., p + q\}$. Note that the matrices A and B are dependent on k; we will say that such systems are time-varying; otherwise we shall call them, as in previous chapters, constant. All matrices and vectors have real components.

In the following section, we study the properties of a system described by (7.8). We follow a similar pattern to the one used in the treatment of the corresponding equations for differential systems in Chapters 3 and 4.

Exercises

1. The classical theory of sampled-data systems is based on the replacement of the samples themselves by impulses of area equal to the value of the sample; the hold then becomes a system whose impulse response is unity for $0 \leqslant t \leqslant T$, and zero

everywhere else. Of course, if we take this approach in the development associated with Figure 7.2, we obtain exactly the same final answer. Assume now that the hold in Figure 7.2 is replaced by a network of impulse response

$$h(t) \quad \begin{aligned} &= e^{-ta}, \quad && 0 \leqslant t \leqslant T, \quad a > 0 \\ &= 0, \quad && \text{otherwise} \end{aligned}$$

Can you obtain a set of state equations for the discrete system with input $\hat{u}$?

2. Consider the above problem in general. If the impulse response of the system is $h(t)$, under what conditions can you obtain a set of state equations for the discrete system? Assume $h(0) = 1$.

3. Discuss the invertibility of the matrix

$$\int_0^T e^{A\mu} B \, d\mu$$

7.2. The Homogeneous System

Consider first the homogeneous system associated with the first equation in (7.8),

$$x(k + 1) = A(k)\, x(k), \qquad x(k_0) = x^0, \quad k \in I \tag{7.9}$$

Our first endeavor will consist in the investigation of existence and uniqueness of the solution of this system. Here we must pause, because a very fundamental difference between the descriptions of differential and difference systems shows up at this stage; this is that the matrices $A(k)$, $k \in I$, must be nonsingular to ensure the uniqueness of the solution of (7.9).

THEOREM 7.1.   The system (7.9) has one and only one solution for all $k \in I$ if and only if the matrices $A(k)$, $k \in I$, are nonsingular. Note that k is *not* assumed to be higher than k_0.

Proof.   Let $A(k)$ be nonsingular, $k \in I$. Let us compute a solution of the system (7.9). If x is a solution,

$$x(k_0 + 1) = A(k_0)\, x^0,$$

$$x(k_0 + 2) = A(k_0 + 1)\, x(k_0 + 1) = A(k_0 + 1)\, A(k_0)\, x^0 \tag{7.10}$$
$$\vdots$$
$$x(k) = A(k - 1)\, A(k - 2) \cdots A(k_0)\, x^0$$

for $k \geqslant k_0$, $k \in I$, and

$$x(k_0 - 1) = A^{-1}(k_0 - 1)\, x^0$$

$$x(k_0 - 2) = A^{-1}(k_0 - 2)\, x(k_0 - 1) = A^{-1}(k_0 - 2)\, A^{-1}(k_0 - 1)\, x^0 \tag{7.11}$$

$$x(k) = A^{-1}(k)\, A^{-1}(k - 1) \cdots A^{-1}(k_0 - 1)\, x^0$$

for $k \leqslant k_0$, $k \in I$. The sufficiency part of the theorem follows by noting that any solution of (7.9) has to satisfy (7.10) and (7.11), and then it is identical to x.

Assume now that there is a unique solution x of (7.9), and that $A(k)$ is singular. Then, if $x(k + 1) = A(k) x(k)$, many vectors $x(k)$ satisfy this equation, so that many solutions will actually coalesce at k. Clearly the solution x is not unique, so that the matrices $A(k)$, $k \in I$, are necessarily nonsingular if the solution is to be unique.

We remark now that a set of vector-valued functions x^1, ..., x^q defined on I, is said to be linearly dependent if there are constants c_1, ..., c_q, not all of which are zero, such that

$$c_1 x^1(k) + \cdots + c_q x^q(k) = 0$$

for *all* $k \in I$. Otherwise, we say it is linearly independent. The following theorem utilizes this concept; it will not be proved since its proof parallels exactly that of Theorem 3.2.

THEOREM 7.2.   The set of all solutions of (7.9) (for different conditions at k_0) forms an n-dimensional vector space over C (if A takes values in a complex space) or over R (if $A(k)$ is real).

A *fundamental set* of solutions of $x(k + 1) = A(k) x(k)$ is, as previously, a set of n solutions which are independent. A matrix with these solutions as columns is a *fundamental matrix* of the system.

The matrix equation

$$X(k + 1) = A(k) X(k) \tag{7.11a}$$

is said to be the *associated matrix equation* of the system (7.9). Again, the following theorem will not be proved because its proof follows, with only some obvious changes, from the proof of Theorem 3.3.

THEOREM 7.3.   Suppose that the matrices $A(k)$, $k \in I$, are nonsingular. Then a solution X of (7.11a) is a fundamental matrix of (7.9) if and only if the determinant of this matrix is nonzero for some $k_0 \in I$. If this is the case, det $X(k)$ is nonzero for all $k \in I$.

Again we must pause here, and consider the implications of what we have learned. Suppose that one matrix, $A(\underline{k})$, $\underline{k} \in I$, is singular. Given $x(\underline{k} + 1)$, it is possible to determine uniquely all the subsequent states (that is, for $k = \underline{k} + 2, \underline{k} + 3,...$)

$$x(k) = A(k - 1) A(k - 2) \cdots A(\underline{k} + 1) x(\underline{k} + 1), \qquad k \geqslant \underline{k} + 1$$

However, it is not possible to determine uniquely the previous states (that is, for $k = \underline{k}, \underline{k} - 1,...$). Suppose that, given $x(\underline{k} + 1)$, we want to determine the state $x(\underline{k})$ which is transformed by $A(\underline{k})$ into the given vector

$$x(\underline{k} + 1) = A(\underline{k})\, x(\underline{k})$$

This expression can be considered as an equation in $x(\underline{k})$. The reader should remember from his studies concerning such systems of algebraic equations that this system has either no solutions at all, or an infinity of them (if $x(\underline{k} + 1)$ is orthogonal to all solutions of the homogeneous algebraic equation $A'(\underline{k})\, y = 0$). In either case, it is not possible to determine uniquely the predecessor of the state $x(\underline{k} + 1)$; this situation is in direct contrast with the properties of differential systems of the type studied in the previous chapters, which always exhibit uniqueness of trajectory in both the forward and backward directions of time; here we find that, even if there is uniqueness of trajectory in the forward direction, there is no uniqueness in the backward direction; some states have many predecessors.

EXAMPLE

Let

$$A(\underline{k}) = \begin{bmatrix} 1 & 2 \\ -1 & -2 \end{bmatrix}$$

Then the equation $x(\underline{k} + 1) = A(\underline{k})\, x(\underline{k})$ has solutions if and only if $x(\underline{k} + 1)$ is orthogonal to the solutions of

$$A'(\underline{k})\, y = \begin{bmatrix} 1 & -1 \\ 2 & -2 \end{bmatrix} \begin{bmatrix} y_1 \\ y_2 \end{bmatrix} = \begin{bmatrix} y_1 - y_2 \\ 2y_1 - 2y_2 \end{bmatrix} = 0$$

The solutions of these equations form a subspace characterized by the equality $y_1 = y_2$. Therefore, $x(\underline{k} + 1)$ should be in a subspace characterized by $x_1 = -x_2$; let then $x(\underline{k} + 1) = [a \ -a]'$. The equation $x(\underline{k} + 1) = A(\underline{k})\, x(\underline{k})$ becomes simply

$$x_1(k) + 2x_2(k) = a$$

this equality defines a subspace of solutions. Note that only some states (those for which $x_1 = -x_2$) have predecessors, and that each of those which do have predecessors have in effect a whole subspace of them.

Of course, we can define, even when all matrices $A(k)$, $k \in I$ are singular, a transition matrix for the system (7.9). If we put $X(k_0) = I_n$ in (7.11a), the solution for $k \geqslant k_0$, to be called $\phi(k, k_0)$, is

$$\phi(k, k_0) = A(k - 1) \cdots A(k_0), \qquad k > k_0, \quad k \in I \qquad (7.12)$$

while $\phi(k_0 , k_0) = I_n$. This solution has the property of generating the solution of (7.9),

$$x(k) = \phi(k, k_0) \, x(k_0) \tag{7.13}$$

$k \geqslant k_0$. This transition matrix shares the following properties with the transition matrix for differential systems

$$\phi(k_0 , k_0) = I_n \tag{7.14}$$

$$\phi(k_2 , k_1) \, \phi(k_1 , k_0) = \phi(k_2 , k_0) \tag{7.15}$$

where $k_0 \leqslant k_1 \leqslant k_2$; note that here we find a departure from the properties of the transition matrices of differential systems, which satisfy $\phi(t_2 , t_1) \, \phi(t_1 , t_0) = \phi(t_2 , t_0)$ *for all* t_0 , t_1 , t_2 .

An important consequence of this fact is that the transition matrix $\phi(k, k_0)$ is not invertible; we cannot, as in the differential case, make $k_0 = k_2 \neq k_1$ in (7.15) and prove the invertibility of the transition matrix; this is of course a consequence of the fact that it is not in general possible to invert the sense of the flow of time for systems such as (7.9).

If the matrices $A(k)$, $k \in I$, *are* invertible, though, then of course we find that

$$\phi(k, k_0) = A^{-1}(k) \cdots A^{-1}(k_0 - 1), \qquad k < k_0 , \quad k \in I \tag{7.16}$$

and then (7.15) is valid for all k_0 , k_1 , k_2 in I, and of course

$$\phi(k_1 , k_2) = \phi^{-1}(k_2 , k_1) \tag{7.17}$$

for all $k_1 , k_2 \in I$. If A is constant and invertible,

$$\phi(k - k_0) = A^{k-k_0} \tag{7.18}$$

and

$$\phi(k + p) = \phi(k) \, \phi(p), \qquad \phi(k) = \phi^{-1}(-k) \tag{7.19}$$

for all integers k, p.

In this case, one can obtain explicit expressions for the transition matrix by the use of the z transform, much as we used the Laplace transform previously. If $x(k + 1) = Ax(k)$, then, if $\hat{x}(z)$ is the z transform of $x(k)$,

$$\hat{x}(z) = (zI_n - A)^{-1} \, zx(0) \tag{7.20}$$

the z transform of the transition matrix is then $(zI_n - A)^{-1} \, z$.

EXAMPLE

$$A = \begin{bmatrix} 0 & 1 \\ -6 & -5 \end{bmatrix}$$

$$(zI_2 - A)^{-1}z = \begin{bmatrix} \dfrac{z(z+5)}{(z+2)(z+3)} & \dfrac{z}{(z+2)(z+3)} \\[2ex] \dfrac{6z}{(z+2)(z+3)} & \dfrac{z^2}{(z+2)(z+3)} \end{bmatrix}$$

so then

$$\phi(k) = \begin{bmatrix} 3(-2)^k - 2(-3^k) & (-2)^k - (-3)^k \\ -6[(-2)^k - (-3)^k] & -2(-2)^k + 3(-3)^k \end{bmatrix}$$

If the matrix A is diagonal with elements c_1, ..., c_m, then

$$A^k = \begin{bmatrix} c_1^{\ k} & & & & \\ & c_2^{\ k} & & & \\ & & \ddots & & \\ & & & \ddots & \\ & & & & c_m^{\ k} \end{bmatrix} \tag{7.21}$$

If A is in Jordan canonical form, as in (4.36) to (4.38), then

$$A^k = \begin{bmatrix} A_1^{\ k} & & & & \\ & A_2^{\ k} & & & \\ & & \ddots & & \\ & & & \ddots & \\ & & & & A_q^{\ k} \end{bmatrix} \tag{7.22}$$

$$A_i^{\ k} = \begin{bmatrix} A_{i1}^k & & & & \\ & A_{i2}^k & & & \\ & & \ddots & & \\ & & & \ddots & \\ & & & & A_{is(i)}^k \end{bmatrix} \tag{7.23}$$

where, remember,

$$A_{ij} = \begin{bmatrix} \lambda_i & & & & & \\ 1 & \lambda_i & & & & \\ & 1 & \ddots & & & \\ & & \ddots & \ddots & & \\ & & & & \lambda_i & \\ & & & & 1 & \lambda_i \end{bmatrix}_{n_{ij} \times n_{ij}}$$

so that

$$
A_{ij}^k =
\begin{bmatrix}
\lambda_i^k & 0 & \cdots & 0 \\
\binom{k}{1} \lambda_i^{k-1} & \lambda_i^k & \cdots & 0 \\
\binom{k}{2} \lambda_i^{k-2} & \binom{k}{1} \lambda_i^{k-1} & \cdots & \vdots \\
\vdots & \vdots & & 0 \\
& & & \lambda_i^k
\end{bmatrix}
\tag{7.24}
$$

where $\binom{k}{1}$, $\binom{k}{2}$, ..., are the coefficients usually defined in connection with the binomial expansion.

Of course, if A is not in Jordan canonical form, we put $Z = S^{-1}X$, where S is the matrix used in the transformation of A into Jordan canonical form, so that the matrix equation $X(k + 1) = AX(k)$, $X(k_0) = I_n$, becomes

$$
Z(k + 1) = S^{-1}ASZ(k), \qquad Z(k_0) = S^{-1}
\tag{7.25}
$$

where the matrix $S^{-1}AS$ is in Jordan canonical form, so that the transition matrix of the system (7.25) is easily computed. It follows that

$$
\phi(k - k_0) = S(S^{-1}AS)^{k-k_0} S^{-1}
\tag{7.26}
$$

EXAMPLE

If

$$
A = \begin{bmatrix} 0 & 1 \\ -6 & -5 \end{bmatrix}
$$

$$
S = \begin{bmatrix} 1 & 1 \\ -2 & -3 \end{bmatrix}, \qquad
S^{-1} = \begin{bmatrix} 3 & 1 \\ -2 & -1 \end{bmatrix}
$$

$$
S^{-1}AS = \begin{bmatrix} 2 & 0 \\ 0 & -3 \end{bmatrix}
$$

$$
\phi(k) = \begin{bmatrix} 1 & 1 \\ -2 & -3 \end{bmatrix}
\begin{bmatrix} (-2)^k & 0 \\ 0 & (-3)^k \end{bmatrix}
\begin{bmatrix} 3 & 1 \\ -2 & -1 \end{bmatrix}
$$

$$
= \begin{bmatrix} (-2)^k & (-3)^k \\ -2(-2)^k & -3(-3)^k \end{bmatrix}
\begin{bmatrix} 3 & 1 \\ -2 & -1 \end{bmatrix}
$$

$$
= \begin{bmatrix} 3(-2)^k - 2(-3)^k & (-2)^k - (-3)^k \\ -6(-2)^k + 6(-3)^k & -2(-2)^k + 3(-3)^k \end{bmatrix}
$$

Exercises

*1. It can be argued that our approach to the homogeneous system does not in fact parallel the treatment given to the differential homogeneous system, and that the following formalism is more appropriate. Instead of (7.9), write

$$\Delta x(k) = x(k + 1) - x(k) = \tilde{A}(k)\, x(k) \qquad (*)$$

where $\tilde{A} = A - I_n$. Here Δ replaces the differential operator, so that we may find now a rather better correspondence between this and the homogeneous differential system. Define and compute the transition matrix of the system (*). Study its properties.

*2. Consider the scalar equation

$$y(k + n) + \alpha_1(k)\, y(k + n - 1) + \cdots + \alpha_{n-1}(k)\, y(k + 1) + \alpha_n(k)\, y(k) = 0$$

and let $y_1, \ldots, y_n$ be linearly independent solutions of it. Show that the following determinant:

$$C(k) = \begin{vmatrix} y_1(k) & y_2(k) & \cdots & y_n(k) \\ y_1(k + 1) & y_2(k + 1) & \cdots & y_n(k + 1) \\ \vdots & \vdots & & \vdots \\ y_1(k + n - 1) & y_2(k + n - 1) & \cdots & y_n(k + n - 1) \end{vmatrix}$$

is nonsingular. This is the *Casorati* of the equation.

*3. Relate the result of the Exercise 2 above with Theorems 7.1, 7.2, and 7.3.

*4. Show that if the Casorati of n solutions $y_1, \ldots, y_n$ is zero for all k, then the solutions are linearly dependent.

*5. Consider the equation $x(\underline{k} + 1) = A(\underline{k})\, x(\underline{k})$. Prove the contention that this equation has a solution if and only if $x(\underline{k} + 1)$ is orthogonal to all solutions of the homogeneous algebraic equation $A'(\underline{k})\, y = 0$; and that, under these conditions, if $A(\underline{k})$ is singular, then it has an infinity of solutions; characterize these solutions. (Hint: The equation has a solution if and only if $x(\underline{k} + 1)$ is in the range of $A(\underline{k})$; show that the null space of $A'(\underline{k})$ equals the orthogonal complement of the range of $A(k)$). Remember that $A(\underline{k})$ is a real matrix.

7.3. The Nonhomogeneous System

We treat now the whole of the system (7.8), and of course we start with the first of these equations

$$x(k + 1) = A(k)\, x(k) + B(k)\, u(k) \qquad (7.27)$$

The input u is a sequence of numbers defined for $k \in I$. If $x(k_0)$ is specified, our aim is to obtain an explicit expression for the solution of (7.27) (it so happens that the solution which will be obtained below is unique), for all $k \geqslant k_0$, $k \in I$, given the values $u(k)$ for $k \geqslant k_0$. Note that, again, we have a preferred direction of the "time."

Let $X(k)$ be a fundamental matrix of the homogeneous system associated with (7.27). Let us look for a solution of (7.27) of the form

$$x(k) = X(k)\, p(k) \tag{7.28}$$

This will be a solution of this equation if and only if p satisfies Equation (7.29) below.

$$x(k + 1) = X(k + 1)\, p(k + 1) = A(k)\, X(k)\, p(k) + B(k)\, u(k)$$

Since $X(k + 1) = A(k)\, X(k)$,

$$X(k + 1)\{p(k + 1) - p(k)\} = B(k)\, u(k)$$

Suppose now that $X(k + 1)$ is invertible. Then

$$p(k + 1) = p(k) + X^{-1}(k + 1)\, B(k)\, u(k) \tag{7.29}$$

The solution of this simple equation can be readily seen to be, if $x(k_0)$ is specified,

$$p(k) = X^{-1}(k_0)\, x(k_0) + \sum_{l=k_0}^{k-1} X^{-1}(l + 1)\, B(l)\, u(l) \tag{7.30}$$

which implies, at last, that

$$x(k) = X(k)\, X^{-1}(k_0)\, x(k_0) + \sum_{l=k}^{k-1} X(k)\, X^{-1}(l + 1)\, B(l)\, u(l)$$

$$= \phi(k, k_0)\, x(k_0) + \sum_{l=k_0}^{k-1} \phi(k, l + 1)\, B(l)\, u(l) \tag{7.31}$$

since $X(k) = \phi(k, k_0)\, X(k_0)$. Note that (7.31) has been derived for $k > k_0$ only. This is the counterpart of Equation (3.59) derived for linear differential systems. We will designate the solution (7.31) by $x(k; k_0, x^0; u)$. The output $y(k; k_0, x^0; u)$ is easily obtained from (7.31) and the second of Equations (7.8).

Note that in the proof of Equation (7.31) we have assumed that the matrices $X(k)$, $k \in I$, are invertible; that is, that the matrices $A(k)$, $k \in I$, are invertible. As a matter of fact, the explicit expression (7.31) satisfies (7.27) even if this condition is not satisfied, as can easily be ascertained by substitution.

If A and B are constant, then

$$x(k; k_0, x^0; u) = \phi(k - k_0)\, x^0 + \sum_{l=k_0}^{k-1} \phi(k - l - 1)\, Bu(l) \tag{7.32}$$

The uniqueness of the solution (7.31) [and then of (7.32)] follows from the fact that the solution of $x(k + 1) = A(k) x(k)$ for $x(k_0) = 0$ is identically equal to zero.

The method of the z transform can be used also for obtaining explicit solutions such as (7.32). If $\hat{x}$, $\hat{u}$ are the z transforms of x, u then, if $k_0 = 0$,

$$\hat{x}(z) = (zI_n - A)^{-1} zx^0 + (zI_n - A)^{-1} B\hat{u}(z) \tag{7.33}$$

EXAMPLE

Consider the equation

$$y(k + 2) + 5y(k + 1) + 6y(k) = u(k)$$

Then

$$x_1(k + 1) = x_2(k)$$

$$x_2(k + 1) = -6x_1(k) - 5x_2(k) + u(k)$$

$$y(k) = x_1(k)$$

that is

$$A = \begin{bmatrix} 0 & 1 \\ -6 & -5 \end{bmatrix}, \qquad B = \begin{bmatrix} 0 \\ 1 \end{bmatrix}, \qquad C = \begin{bmatrix} 1 & 0 \end{bmatrix}$$

As we have already computed the transition matrix of this system, we will simply show how to use the second expression in the right-hand side of (7.33) to compute the corresponding part of the output. Suppose u is a scalar, $u(k) = 1$, $k \geqslant 0$. Then, if $x(0) = 0$,

$$\hat{y}(z) = C\hat{x}(z) = C(zI_n - A)^{-1}B \, \frac{z}{z - 1}$$

$$= [1 \ 0] \begin{bmatrix} \dfrac{z(z + 5)}{(z + 2)(z + 3)(z - 1)} & \dfrac{z}{(z + 2)(z + 3)(z - 1)} \\[3mm] \dfrac{6z}{(z + 2)(z + 3)(z - 1)} & \dfrac{z^2}{(z + 2)(z + 3)(z - 1)} \end{bmatrix} \begin{bmatrix} 0 \\ 1 \end{bmatrix}$$

$$= \frac{z}{(z + 2)(z + 3)(z - 1)}$$

which corresponds to

$$y(k) = \tfrac{1}{12} - \tfrac{1}{3}(-2)^k + \tfrac{1}{4}(-3)^k, \qquad k \geqslant 0$$

In the next section we summarize the aspects of the solutions just developed which have relevance to our goal of eventually obtaining an abstract class of systems, the dynamical systems, from the properties of single differential and difference systems which we have been studying in these first chapters.

Exercises

1. Let a system be described by

$$x(k + 1) = Ax(k) + bu(k), \qquad x(0) = a$$
$$y(k) = Cx(k)$$

(1)

Suppose u satisfies the difference equation

$$u(k + 1) = Fu(k), \qquad u(0) = b$$

(2)

   a. Construct a homogeneous system equivalent to (1) and (2), in the sense that the solution of this homogeneous system can be used to obtain the solution of (1) and the solution of (2).
   b. Obtain the solution of the homogeneous system and then the solution of (1) and the solution of (2).

2. Consider a system as above, with

$$A = \begin{bmatrix} 1 & -1 \\ 0 & 1 \end{bmatrix}, \qquad B = \begin{bmatrix} 1 \\ 1 \end{bmatrix}, \qquad C = [1\ 1]$$

(1)

$$x(0) = [1\ \ 1]'$$

Suppose u, a scalar input, satisfies

$$u(k + 2) - u(k + 1) - u(k) = 0, \qquad u(0) = 1, \quad u(1) = -1$$

(2)

Find a homogeneous system equivalent to (1) and (2) in the same sense as in Exercise 1. Obtain the solution of this system, and then of (1) and of (2).

*3. Define an adjoint system for (7.9), and use this system for the derivation of (7.31).

7.4. A Summary

We paraphrase Section 3.7. The following elements are of importance in the construction of the solution of the system (7.8):

i. The *state space* R^n, where the solutions of the first of Equations (7.8) take values.

ii. A subset I of the space of all integers.

iii. A space U of functions with domain consisting of sets of integers in I of the form $\{p, p + 1, ..., p + q\}$, and range in R^r. These are r-vector-valued control functions with finite entries.

iv. A space Y of outputs of the system.

v. A *transition function* $x(\cdot, \cdot, \cdot, \cdot)$ with domain the product space $I \times I \times R^n \times U$ and with range in R^n, which defines the state $x(k; k_0, x^0; u)$ for any initial $k_0 \in I$, any initial state $x^0 \in R^n$ and any input defined on a set $\{k_0, ..., k - 1\}$, $k > k_0$. This function is defined by

$$x(k; k_0, x^0; u) = \phi(k, k_0) x(k_0) + \sum_{l=k_0}^{k-1} \phi(k, l+1) B(l) u(l) \qquad (7.34)$$

vi. An *output function* $y(\cdot, \cdot, \cdot)$ with domain the product space $I \times R^n \times U$, which assigns to each $k \in I$, each $x \in R^n$ and each $u \in U$ an output

$$y(k, x, u) = C(k)x + D(k) u(k)$$

The transition function $x(\cdot, \cdot, \cdot, \cdot)$ has the following properties:

i. $x(k_0 ; k_0, x^0; u) = x^0$ for all $u \in U$, $k_0 \in I$, $x^0 \in R^n$.

ii. It satisfies Equation (3.62), with some obvious modifications Note that the requirement $k_2 \geqslant k_1 \geqslant k_0$ is crucial here because it is not possible in this case to go back in time unless the matrices $A(k)$ happen to be nonsingular.

iii. We will refer to the continuity of functions such as this in a general way in Chapter 9.

Other remarks which parallel the more extensive treatment of Section 3.7 are:

i. The transition function is linear with respect to x^0 when u is zero and linear with respect to u when x^0 is zero, as in (3.63).

ii. We can, again, separate the output into a *zero-input* response and a *zero-state* response.

iii. If the matrices A, B, C, and D are constant, then a shift in the time origin changes neither the values of the state nor of the output, provided that the input is shifted accordingly. The system is said to be *time invariant* or *constant*.

7.5. Controllability, Observability

We shall study in this section only the controllability and observability of constant difference systems whose matrix A is nonsingular. We start with controllability. Let $x(k_0) = 0$. Then, from (7.32) it follows quite simply that

$$x(k_0 + p) = \sum_{l=1}^{p} A^{p-l} Bu(k_0 + l - 1)$$

$$= Bu(k_0 + p - 1) + ABu(k_0 + p - 2) + \cdots$$

$$+ A^{p-2} Bu(k_0 + 1) + A^{p-1} Bu(k_0) \qquad (7.35)$$

Let us consider the set of all vectors $x(k_0 + p)$ when the input u takes all possible values at each k, $k = k_0$, $k_0 + 1$, ..., $k_0 + p - 1$. Clearly this set is a subspace of R^n, and is spanned by the columns of the matrices B, AB, ..., $A^{p-1}B$.

Suppose now that the following matrix

$$Q = [B \ AB \ \cdots \ A^{n-1}B] \tag{7.36}$$

has rank n. Then the set of all vectors in R^n which can be reached (in n steps) equals the whole of R^n, since in this case there would be n independent columns in Q. On the other hand, assume that every state in R^n can be reached after a finite number of steps, but that Q has rank less than n. This implies that every matrix $[B \ AB \ \cdots \ A^{N-1}B]$, for any N, has rank less than n, so that the set of all vectors which can be reached from the origin has dimension less than n, which contradicts our assumption that this set is identical with R^n.

It seems natural, then, to define a system $x(k + 1) = Ax(k) + Bu(k)$ as *controllable* if every state can be reached from the origin in a finite number of steps. We have then proved:

THEOREM 7.4. The system described by $x(k + 1) = Ax(k) + Bu$, with A and B constant matrices, is controllable if and only if the matrix Q has rank n.

We say that the system described by $x(k + 1) = Ax(k)$, $y(k) = Cy(k)$, with A and C constant matrices, is *observable* if the initial conditions $x(k_0)$ can be determined from a finite sequence of measurements, $y(k_0)$, ..., $y(k_0 + p)$. In a similar way as in Chapter 5, we can show by the use of the adjoint system, that the following theorem holds.

THEOREM 7.5. The system described by $x(k + 1) = Ax(k)$, $y(k) = Cx(k)$, A and C constant, is observable if and only if the matrix P has rank n, where

$$P = [C' \ A'C' \ \cdots \ A'^{(n-1)} C'] \tag{7.37}$$

Note that the conditions for controllability and observability take the same form here as in the case of differential systems. However, observe that the matrix A is, in this instance, nonsingular.

At last, we show the applicability of Theorems 5.18 and 5.28, which are concerned with the controllability and observability of differential systems whose matrix A is in Jordan canonical form, to the present situation. We assume, then, that the matrix A of a difference system is in Jordan canonical form; we use the same partition of the matrices as in Theorems 5.18 and 5.28, and prove

THEOREM 7.6 [*1*].   Consider

$$x(k + 1) = Ax(k) + Bu(k)$$
$$y(k) = Cx(k)$$
(7.38)

where A is in Jordan canonical form and B and C are partitioned as in Theorems 5.18 and 5.28. Then:

i. The system (7.38) is controllable if and only if for each $i = 1, ..., q$, the set of $s(i)$ r-dimensional row vectors

$$b_{fi1}, b_{fi2}, ..., b_{fis(i)}$$
(7.39)

is a linearly independent set.

ii. The system (7.38) is observable if and only if for each $i = 1, ..., q$ the set of $s(i)$ m-dimensional column vectors

$$c_{li1}, c_{li2}, ..., c_{lis(i)}$$

is a linearly independent set.

Proof.   We will only prove the first part of this theorem; the second can be proved in a similar way. The core of the proof will consist in showing that the matrix Q in (7.36) has rank n if and only if the contention in the part i of the theorem is true. It should be clear, then, that this proof could also be used to prove Theorem 5.18 (since we will not use the condition that A is nonsingular), while the proof given for Theorem 5.18, which is based upon the exponential nature of the transition matrix of a constant differential system, cannot be used to study the controllability of a constant difference system.

It will be convenient to refer to Q in (7.36) as $Q(A, B, n)$. We first prove necessity, that is, that the rank of $Q(A, B, n)$ is n only if i holds. Assume that $\operatorname{rank} Q = n$, but that i does not hold, say for $i = k$. Observe first that

$$\operatorname{rank} Q(A, B, n) = \operatorname{rank} Q(A - \lambda I, B, n)$$
(7.40)

for any λ because $Q(A - \lambda I, B, n)$ can be obtained from $Q(A, B, n)$ by means of a finite sequence of elementary column operations (see Appendix).

If A and B are partitioned as in Theorem 5.18, then,

$$Q(A - \lambda_k I, B, n) = \begin{bmatrix} Q(A_{11} - \lambda_k I, B_{11}, n) \\ \vdots \\ Q(A_{qs(q)} - \lambda_k I, B_{qs(q)}, n) \end{bmatrix}$$
(7.41)

where we have used the same symbol I for identity matrices of various dimensions.

The submatrix $Q(A_{kj} - \lambda_k I, B_{kj}, n)$ is

$$Q(A_{kj} - \lambda_k I, b_{kj}, n) = \begin{bmatrix} 0 & \cdots & & 0 \\ b_{fkj} & 0 & & 0 \\ b_{2kj} & b_{fkj} & & 0 \\ \vdots & & & \vdots \end{bmatrix} \qquad (7.42)$$

note that the second row of (7.42) is $[b_{fkj} 0 \cdots 0]$, so that, since the vectors

$$b_{fk1}, b_{fk2}, ..., b_{fks(k)}$$

are, by assumption, linearly dependent, then by (7.41) there are rows of $Q(A - \lambda_k I, B, n)$ which are linearly dependent, which implies that the rank of $Q(A, B, n)$ is not n, which contradicts our assumption that this rank is actually equal to n. The necessity of i follows.

We present only a sketch of the proof of sufficiency, the reader is referred to [1] for the details. The first step consists in showing that the rank of $Q(A_i, B_i, n)$ is n_i ; the matrices A_i, B_i and the integer n_i are defined in (5.24). This of course implies that all the rows in $Q(A_i, B_i, n)$ are linearly independent. The sufficiency of i follows from the proof that the rows of $Q(A_i, B_i, n)$ and $Q(A_j, B_j, n)$ are linearly independent for $i \neq j$.

We end this section, and this chapter, by noting that we shall not treat the synthesis problem for difference systems. We shall return to the study of these systems, however, in the next chapter, when we treat the stability of linear systems.

Exercise

1. Suppose a discrete system is obtained from a continuous one by means of a sample-and-hold operation as in Figure 7.2. Can you relate the controllability properties of the discrete system with the corresponding characteristics of the continuous system from which it is derived?

SUPPLEMENTARY NOTES AND REFERENCES

Two good monographs on subjects related to the material presented in this chapter are

K. S. Miller, "An Introduction to the Calculus of Finite Difference and Difference Equations." Dover, New York, 1966.

K. S. Miller, "Linear Difference Equations." Benjamin, New York, 1968.

The following text presents well several aspects of linear difference systems:

H. Freeman, "Discrete-Times Systems." Wiley, New York, 1965.

REFERENCE

1. C. Chen and C. A. Desoer, A proof of controllability of Jordan form state equations. *IEEE Trans. Automatic Control* **AC-13** (1968), 195–196.

8

Stability

8.1. Introduction

In this chapter we end our study of finite-dimensional systems by considering again the homogeneous differential system of Chapter 3 $\dot{x}(t) = A(t)\, x(t)$ as well as, later on, the homogeneous difference system of Chapter 7, $x(k + 1) = A(k)\, x(k)$. Our interest is now centered on the behavior of the solutions of $\dot{x}(t) = A(t)\, x(t)$ as the time gets large (and on the behavior of the solutions of $x(k + 1) = A(k)\, x(k)$ as k gets large). Of course, we must assume that the interval of definition of the matrix A is a set of the type $[t_0 , \infty)$, in the case of the differential system, or a set $[k_0 , \infty)$, in the case of the difference system.

We concentrate hereafter on the homogeneous differential system, and present first an example of systems which exhibit several interesting types of behavior as the time grows.

EXAMPLE

a. Consider first the homogeneous system described by

$$A = \begin{bmatrix} -1 & 0 \\ 0 & -2 \end{bmatrix}$$

Here $x_1(t) = x_1{}^0 e^{-(t-t_0)}$, $x_2(t) = x_2{}^0 e^{-2(t-t_0)}$. As t gets large, $t - t_0$ gets large too and $x_1(t)$ and $x_2(t)$ tend to the origin of the state space, regardless of the actual value of the initial condition vector. We say that the origin is a *point of equilibrium* of this system, and that this point of equilibrium is *asymptotically stable*, to indicate that every trajectory

239

of the system has points in every neighborhood of the origin, no matter how small the radius of the neighborhood.

b.   Consider now

$$A = \begin{bmatrix} 0 & 1 \\ -1 & 0 \end{bmatrix}$$

here

$$x_1(t) = x_1{}^0 \cos(t - t_0) + x_2{}^0 \sin(t - t_0)$$
$$x_2(t) = -x_1{}^0 \sin(t - t_0) + x_2{}^0 \cos(t - t_0)$$

The origin is not an asymptotically stable point of equilibrium for this system since the trajectories for which $x^0 \neq 0$ do not get arbitrarily near the origin as the time increases. Note that the origin is still a point of equilibrium, in the sense that, if $x^0 = 0$, then the whole trajectory is zero; as a matter of fact, this property was proved for a general homogeneous system in Chapter 3.

The origin is said to be a *stable* point of equilibrium of this system. Consider the norm of $x(t)$,

$$\| x(t) \| = (x_1{}^2(t) + x_2{}^2(t))^{1/2} = \| x^0 \|$$

it happens to be equal to the initial norm. It follows that, given any $\epsilon > 0$, a $\delta(\epsilon) > 0$ [in this case defined by $\delta(\epsilon) = \epsilon$] can be found such that if $\| x^0 \| < \delta(\epsilon)$, then $\| x(t) \| < \epsilon$, for $t \geqslant t_0$. If the motion starts sufficiently near the origin, it never gets very far from it.

c.   At last, consider

$$A = \begin{bmatrix} 1 & 0 \\ 0 & 2 \end{bmatrix}$$

Here $x_1(t) = x_1{}^0 e^{t-t_0}$, $x_2(t) = x_2{}^0 e^{2(t-t_0)}$. The origin is said to be in this case an *unstable* point of equilibrium; as $(t - t_0)$ increases, the norm of the trajectory also increases, and, given a number $N > 0$, no matter how large, a value of t can be found such that $\| x(t) \| > N$.

Note that in these examples, we have considered $t \geqslant t_0$, giving the time a preferred sense of direction; this is in contrast with the approach taken in Chapter 3. As a matter of fact, during this whole chapter we will give the time this same, preferred sense of direction.

In this chapter, we will consider the general homogeneous system

$$\dot{x}(t) = A(t)\, x(t), \qquad t \in [t_0, \infty) \tag{8.1}$$

where the state vector $x \in R^n$; a more general assumption (that $x \in C^n$) could be made, and the resulting theory would be essentially identical. The entries of the matrix function A will be considered to be continuous, *bounded*, real-valued functions defined on $[t_0, \infty)$.

We say that the origin is a point of *equilibrium* (or *singular* point) of the system (8.1); if $x(t_1) = 0$, $t_1 \in [t_0, \infty)$, then $x(t) = 0$, $t \in [t_1, \infty)$. Clearly, if the system somehow reaches the origin, it stays there. The system (8.1) may have other points of equilibrium; see Exercises 1 and 2 at the end of this section. Most of this chapter is devoted to the study of the stability properties of the origin. We define three types of behavior of systems such as (8.1). It is convenient to define first the type of stability illustrated by part b of the example above.

Definition.[†] The origin is said to be a *stable* point of equilibrium of (8.1) if, given t_0, a number $\delta(\epsilon, t_0)$ can be found such that if $\| x(t_0)\| < \delta(\epsilon, t_0)$, then $\| x(t)\| < \epsilon$ for $t \geqslant t_0$.

If the number $\delta(\epsilon, t_0)$ does not depend on t_0, the origin is said to be *uniformly* stable.

Definition. The origin is said to be an *asymptotically stable* point of equilibrium of (8.1) if it is stable and if every trajectory of this system, regardless of the initial condition, satisfies

$$\lim_{t \to \infty} \| x(t) \| = 0 \tag{8.2}$$

that is, given $x(t_0) = x^0$, and $\epsilon > 0$, a number $N(x^0, t_0, \epsilon)$ can be found such that $\| x(t)\| < \epsilon$ for $t > N(x^0, t_0, \epsilon)$. Note that we have to assume explicitly that the system is stable; in other words (8.2) in general does not imply stability in the sense of the definition above.

If the number $N(x^0, t_0, \epsilon)$ does not depend on t_0, the origin is said to be *uniformly* asymptotically stable.

Definition. The origin is said to be an *unstable* point of equilibrium of (8.1) if every trajectory with nonzero initial condition has the following property: Given $N > 0$, no matter how large, a time $\tilde{t}(N, x^0, t_0)$ can be found such that $x(\tilde{t}) > N$.

In the engineering literature it is said sometimes that the *system* (8.1) is stable, asymptotically stable, or unstable whenever the origin has the corresponding property.

The stability of constant homogeneous systems is the subject of the next section, where we relate the stability of the origin to properties of the matrix A.

[†] In this chapter we use in general an arbitrary norm for vectors in R^n, and, when necessary, the matrix norm induced by the vector norm. In Theorems 8.2 and 8.6, and in Section 8.4, we use, however, the norm (2.32) generated by the inner product operation in R^n. Note that the stability properties of the origin do not depend on the particular norm used in the corresponding definitions.

Exercises

*1.   Let A be a constant matrix. Show that the homogeneous system $\dot{x} = Ax$ has either one or an infinite number of points of equilibrium.

2.   Give an example in which there are an infinite number of points of equilibrium. In which sense can you say that no system such as $\dot{x} = Ax$ can have more than one *isolated* point of equilibrium?

*3.   Show by means of a counterexample that asymptotic stability does not imply stability.

*4.   Show that if there is stability for t_0 there is stability for any other initial time t_1.

5.   Give two examples of homogeneous systems for which the origin is asymptotically stable but not uniformly asymptotically stable.

*6.   Show that, for constant systems, stability (or asymptotic stability) of the origin implies uniform stability (or uniform asymptotic stability).

*7.   Show that if the origin is an asymptotically stable point of equilibrium in the sense of the definition above, then the origin is the only point of equilibrium of the system.

8.2. The Stability of Constant Systems

Our first result relates asymptotic stability to the eigenvalues of the matrix A.

THEOREM 8.1.   The origin is an asymptotically stable point of equilibrium for the system

$$\dot{x} = Ax \tag{8.3}$$

with A a constant $n \times n$ matrix if and only if the eigenvalues of A have negative real parts.

Proof.   This theorem follows by writing

$$e^{tA} = Se^{t\tilde{A}}S^{-1}$$

where $\tilde{A}$ is the Jordan canonical form of A. If all the eigenvalues of A have negative real parts, every component of $e^{(t-t_0)A} x(t_0)$ for an arbitrary $x(t_0)$ will tend to zero as t tends to infinity. On the other hand, suppose that all components of this vector do tend to zero as t tends to infinity, but that an eigenvalue λ_l has a nonnegative real part. Then there will be components of e^{tA} which do not tend to zero as t tends to infinity, and then components of $e^{(t-t_0)A} x(t_0)$ for some $x(t_0)$ which do not converge to zero as t tends to infinity. The theorem follows.

Similar theorems can easily be proved regarding conditions on the eigenvalues of A which imply (and are implied by) stability or instability of the origin; we present those theorems as exercises for the reader. We shall now leave these matters, and approach the study of the asymptotic stability of system (8.3) from a rather different side, related to the *Lyapunov functions*. We remark here that this method, the *Lyapunov second method*, is an extremely powerful tool, especially suited for the analysis of the stability of nonlinear systems. Our needs in this book are quite modest, and thus we shall not make use of the whole full-dress apparatus of the method, so that our treatment will be confined to the few special cases necessary for the study of the stability of linear systems. Incidentally, we will encounter a type of matrix equation of interest not only in stability studies but also in many other fields.

Consider the system (8.3), and the quadratic form (x, Fx), where F is a symmetric, constant, positive definite matrix. Then, if x is a solution of (8.3) defined for all t in R,

$$\frac{d}{dt}(x(t), Fx(t)) = (\dot{x}(t), Fx(t)) + (x(t), F\dot{x}(t))$$

$$= (Ax(t), Fx(t)) + (x(t), FAx(t))$$

$$= (x(t), (A'F + FA)x(t)) \qquad (8.4)$$

Let us assume now that the following equation

$$A'F + FA = -G \qquad (8.5)$$

where G is any positive definite, $n \times n$, constant matrix (note that $G = G'$) has a positive definite solution F. Note as well that this is purely an assumption, and that surely some $n \times n$ constant matrices A are such that the corresponding equations (8.5) do not have positive definite solutions; for instance, if $A = I_n$, (8.5) becomes $2F = -G$, which of course does not have a positive definite solution F (since G is positive definite itself). If A is such that (8.5) does have a positive definite solution, (8.4) becomes

$$\frac{d}{dt}(x(t), Fx(t)) = -(x(t), Gx(t)) \qquad (8.6)$$

for any solution x of the system (8.3). It so happens that, under these conditions [that is, if A is such that (8.5) has a positive definite solution F] the origin is an asymptotically stable point of equilibrium. This is shown by considering that:

i. The origin is stable. The quadratic form (x, Fx), which is positive if x is nonzero, can be shown to satisfy a stronger relationship: There is a positive number μ such that

$$(x, Fx) \geqslant \mu \| x \|^2 \qquad (8.7)$$

see Exercise 2 at the end of Section 5.3. Let ϵ be given, and choose a vector x^0 such that $\| x^0 \| < \epsilon$ and $(x^0, Fx^0) < \mu\epsilon^2$; such a vector certainly exists, since (x, Fx) is zero at the origin and is a continuous function of x, so that inside the set $\{x : \| x \| < \epsilon\}$ there are bound to be many points for which (x, Fx) is less than $\mu\epsilon^2$. Consider now the trajectory with values $x(t; t_0, x^0)$. Since G is positive definite, by (8.6),

$$\frac{d}{dt}(x(t; t_0, x^0), Fx(t; t_0, x^0)) < 0 \tag{8.8}$$

so that $(x(t_1; t_0, x^0), Fx(t_1; t_0, x^0)) < (x^0, Fx^0) < \mu\epsilon^2$, for any $t_1 > t_0$. This implies that $\| x(t_1; x^0, t_0) \| < \epsilon$ for any $t_1 > t_0$, since if there were a value $\hat{t} > t_0$ such that $\| x(\hat{t}; x^0, t_0) \| = \epsilon$ then, by (8.7),

$$(x(\hat{t}, t_0, x^0), Fx(\hat{t}; t_0, x^0)) \geqslant \mu\epsilon^2 \tag{8.9}$$

which contradicts our previous finding that the quadratic form in (8.9) is actually less than $\mu\epsilon^2$. The origin is stable; in this case, the number $\delta(\epsilon, t_0)$ defined in Section 8.1 equals simply ϵ.

ii. Every trajectory tends to the origin as the time tends to infinity. If x is a trajectory, $(x(t; t_0, x^0), Fx(t; t_0, x^0))$ decreases monotonically as the time increases, since its derivative, by (8.7), is always negative. Since $(x(t; t_0, x^0), Fx(t; t_0, x^0))$ is always nonnegative, its limit exists and is nonnegative,

$$\lim_{t \to \infty} (x(t; t_0, x^0), Fx(t; t_0, x^0)) \geqslant 0 \tag{8.10}$$

We show that this limit is actually equal to zero. Assume otherwise, that is, that it equals a number $\alpha > 0$. Since G is positive definite, there is a positive constant ν such that $(x, Gx) \geqslant \nu \| x \|^2$. On the other hand, by using the Cauchy–Schwarz inequality, we prove that

$$(x, Fx) \leqslant \| F \| \cdot \| x \|^2 = k \| x \|^2 \tag{8.11}$$

where we put $k = \| F \|$. If the limit in (8.10) equals $\alpha > 0$, then $\| x(t; t_0, x^0) \| \geqslant \beta = (\alpha/k)^{1/2}$, since $(x(t; t_0, x^0), Fx(t; t_0, x^0))$ decreases monotonically. Then, by (8.6),

$$\frac{d}{dt}(x(t; t_0, x^0), Fx(t; t_0, x^0)) \leqslant \nu\beta^2 \tag{8.12}$$

from which we obtain

$$(x(t; t_0, x^0), Fx(t; t_0, x^0)) \leqslant (x^0, Fx^0) - (t - t_0)\nu\beta^2, \qquad t > t_0 \tag{8.13}$$

which implies, since v is positive, that for sufficiently large t the quadratic form $(x(t; t_0, x^0), Fx(t; t_0, x^0))$ is actually negative, which is a contradiction since this function is always positive. Therefore the limit in (8.10) is zero. We have proved:

THEOREM 8.2. The origin is an asymptotically stable point of equilibrium for the system (8.3) if the equation (8.5) has a positive definite solution F.

The quadratic form (x, Fx) is known as a *Lyapunov function* for the system (8.5).

The rest of this section will be spent studying properties and methods of solution of the matrix equation (8.5). Among other results, we will strenghten Theorem 8.2, and show that if the system (8.3) does have the origin as an asymptotically stable point of equilibrium, then Equation (8.5) does have a unique positive definite solution.

We establish first conditions as to whether Equation (8.5) does have a unique solution. Let u, v be eigenvectors of A' associated with the eigenvalues λ and μ respectively. Then,

$$A'uv' + uv'A = (\lambda + \mu) uv' \qquad (8.14)$$

since $A'u = \lambda u$, $(v'A)' = A'v = \lambda v$, so that $v'A = \lambda v'$, so that $\lambda + \mu$ is an eigenvalue of the operator defined by the left-hand side of (8.5). Since this equation has a unique solution if and only if this operator is nonsingular, that is, if none of its eigenvalues is zero, it follows that (8.5) has a unique solution if and only if the sums of the type $\lambda + \mu$ are not zero. Since the eigenvalues of A and those of A' are identical, we have, at last

THEOREM 8.3. Equation (8.5) has a unique solution F if and only if none of the eigenvalues $\lambda_1, ..., \lambda_n$, and none of the sums $\lambda_i + \lambda_j$, $i \neq j$, $i, j = 1, ..., n$, are zero.

See also [1] for a different proof of this theorem. Note that, if all eigenvalues of A have negative real parts, that is, if the origin is asymptotically stable, then the conditions of Theorem 8.3 are satisfied. We show now:

THEOREM 8.4. Let A have eigenvalues with negative real parts. Then the (unique) solution of Equation (8.5) is positive definite.

Proof. We shall use for this proof an explicit expression for the solution of (8.5). Consider the equation

$$\dot{Z} = A'Z + ZA, \qquad Z(0) = -G \qquad (8.15)$$

The solution to this equation is simply

$$Z(t) = -e^{A't}Ge^{At} \qquad (8.16)$$

Since the eigenvalues of A have negative real parts, $Z(t)$ tends to zero as t tends to infinity. If (8.15) is integrated between zero and infinity,

$$\int_0^\infty \dot{Z}(t)\, dt = -Z(0) = G = A'\left[\int_0^\infty Z(t)\, dt\right] + \left[\int_0^\infty Z(t)\, dt\right] A \quad (8.17)$$

so that the solution of (8.5) is actually

$$F = -\int_0^\infty Z(t)\, dt = \int_0^\infty e^{A't}Ge^{At}\, dt \qquad (8.18)$$

We show that F is positive definite by computing (x, Fx)

$$(x, Fx) = \int_0^\infty (x, e^{A't}Ge^{At}x)\, dt = \int_0^\infty (e^{At}x, Ge^{At}x)\, dt$$

this is positive whenever x is not zero because G is positive definite, so that we conclude that F is positive definite. The theorem follows.

This theorem represents the previously announced extension of Theorem 8.2: If the origin is asymptotically stable, then A has only eigenvalues with negative real parts, and then Equation (8.5) has a unique positive definite solution.

We have acquired a rather complete picture of the properties of the matrix Equation (8.5). We shall study now some of its computational aspects. The general approach to the determination of the asymptotic stability of the system (8.3) consists simply in computing F from (8.5), and testing whether this matrix is or is not positive definite; we show numerical methods for doing this in the Appendix. In the rest of this section we show a method to compute the matrix F from Equation (8.5). We assume that the conditions of Theorem 8.3 are satisfied; we follow [2].

It should be recognized that since F is a symmetric matrix, Equation (8.5) represents $\frac{1}{2}n(n+1)$ linear equations for the $\frac{1}{2}n(n+1)$ different elements of F. This system of linear equations can of course be solved numerically; however, it is possible to transform this system into another, Equation (8.21) below, which consists only of $\frac{1}{2}n(n-1)$ equations and unknowns, so that the number of these has been reduced by n. Indeed, write (8.5) as

$$(FA)' + \tfrac{1}{2}G = -FA - \tfrac{1}{2}G \qquad (8.19)$$

and put $FA + \tfrac{1}{2}G = S$; (8.19) shows that S is a skew-symmetric matrix. Of course, since A is invertible

$$F = (S - \tfrac{1}{2}G)\, A^{-1} \qquad (8.20)$$

and the condition that F be symmetric gives us our new equation (in S) to replace (8.5):

$$A'S + SA = \tfrac{1}{2}(A'G - GA) \tag{8.21}$$

Since S and both sides of (8.21) are skew-symmetric, this equation consists of only $\tfrac{1}{2}n(n - 1)$ equations and unknowns.

A further simplification is possible if it is not necessary to determine F itself, but only whether this matrix is or is not positive definite; in this case it is not necessary to compute F from (8.20), avoiding the need for inverting the matrix A. This observation is based on the fact that F is positive definite if and only if $A'FA$ is positive definite, provided A is nonsingular; since

$$A'SA = A'(S - \tfrac{1}{2}G) \tag{8.22}$$

it is necessary to test only this matrix for positive definiteness once S has been computed.

It is possible to connect this method with a rather different approach due to Smith [3]. Suppose that the characteristic polynomial of A is known; we write it as

$$\det(\lambda I_n - A) = \lambda^n + k_1\lambda^{n-1} + \cdots + k_n \tag{8.22}$$

and define an $n \times n$ matrix Q by

$$Q = \begin{bmatrix} 0 & 1 & 0 & & 0 \\ \cdot & 0 & 1 & & \cdot \\ \cdot & \cdot & \cdot & & \cdot \\ 0 & 0 & 0 & & 1 \\ -k_n & -k_{n-1} & -k_{n-2} & \cdots & -k_1 \end{bmatrix} \tag{8.23}$$

If $U = (u_{ij})$ is the solution of

$$Q'U + UQ = -C \tag{8.24}$$

where we assume that the eigenvalues of Q (which are the eigenvalues of A) are such that (8.24) has a unique solution, and where

$$C = \begin{bmatrix} 1 & & & \\ & 0 & & \\ & & \cdot & \\ & & & \cdot \\ & & & & 0 \end{bmatrix} \tag{8.25}$$

then the matrix F is [3]

$$F = \sum_i \sum_j u_{ij}(A')^{i-1}\, GA^{j-1} \tag{8.26}$$

we have transformed the problem of solving (8.5) into the (much simpler) problem of solving (8.24); we use the same type of transformation to simplify (8.24) as in Equations (8.19)–(8.22)

$$U = (S_2 - \tfrac{1}{2}C)Q^{-1} \tag{8.27}$$

where S_2 is the skew-symmetric solution of

$$Q'S_2 + S_2Q = \tfrac{1}{2}(Q'C - CQ) \tag{8.28}$$

The right-hand side of this equation is just

$$\tfrac{1}{2}(Q'C - CQ) = \frac{1}{2} \begin{bmatrix} 0 & 1 & 0 & \cdots & 0 \\ -1 & 0 & 0 & \cdots & 0 \\ 0 & 0 & 0 & \cdots & 0 \\ 0 & 0 & 0 & \cdots & 0 \end{bmatrix} \tag{8.29}$$

If we define a column vector s of dimension $\tfrac{1}{2}n(n-1)$ formed with the elements above the principal diagonal of S_2, so that

$$s' = (s_{12} \ s_{13} \ \cdots \ s_{n-1 \ n}) \tag{8.30}$$

then (8.28) can be written as

$$Hs = g \tag{8.31}$$

where $g' = (\tfrac{1}{2} 0 \cdots 0)$ and H is a square matrix of order $\tfrac{1}{2}n(n-1)$. The elements of H are given by Barnett and Storey in [2] as

$$h_{ij} = 1 \qquad \qquad i = \tfrac{1}{2}J(J-1) + I, \quad j = \tfrac{1}{2}(J-1)(J-2) + I$$

$$I = 1, 2, ..., J-1; \quad J = 2, 3, ..., n-1$$

$$i = j+1, \quad j = \tfrac{1}{2}(J-1)(J-2) + I$$

$$I = 1, 2, ..., J-2; \quad J = 3, 4, ..., n$$

$$h_{ij} = k_{n-p+1}, \qquad i = \tfrac{1}{2}(I-1)(I-2) + p$$

$$j = \tfrac{1}{2}(n-1)(n-2) + I$$

$$p = 1, 2, ..., I-1$$

$$I = 1, 2, ..., n-1$$

$$h_{ij} = -k_{n-r+1}, \qquad j = \tfrac{1}{2}(r-1)(r-2) + I$$

$$j = \tfrac{1}{2}(n-1)(n-2) + I$$

$$r = I+1, ..., n$$

$$I = 1, 2, ..., n-1$$

All other $h_{ij} = 0$. The matrix H is very sparse, no row having more than four nonzero elements. The vector s (and then the matrix S_2) should be simple to compute; the computations of U and then of F present no difficulty.

In the next section we study another type of stability, exponential stability, which is of interest in connection with the study of time-varying systems.

Exercises

*1. Relate stability and instability of the origin to the position of the eigenvalues of the matrix A.

2. Suppose A is not invertible, so that it has zero eigenvalues. Describe the behavior of the state vector as t increases.

3. Prove Theorem 8.3 by applying a similar kind of reasoning to that used in analyzing the observer equation.

4. Verify that (8.18) is a solution of (8.5) by direct substitution.

8.3. Exponential Stability

If the origin is an asymptotically stable point of equilibrium for the system (8.1), we know that all solutions tend to the origin as t increases. It is of interest sometimes to say more about the way this convergence occurs, simply by comparing the norm of the solution under consideration with a suitable function, as in the following definition.

Definition. The origin is said to be an *exponentially stable* point of equilibrium for (8.1) if there exist two constants $\alpha > 0$ and $M > 0$ independent of t_0, such that

$$\| x(t; t_0, x^0) \| \leqslant M e^{-\alpha(t-t_0)} \tag{8.32}$$

for all $t \geqslant t_0 \geqslant 0$.

Of course, if the origin is exponentially stable it is also asymptotically stable. This definition is of interest mostly for time-varying systems because if the origin is asymptotically stable for a constant system it is also exponentially stable; this can be proved directly from the theorems of the previous section, and also follows from the theorems on exponential stability in this section.

We present first a simple lemma, whose proof we leave as an exercise for the reader.

LEMMA.   Let $\phi(t, \tau)$ be the transition matrix of (8.1). Condition (8.32) is implied by and implies

$$\| \phi(t, t_0) \| \leqslant N e^{-\alpha(t-t_0)}, \qquad t \geqslant t_0 \geqslant 0 \tag{8.33}$$

where N and α are positive constants independent of t_0 ; α is the same constant α which appears in (8.32). The norm of the matrix $\phi(t, t_0)$ in (8.33) is the matrix norm generated by the vector norm in (8.32). In the rest of this section we assume that t_0 is a nonnegative number.

It should be clear, then, that we could have defined exponential stability by means of the condition (8.33) instead of (8.32). Before going on to establish some theorems on exponential stability, we must clarify the concept of a matrix A in (8.1) with *bounded* entries. We say that $f: [t_0 , \infty) \rightarrow R$ is bounded if there is a constant K such that $| f(t) | < K$ for $t \in [t_0 , \infty)$. Likewise, the matrix A is said to be bounded if all its entries are bounded, or, alternatively, if there is a constant K such that $\| A(t) \| < K$ for all $t \in [t_0 , \infty)$.

We prove now our first theorem on exponential stability [4]:

THEOREM 8.5.   Let the matrix A in (8.1), defined on $[t_0 , \infty)$, be bounded. Then the origin is an exponentially stable point of equilibrium of (8.1) if and only if a constant $\tilde{N}$, independent of both t and t_0 , exists such that

$$\int_{t_0}^{t} \| \phi(t, \tau) \| \, d\tau \leqslant \tilde{N} \tag{8.34}$$

for all $t \geqslant t_0$. The conditions (8.34) and (8.33) are, in this case, equivalent.

Proof.   It follows simply by substitution that (8.33) implies (8.34):

$$\int_{t_0}^{t} \| \phi(t, \tau) \| \, d\tau \leqslant N \int_{t_0}^{t} e^{-\alpha(t-\tau)} \, d\tau$$

$$= (N/\alpha)[1 - e^{-\alpha(t-t_0)}] \leqslant N/\alpha = \tilde{N} \tag{8.35}$$

To show the converse statement, we prove first that (8.34) and the boundedness of A imply that there is a constant $\mu > 0$, which is independent of t, such that $\| \phi(t, \tau) \| \leqslant \mu$ for $t \geqslant \tau \geqslant t_0$. Indeed, it was shown in Chapter 4 that

$$\frac{d}{d\tau} [\phi(t, \tau)] = -\phi(t, \tau) A(\tau) \tag{8.36}$$

so that, if $t \geqslant \lambda \geqslant t_0$,

$$\left\| \int_{\lambda}^{t} \frac{d}{d\tau} \phi(t, \tau) \, d\tau \right\| = \| \phi(t, \lambda) - I_n \|$$

$$= \left\| \int_{\lambda}^{t} \phi(t, \tau) \, A(\tau) \, d\tau \right\|$$

$$\leqslant \int_{\lambda}^{t} \| \phi(t, \tau) \| \, \| A(\tau) \| \, d\tau \leqslant \tilde{N} K \qquad (8.37)$$

from which it follows that $\| \phi(t, \tau) \|$ should be bounded for $t \geqslant \tau \geqslant t_0$ by a constant μ which is independent of t because $\tilde{N}$ and K are independent of t.

Note that this result implies that $\| \phi(t, \tau) \|$ is bounded by μ in the shaded region of the (t, τ) plane shown in Figure 8.1 so that, for instance,

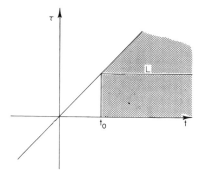

FIGURE 8.1. The matrix $\| \phi(t, \tau) \|$ is bounded by a constant μ in the shaded region of the (t, τ) plane shown above.

we can say the $\| \phi(\tau, t_0) \|$ is bounded by μ for $t_0 \leqslant \tau \leqslant t < \infty$, that is, $\| \phi(t, \tau) \|$ is bounded by μ along the line L in Figure 8.1. Let us compute now, for $t \geqslant \tau \geqslant t_0$,

$$\left\| \int_{t_0}^{t} \phi(t, \tau) \, \phi(\tau, t_0) \, d\tau \right\| \leqslant \int_{t_0}^{t} \| \phi(t, \tau) \| \, \| \phi(\tau, t_0) \| \, d\tau$$

$$\leqslant \tilde{N} \mu \qquad (8.38)$$

while the norm in the left-hand side of this expression actually equals, by the group property of the transition matrix, $(t - t_0) \| \phi(t, t_0) \|$, so that we have proved

$$(t - t_0) \| \phi(t, t_0) \| \leqslant \tilde{N} \mu \qquad (8.39)$$

We use this estimate in the following manner:

$$\left\| \int_{t_0}^{t} \phi(t, \tau)[(\tau - t_0)\, \phi(\tau, t_0)]\, d\tau \right\| = \| \phi(t, t_0) \| \frac{(t - t_0)^2}{2!}$$

$$\leqslant \int_{t_0}^{t} \| \phi(t, \tau) \| \, \|(\tau - t_0)\, \phi(\tau, t_0) \| \, d\tau \leqslant N^2 \mu \tag{8.40}$$

so that

$$\frac{(t - t_0)^2}{2!} \| \phi(t, t_0) \| \leqslant \tilde{N}^2 \mu \tag{8.41}$$

and then, by induction,

$$\frac{(t - t_0)^n}{n!} \| \phi(t, t_0) \| \leqslant \tilde{N}^n \mu \tag{8.42}$$

for $t \geqslant t_0 \geqslant 0$, $n = 0, 1, 2, \ldots$. If we make $0 < \alpha < \tilde{N}^{-1}$, multiply the first equation in (8.42) by α^0, the second by α, ..., the nth by α^{n-1}, ..., and add all the resulting expression, we obtain

$$e^{\alpha(t-t_0)} \| \phi(t, t_0)\| \leqslant \mu \sum_{n=0}^{\infty} (\alpha \tilde{N})^n = \frac{\mu}{1 - \alpha \tilde{N}} \tag{8.43}$$

from which it follows that

$$\| \phi(t, t_0)\| \leqslant \frac{\mu}{1 - \alpha \tilde{N}} \, e^{\alpha(t-t_0)} \tag{8.44}$$

for all $t \geqslant t_0 \geqslant 0$, which proves that (8.34) implies (8.33). The theorem follows.

In preparation for the following, and last, theorem in this section, we define a quadratic form $(x, V(t)x)$ to be positive definite if there exists a constant $a > 0$ such that

$$(x, V(t)x) \geqslant a \| x \|^2 \tag{8.45}$$

for all t. If the elements of V are bounded on $[t_0, \infty)$, which is equivalent to requiring that the norm of $V(t)$ be bounded on $[t_0, \infty)$, there is a constant $b > 0$ such that

$$b \| x \|^2 \geqslant (x, V(t)x) \tag{8.46}$$

on $[t_0, \infty)$. We can prove now:

THEOREM 8.6 [5].   Assume that the matrix A is bounded on $[t_0, \infty)$. Then condition (8.33) holds if there is a positive definite quadratic form $(x, V(t)x)$, with coefficients that are continuously differentiable and bounded functions of t on $[t_0, \infty)$ and which satisfies for some constant c

$$\frac{d}{dt}(x(t; t_0, x^0), V(t)x(t; t_0, x^0)) \leqslant -c\|x(t; t_0, x^0)\|^2 \qquad (8.47)$$

where $x(t; t_0, x^0)$ is any solution of (8.1).

Proof.   The quadratic form $(x, V(t)x)$ is said to be a Lyapunov function for the system (8.1). Let k be a positive number to be selected later. Put

$$y(t; t_0, x^0) = e^{[k(t-t_0)]}x(t; t_0, x^0)$$

so that y is a solution of the system

$$\dot{y} = (A + kI_n)y \qquad (8.48)$$

which satisfies $y(t_0; t_0, x^0) = x^0$. Then, using the properties of $(x, V(t)x)$ we have, from (8.47) and (8.48),

$$\frac{d}{dt}(y(t; t_0, x^0), V(t)y(t; t_0, x^0))$$

$$\leqslant -c\|y(t; t_0, x^0)\| + 2k(y(t; t_0, x^0), V(t)y(t; t_0, x^0))$$

$$\leqslant -\|y(t; t_0, x^0)\|(c - 2ka) \qquad (8.49)$$

By taking k sufficiently small, we can make $(c - 2ka)$ positive, so that the derivative in the left-hand side of (8.49) is negative; this implies that the quadratic form $(y(t; t_0, x^0), V(t)y(t; t_0, x^0))$ is nonincreasing with the time, that is

$$(y(t; t_0, x^0), V(t)y(t; t_0, x^0)) \leqslant (y(t_0; t_0, x^0), V(t)y(t_0; t_0, x^0))$$

$$= (x^0, V(t)x^0) \leqslant b\|x^0\|^2$$

which implies that $\|y(t; t_0, x^0)\| \leqslant N$ for all $t \geqslant t_0 \geqslant 0$, where N is a positive constant independent of t_0. Then the theorem follows, since

$$\|x(t; t_0, x^0)\| \leqslant Ne^{-k(t-t_0)} \qquad (8.50)$$

The necessity of the contention of this theorem is proved in [5].

It should be clear that constant systems for which the origin is asymptotically stable satisfy (8.34), so that these systems are exponentially stable as well. The same result can be derived by considering that, by Theorem 8.2, these systems do have Lyapunov functions which satisfy exactly the conditions of Theorem (8.6).

In the next section we return to considering input–output problems, and talk about *external stability* and its relationships with some types of internal stabilities studied so far.

Exercises

*1. Show that the origin is a stable point of equilibrium of the homogeneous system $\dot{x} = Ax$ if and only if there exists a constant M, which may depend on t_0, such that

$$\| \phi(t, t_0)\| \leqslant M \qquad \text{for all } t \geqslant t_0$$

*2. Show that the origin is asymptotically stable if and only if there is a constant M, such that

$$\| \phi(t, t_0)\| \leqslant M \qquad \text{for all } t \geqslant t_0$$

$$\lim_{t \to \infty} \| \phi(t, t_0)\| = 0, \qquad \text{for all } t_0$$

3. Show that if $\lim_{t \to \infty} \| \phi(t, t_0)\| = 0$ for one value of t_0, then this equality applies for every t_0.

*4. Show in detail that if the origin is asymptotically stable for a constant system, then it is exponentially stable.

*5. Show by means of a counterexample that a function which takes finite values on $[0, \infty)$ is not necessarily bounded there.

*6. Show by means of a counterexample that a function which takes finite values on $(0, 1)$ is not necessarily bounded there.

8.4. External Stability, Controllability, Observability and Exponential Stability

In previous chapters we have encountered two kinds of descriptions of linear systems. The internal structure of the system is characterized by the state representation, while an input–output description, in the form of a transfer function or an impulse response matrix, may fail to convey information about the internal structure if the system is not controllable and observable. We may expect that systems which are unstable would behave quite wildly from an input–output standpoint, and that, if the state vector is not available, one would be able to diagnose instability, stability, asymptotic stability, etc., as the case may be, from a series of input–output measurements. This is only partially true; if the system is not controllable and observable the behavior of the whole system simply does not get reflected in the input–output characteristics of it, so that it may well be that an unstable system has a very nice, orderly input–output behavior. As a matter of fact, we shall have to strengthen our definitions of controllability and observability in the general time-varying case so as to obtain meaningful relationships between input–output behavior and, as it will turn out, exponential stability.

In this section we will describe first in a meaningful way the input–output stability of linear systems; the relationship between this kind of stability, controllability, observability, and exponential stability will then be established.

The input–output behavior of a linear system with respect to stability will be characterized by whether every bounded input does or does not give rise to a bounded output when the initial state is zero. We shall call this *external stability* (also known as *bounded-input, bounded-output, BIBO, stability*). We define formally,

Definition. Let the inputs of a linear system be defined over $[t_0, \infty)$ and let $x(t_0) = 0$. We say that this system is *externally stable* if every bounded input gives rise to a bounded output, that is, if the inequality

$$\| u(t) \| < c_1, \qquad t \in [t_0, \infty) \tag{8.51}$$

implies $\| y(t; t_0, 0; u) \| < c_2$, for some c_2 (which may depend on t_0 and on c_1) and $t \in [t_0, \infty)$. Note that, as usual, we are using the same notations for norms in R^r and R^n.

If c_2 is independent of t_0, but dependent only on c_1, we say that the system is *uniformly* externally stable.

It is necessary to strengthen now our definition of controllability and observability. We consider the system

$$\dot{x} = Ax + Bu, \qquad y = Cx \tag{8.52}$$

where A, B, and C are defined for all t, their entries are continuous and they are bounded, that is, there is a constant K such that

$$\| A(t) \| < K, \qquad \| B(t) \| < K, \qquad \| C(t) \| < K \tag{8.53}$$

for all t. We define:

Definition [6]. The system (8.52) is said to be *uniformly controllable* if and only if there exists $\delta > 0$ such that for every state $\xi \in R^n$ and for any time t, there exists an input u, defined on $[t - \delta, t)$ such that if $x(t - \delta) = 0$, then $x(t) = \xi$ and $\| u(\tau) \| \leqslant \gamma(\delta, \| \xi \|)$ for all $\tau \in [t - \delta, t)$.

Note that a number δ, not necessarily small, is associated with each uniformly controllable system so that on each half-open interval of length δ the control u defined above is bounded in norm by a number which depends on δ (and on the norm of the final state) only, but not on the value of the time t; hence the name uniform. It should be clear that a controllable system satisfies all the requirements of the definition of uniform controllability, except the one that the bound on $\| u \|$ should not depend on t.

Before showing an example which demonstrates the existence of controllable but not uniformly controllable systems, thus justifying the definition above, we prove a theorem which provides with what is in fact an alternative definition of uniform controllability. If A, B are symmetric $n \times n$ matrices, we say that $A > B$ if $A - B$ is positive definite; we say that $A \geqslant B$ if $A - B$ is positive semidefinite.

THEOREM 8.7. The system (8.52) is uniformly controllable if and only if there exists a number δ such that, for all t,

$$W(t - \delta, t) \geqslant \alpha_1(\delta) \, I_n > 0_n \qquad (8.54)$$

where W is the matrix function defined in Equation (5.6):

$$W(t_0, t_1) = \int_{t_0}^{t_1} \phi(t_1, t) \, B(t) \, B'(t) \, \phi'(t_1, t) \, dt$$

and $\alpha_1 : R_+ \to R_+$. Recall that R_+ is the set of all positive real numbers.

Proof. Suppose (8.54) holds for some function α_1. Since W is invertible, we can define u as

$$u(\tau) = B'(\tau) \, \phi'(t, \tau) \, W^{-1}(t - \delta, t)\xi \qquad (8.55)$$

This control transfers the system from the zero state at time $t - \delta$ to the state ξ at time t; the proof is essentially the same as that in Theorem 5.5. We only have to show that the control u is bounded in norm in the way described in the definition of uniform controllability. If A is bounded as in (8.53), it can be shown quite simply that

$$\| \phi(t, \tau) \| \leqslant e^{K|t-\tau|} \qquad (8.56)$$

here K is the number which bounds the norm of A; see Exercise 1 of Section 3.2. Since B is also bounded, the boundedness of $\| u(\tau) \|$ as in the definition above will follow quite simply if we show that the norm of the inverse of W is bounded in an appropriate way. Indeed, it was suggested as an Exercise in Chapter 5, and is easily proved, that if V and U are two symmetric matrices and $V \geqslant U$, then $U^{-1} \geqslant V^{-1}$, provided that these two matrices have inverses. In our case, then,

$$(W^{-1}(t - \delta, t)x, x) \leqslant \frac{1}{\alpha_1(\delta)} \| x \|^2 \tag{8.57}$$

which implies that

$$\| W^{-1}(t - \delta, t) \| \leqslant \frac{1}{\alpha_1(\delta)} \tag{8.58}$$

since, as shown in Chapter 2, the norm of W^{-1} is the least number k for which $\| W^{-1}x \| \leqslant k \| x \|$, and clearly $1/\alpha_1(\delta)$ is one of such numbers. Therefore,

$$\| u(\tau) \| \leqslant K e^{-K\delta} \frac{1}{\alpha_1(\delta)} \| \xi \| \tag{8.59}$$

for all $\tau \in [t - \delta, t)$ and for all t. The bound (8.59) does not depend on t, and the system (8.52) is, under the hypothesis of the theorem, uniformly controllable.

Suppose now that the system (8.52) is uniformly controllable, but (8.54) does not hold for any function α_1 from R_+ into R_+. Then for each $\delta > 0$ and for any such function α_1 there is a vector $\eta \in R^n$, with $\| \eta \| = 1$, such that, for some t,

$$(W(t - \delta, t)\eta, \eta) < \alpha_1(\delta)$$

or, since α_1 is any arbitrary function from the positive reals to the positive reals, and δ is arbitrary,

$$\int_{t-\delta}^{t} (\phi(t, \tau) B(\tau) B'(\tau) \phi'(t, \tau)\eta, \eta) \, d\tau = \int_{t-\delta}^{t} \| \eta'\phi(t, \tau) B(\tau) \|^2 \, d\tau < \alpha \tag{8.60}$$

for any $\alpha > 0$ and some t.

Since the system is uniformly controllable, a control bounded in the manner described in the definition above exists which drives the zero state at time $t - \delta$ to the state η at time t; that is,

$$\eta = \int_{t-\delta}^{t} \phi(t, \tau) B(\tau) u(\tau) \, d\tau \tag{8.61}$$

then the Cauchy–Schwarz inequality implies that

$$\| \eta \|^2 \leqslant \left[\int_{t-\delta}^{t} \| \eta'\phi(t, \tau) B(\tau) \|^2 \, d\tau \right]^{1/2} \left[\int_{t-\delta}^{t} \| u(\tau) \|^2 \, d\tau \right]^{1/2} \tag{8.62}$$

Since $\| u(\tau) \| \leqslant \gamma(\delta, 1)$ for some function γ, then (8.60) and (8.62) imply that

$$\gamma(\delta, 1)(\alpha\delta)^{1/2} \geqslant \| \eta \| = 1 \qquad (8.63)$$

which is a contradiction since α can be made arbitrarily small. Uniform controllability implies (8.54), and the theorem follows.

We can show now quite simply that uniform controllability is a concept which is relevant only to time-varying systems because all controllable constant systems are uniformly controllable as well. Indeed, if a system is controllable, $W(t - \delta, \delta)$ is positive definite, which implies that an inequality such as (8.54) is satisfied with the provision that α_1 depends generally on t, while for uniformly controllable systems it does not depend on the time. For a constant system, however, the matrix $W(t - \delta, \delta)$ does not depend on t, as can easily be seen from (5.6) using the fact that B is constant and that $\phi(t, \lambda)$ depends only on $t - \lambda$, so that it satisfies (8.54) for a function α_1 independent of t. The contention follows.

There are, however, time-varying controllable systems which are not uniformly controllable, as shown by the following example.

EXAMPLE

Consider the scalar system $\dot{x}(t) = t\, u(t)$. Here

$$W(t - \delta, t) = \int_{t-\delta}^{t} \tau^2 \, d\tau = \tfrac{1}{3}[t^3 - (t - \delta)^3]$$

$$= \tfrac{1}{3}[3t^2\delta - 3t\delta^2 + \delta^3] \geqslant \tfrac{1}{4}t^2\delta > 0$$

so that this system is controllable (indeed, totally controllable, since the above inequality is valid for any $\delta > 0$), but is not uniformly controllable; a little thought will convince the reader that no function of δ alone can be found such that for all t and a fixed δ,

$$\tfrac{1}{4}t^2\delta > \alpha_1(\delta) > 0$$

since by making t small enough $\tfrac{1}{4}t^2\delta$ can be made less than any number $\alpha_1(\delta)$.

At last, theorems can be established which relate external and exponential stability. The first one refers to the behavior of the first of the equations in (8.52), and its contention is an intermediate result, in the sense that it relates exponential stability with the behavior of the state vector x, not with the output vector; a further connection, between state and output vectors, will appear in the following theorem.

We define now a class of external stability which describes properties of the first equation in (8.52); we say that the system (8.52) is *state-*

externable stable if every bounded input defined on an interval such as $[t_0 , \infty)$ gives rise to a bounded state vector. We prove

THEOREM 8.8 [6]. Let the system (8.52) be uniformly controllable. Then this system is state-externally stable if and only if the origin is an exponentially stable equilibrium point for this system (or, as we also say, if the system is exponentially stable).

Proof. If (8.52) is exponentially stable, then (8.33) and (8.34) are satisfied by the norm of the transition matrix and its integral respectively. This implies that, if $\| u(t) \| \leqslant c_1$,

$$\| x(t; t_0 , x^0; u) \| \leqslant \| \phi(t, t_0) \| \, \| \, x^0 \| + \int_{t_0}^t \| \phi(t, \lambda) \| \, \| \, B(\lambda) \| \, \| \, u(\lambda) \| \, d\lambda$$

$$\leqslant N \| x^0 \| + K c_1 \tilde{N} \tag{8.64}$$

where K is the bound on $\| B(t) \|$ defined in (8.53). The first part of the theorem follows, since (8.64) implies that the state vector is bounded on $[t_0 , \infty)$.

Let η be a vector in R^n with norm equal to unity. If (8.52) is uniformly controllable, there is a $\delta > 0$ such that for all t there is an input u such that

$$\eta = \int_{t-\delta}^t \phi(t, \tau) \, B(\tau) \, u(\tau) \, d\tau$$

and $\| u(\tau) \| \leqslant \gamma(\delta, 1)$ for $\tau \in [t - \delta, t)$. We multiply both sides of this equation by $\phi(t_1 , t)$, and integrate the norm of the result between t_0 and t_1 :

$$\int_{t_0}^{t_1} \| \phi(t_1 , t)\eta \| \, dt \leqslant \gamma(\delta, 1) \int_{t_0}^{t_1} \left\{ \int_{t-\delta}^t \| \phi(t, \tau) \, B(\tau) \| \, d\tau \right\} dt \tag{8.65}$$

If we make $r = \tau - t + \delta$, we obtain

$$\int_{t_0}^{t_1} \| \phi(t_1 , t)\eta \| \, dt \leqslant \gamma(\delta, 1) \int_0^\delta \left\{ \int_{t_0+r-\delta}^{t_1+r-\delta} \| \phi(t, \tau) \, B(\tau) \| \, d\tau \right\} dr \tag{8.66}$$

and for $0 \leqslant r \leqslant \delta$,

$$\int_{t_0+r-\delta}^{t_1+r-\delta} \| \phi(t_1 , \tau) \, B(\tau) \| \, d\tau \leqslant \int_{-\infty}^{t_1} \| \phi(t_1 , \tau) \, B(\tau) \| \, d\tau \tag{8.67}$$

Since we are assuming that the norm of the state is bounded when the input is bounded, then the right-hand side of (8.67) is bounded by a constant c, so that, since (8.65) to (8.67) are valid for any t_0,

$$\int_{-\infty}^{t_1} \| \phi(t_1, t)\eta \| \, dt \leqslant \gamma(\delta, 1)\, \delta c \qquad (8.68)$$

for all t_1. The norm of $\phi(t_1, t)$ is

$$\| \phi(t_1, t) \| = \sup_{\|\eta\|=1} \| \phi(t_1, t)\eta \|$$

so that from (8.68),

$$\int_{-\infty}^{t_1} \| \phi(t_1, t) \| \, dt \leqslant \gamma(\delta, 1)\, \delta c \qquad (8.69)$$

which implies, by Theorem (8.5), the exponential stability of the origin. The theorem follows.

As announced above, we will now relate state-external stability with external stability itself. Of course, state-external stability implies, by the second of Equations (8.52) and the boundedness of C, external stability; for the proof of the converse statement, however, we need a concept closely related to uniform controllability, *uniform observability*.

Definition.   The system described by

$$\dot{x} = Ax, \qquad y = Cx \qquad (8.70)$$

is *uniformly observable* if the system $\dot{z} = -A'z + C'u$ is uniformly controllable.

We have, then, from Theorem 8.8,

THEOREM 8.9.   The system (8.70) [or (8.52)] is uniformly observable if and only if the matrix function M defined in Theorem 5.22 satisfies

$$M(t - \delta, t) \geqslant \alpha_2(\delta)\, I_n > 0_n$$

for some function α_2 from the positive reals to the positive reals.

We establish [6]:

THEOREM 8.10.   If the system (8.52) is uniformly observable, then it is externally stable if and only if it is state-externally stable.

Proof.   If (8.52) is state-externally stable, it is externally stable, since the matrix C is bounded. We only have to show therefore that if (8.52) is externally stable and uniformly observable, then it is state-externally stable. Suppose otherwise, that is, that there is a bounded input u which produces both a bounded output and an unbounded state. Then, corresponding to each positive number N there is a value of time $t - \delta$ for which $\| x(t - \delta) \| > N$. If the input is zero over $[t - \delta, t)$, then

$$y(\tau) = C(\tau)\phi(\tau, t - \delta) x(t - \delta), \qquad \tau \in [t - \delta, t)$$

Then,

$$\int_{t-\delta}^{t} \| y(\tau) \|^2 \, d\tau = \int_{t-\delta}^{t} x'(t - \delta) \phi'(\tau, t - \delta) C'(\tau) C(\tau) \phi(\tau, t - \delta) x(t - \delta) \, d\tau$$

$$= x'(t - \delta) M(t - \delta, t) x(t - \delta) \geqslant \alpha_2(\delta) \| x(t - \delta) \|^2 \geqslant \alpha_2(\delta) N^2$$

$$(8.71)$$

by (8.70). At some point in $[t - \delta, t)$, the norm of the output has to be larger than $N[\alpha_2(\delta)/\delta]^{1/2}$, since otherwise the integral of this norm over this interval would be less than $\alpha_2(\delta) N^2$. But since N is arbitrary while u is bounded, this contradicts the external stability of (8.52). No such control u exists, and the theorem follows.

Theorems 8.8 and 8.10 imply the following theorem.

THEOREM 8.11.   If the system (8.52) is uniformly controllable and observable, then it is externally stable if and only if it is exponentially stable.

If (8.52) is a constant system, we can specialize this theorem as follows.

THEOREM 8.12.   If the system (8.52) is constant, controllable, and observable, then it is externally stable if and only if it is exponentially stable (or asymptotically stable).

We have succeeded in relating internal and external types of stability for systems which are, in a uniform sense, controllable and observable. In the case of linear constant systems, both types of stability are equivalent provided that, logically enough, there are no parts of the system which are not influenced by (or influence) the input (or output), that is, provided the system is controllable and observable. Otherwise it may happen that some very wild behavior of some of these parts does not get reflected in the behavior of the output.

Exercises

*1.   Let $f: J \rightarrow R$, $J \subset R$, be continuous. Find a sufficient condition on J so that f is bounded over this set. Of course, f is finite for each point $x \in J$.

2. Find a (time-varying) system which, because either or both of its matrices B and C are not bounded, fails to satisfy Theorem 8.11.
3. Find a two-dimensional system which is controllable but not uniformly controllable.
*4. Show in detail that all controllable constant systems are uniformly controllable.

8.5. The Stability of Difference Systems

Many of the concepts developed in conjunction with the study of the stability of differential systems can be obviously (and sometimes trivially) extended to the treatment of the stability of difference systems.

We say that $\hat{x}$ is a point of equilibrium of $x(k + 1) = A(k) x(k)$, where we assume A is a matrix function defined for all $k \geq k_0$, if $A(k)\hat{x} = 0$ for some $k \geq k_0$.

We could, but we will not, paraphrase the definitions of stability, asymptotic stability, and instability made in Section 8.1 in order to define similar concepts for difference systems; we leave this task to the reader; it is only necessary to replace t by k. For constant difference systems, we prove

THEOREM 8.13.   The system

$$x(k + 1) = Ax(k) \qquad (8.72)$$

is asymptotically stable if and only if the eigenvalues of the matrix A have an absolute value less than unity.

Proof.   The assertion follows from (7.13), (7.24) and the fact that $k^n \lambda^k$ tends to zero as k tends to infinity provided that n is fixed and $| \lambda | < 1$.

Doubtless, many theorems and ideas from the previous sections can be extended to the study of related ideas for difference systems, sometimes painlessly, on occassions less so; lack of space will not permit us to undertake this task. We will end this section, and this chapter, with a theorem reminiscent of Theorem 8.2; in the case of difference systems, however, the role of the derivative (8.4) is played by the difference $(x(k + 1), Fx(k + 1)) - (x(k), Fx(k))$, so that

$$(x(k + 1), Fx(k + 1)) - (x(k), Fx(k)) = (Ax(k), FAx(k)) - (x(k), Fx(k))$$

$$= (x(k), (A'FA - F) x(k)) \qquad (8.73)$$

We have therefore:

THEOREM 8.14.   The origin is an asymptotically stable point of equilibrium for the system (8.72) if and only if given any positive definite matrix G there is a positive definite matrix F which is the unique solution of the equation

$$A'FA - F = -G \qquad (8.74)$$

A connection can be established between properties of this equation and the eigenvalues of A much as it was done for the corresponding equation in the differential case.

SUPPLEMENTARY NOTES AND REFERENCES

§8.1. There are many texts which treat the general subject of stability in great depth and breadth. The following treatises are quite advanced:

W. Hahn, "Stability Theory." Springer, Berlin, 1967.
J. L. Massera and J. J. Schäffer, "Linear Differential Equations and Function Spaces." Academic Press, New York, 1966.
N. N. Krasovski, "Stability of Motion." Stanford Univ. Press, Stanford, California, 1963.

Surveys of the literature on the subject of stability abound; see for instance,

W. Hahn, "Theory and Application of Lyapunov's Direct Method." Prentice-Hall, Englewood Cliffs, New Jersey, 1963.
L. Cesari, "Asymptotic Behavior and Stability Problems in Ordinary Differential Equations." Academic Press, New York, 1963.
R. E. Kalman and J. E. Bertram, Control system design via the "second method" of Lyapunov—Pt. I: Continuous-time systems, Pt. II: Discrete-time systems, *J. Basic Engineering Trans. ASME* **82** (1960), 371–400.
R. W. Brockett, The status of stability theory for deterministic systems, *IEEE Trans. Automatic Control* **AC-11** (1966), 596–606.

§8.2. The matrix Equation (8.5) is related to the matrix Equation (6.105) encountered when we studied the observer in Chapter 6. In fact, this equation is found also in connection with elasticity studies. Further treatment of this equation can be found in

Er-Chieh Ma, A finite series solution of the matrix equation $AX - XB = C$, *SIAM J. Appl. Math.* **14** (1966), 490–495.
R. A. Smith, Matrix equation $XA + BX = C$, *SIAM J. Appl. Math.* **16** (1968), 198–201.
A. Jameson, Solution of the equation $AX + XB = C$ by inversion of an $M \times M$ or $N \times N$ matrix, *SIAM J. Appl. Math.* **16** (1968), 1020–1023.

An interesting development, which relates the observability of a linear constant system to its asymptotical stability and to the existence of a quadratic Lyapunov function whose derivative is *not* negative definite, appears in

R. E. Kalman, Lyapunov functions for the problem of Lur'e in automatic control, *Proc. Nat. Acad. Sci. U.S.A.* **49** (1963), 201–205.

§8.4. There are many theorems which relate external and exponential stability under a variety of assumptions. See the paper by Kalman and Bertram and the book by Massera and Schäffer cited above, and also

W. Kaplan, "Operational Methods for Linear Systems," Addison-Wesley, Reading, Massachusetts, 1962.

D. C. Youla, On the stability of linear systems, *IEEE Trans. Circuit Theory* **CT-10** (1963), 276–279.

The original definition of uniform controllability appears in

R. E. Kalman, Contributions to the theory of optimal control, *Bol. Soc. Mat. Mexicana* **5** (1960), 102–119.

REFERENCES

1. R. Bellman, Kronecker products and the second method of Lyapunov, *Math. Nachr.* **17** (1959), 17–19.
2. S. Barnett and C. Storey, Analysis and synthesis of stability matrices, *J. Differential Equations* **3** (1967), 414–422.
3. R. A. Smith, Matrix calculations for Lyapunov quadratic forms, *J. Differential Equations* **2** (1966), 208–217.
4. T. F. Bridgland, Stability of linear signal transmission systems, *SIAM Rev.* **5** (1963), 7–32.
5. H. A. Antonsiewicz and P. Davis, Some implications of Lyapunov's conditions for stability, *J. Rational Mech. Anal.* **3** (1954), 447–457.
6. L. M. Silverman and B. D. O. Anderson, Controllability, observability and stability of linear systems, *SIAM J. Control* **6** (1968), 121–130.

9

Infinite Dimensions

9.1. Introduction

It was mentioned in Section 1.4 that the introduction of the concept of state should be considered as an important step forward in the development of a theory of systems. The time has come to take another such step, when we no longer consider finite-dimensional state spaces, but rather focus our attention on a larger class of systems which do not have such simple descriptions, that is, systems whose state spaces can be said to be infinite dimensional. The reader will surely remember that examples of these systems were given at the end of Chapter 1.

The material to follow reaches a more profound intellectual level than the previous chapters. As a matter of fact, even if the treatment will be mostly self-contained, the reader should have, in order really to profit from it, some background in functional analysis. If the interest of the student is to become creative in the subject matter of this chapter, then a thorough knowledge of functional analysis becomes imperative.

We shall start by going deeper than before into some aspects of the theory of linear normed spaces, so that we can subsequently define the entity known as a *dynamical system*. The axioms which constitute this definition will be used to construct a fairly general theory of the behavior of these systems, simply by deriving several properties which are implied by these axioms; central to this theory is the concept of a strongly continuous semigroup of operators.

The theory will be applied to analyze from a dynamical standpoint *one* system, described by the *diffusion equation*

$$\frac{\partial}{\partial t} x(t, z) = \alpha \frac{\partial^2}{\partial z^2} x(t, z) + u(t, z)$$

The state variable x takes values $x(z, t)$, which can be thought of as being the temperature of a bar at a distance z from the origin at time t. Suppose first that this bar is infinite; then an interesting control problem is the devising of a *distributed* control function u such that the temperature of the bar takes a desired path in time and space. Of course, a dynamical description of the solutions of the diffusion equation and an understanding of the properties of these solutions are necessary before attempting to influence them; this is, essentially, the task which we are attempting in this chapter.

Of course, the bar may be finite in length; control of the temperature in this case may be effected by varying the temperatures at the *boundaries* of the bar, as well as by some sort of distributed control. Again, it is important to reach in this case an understanding of the basic fabric of the solutions, similar to the understanding gained of the solutions of systems with finite-dimensional state spaces.

A word of caution is needed here, because we will not be able to offer a comprehensive theory of controllability of these systems, and we shall not even touch upon the subjects of observability, synthesis, or stability. It happens that a very extensive background in mathematics is needed to master these subjects; the main objective of this chapter is to motivate the student into acquiring such a background, so that he can subsequently deepen his knowledge far beyond the elements given here.

9.2. Some Further Mathematical Topics

The concept of a linear normed space was defined in Chapter 2, where we went so far as giving some examples of spaces, such as $C[0, 1]$ or $L_p[0, 1]$, which are not finite dimensional. We will now define an important subclass of these spaces, the *complete* linear normed, or *Banach*, spaces; the basic setting of our axiomatics for dynamical systems will be built around this concept.

Definition.   Let X be a linear normed space. A point $x \in X$ is said to be the *limit* of a sequence of elements $x^1, x^2, ..., x^n, ...$ of X if the sequence of real numbers $\| x^n - x \|$ tends to zero as n tends to infinity.

It is easily proved that the limit of a sequence is unique, that is, if $x^n \to x$, and $x^n \to \bar{x}$, then $x = \bar{x}$.

Let X be the space $C[a, b]$. Convergence in this space has some interesting characteristics which we shall need later on. Suppose $f^n \to f, f^n, f$ in $C[a, b]$. This implies that

$$\max_{t \in [a,b]} |f^n(t) - f(t)| \to 0 \tag{9.1}$$

as n tends to infinity. We see that convergence in this space corresponds to *uniform* convergence of functions, since the expression (9.1) above implies that given $\epsilon > 0$, a number $N(\epsilon)$ can be found such that for $n > N(\epsilon)$

$$|f^n(t) - f(t)| < \epsilon$$

for all $t \in [a, b]$. Note that $N(\epsilon)$ does not depend on t, which is exactly the meaning of uniform convergence.

We introduce now another basic concept.

Definition. Let $x^n \in X$, a linear normed space, for $n = 1, 2, \ldots$. This sequence is said to be a *fundamental*, or Cauchy, sequence if $\| x^n - x^m \|$ tends to zero as n, m tend to infinity.

It is not difficult to show that a convergent sequence is fundamental. Indeed, let x^n tend to x as n tends to infinity. Then,

$$\| x^n - x^m \| \leqslant \| x^n - x \| + \| x^m - x \|$$

Since each of the two norms in the right-hand side of this expression can be made less that $\epsilon/2$ by taking n and m larger than a number $N(\epsilon/2)$, the assertion follows.

It happens that the converse of this assertion is generally false; there are linear normed spaces in which one can construct fundamental sequences which do not converge to a member of the space. This will be illustrated by means of an example.

EXAMPLE

Let X be the space of continuous, real-valued functions defined on [0, 1] with the norm

$$\|f\| = \int_0^1 |f(t)| \, dt$$

Consider the sequence $\{f^n\}$, with

$$
f^n(t) = \begin{cases} 0, & 0 \leqslant t \leqslant \dfrac{1}{2} \\[2mm] n\left(t - \dfrac{1}{2}\right), & \dfrac{1}{2} \leqslant t \leqslant \dfrac{1}{2} + \dfrac{1}{n} \\[2mm] 1, & \dfrac{1}{2} + \dfrac{1}{n} \leqslant t \leqslant 1 \end{cases}
$$

This is a fundamental sequence because

$$
\| f^n - f^m \| = \frac{1}{2}\left| \frac{1}{n} - \frac{1}{m} \right|
$$

which tends to zero as $n, m \to \infty$. However, the limit function of f^n is given by

$$
f(t) = \begin{cases} 0, & 0 \leqslant t \leqslant \tfrac{1}{2} \\ 1, & \tfrac{1}{2} < t \leqslant 1 \end{cases}
$$

which is not continuous, and consequently does not belong to the original space.

The only spaces which are of interest to us are those which do have the property that every fundamental sequence converges to an element of the space; this requirement prompts us to make the following

Definition. A linear normed space is said to be *complete* if every fundamental sequence in it has a limit which belongs to it. A complete linear normed space is said to be a *Banach* space.

The spaces $C[a, b]$, and the spaces $L_p[a, b]$, $p \geqslant 1$, are complete [1]. The proof of this assertion is not trivial.

In the development to follow we shall require that the state spaces of dynamical systems be Banach spaces. This requirement of completeness is not, however, so restrictive as it may seem because it is possible to enrich an incomplete linear normed space by simply adding to it new elements, in such a way that the new linear normed space (with the same norm as before) is complete. The description and justification of this process of *completion* [1] is rather beyond our reach, so that we will limit our treatment to a simple example. The space of all rational real numbers with the usual norm is not complete, since the (fundamental) sequence

$$
\left\{ 1, 1 + \frac{1}{2!}, 1 + \frac{1}{2!} + \frac{1}{3!}, \ldots \right\} \tag{*}
$$

does not converge to a rational number. However, this space can be completed by adding to it all irrational numbers to form R, the set of all real numbers; every fundamental sequence in it does converge to a member of the space; the sequence (*) converges to e, an irrational number.

We pass now to study another topic which is a prerequisite for the proper definition of dynamic systems, that of the *continuity* of transformations.

Let X, Y be Banach spaces, and let $T: X \to Y$ be a transformation. We say that T (not necessarily linear) is *continuous at* $x \in X$ if $x^n \to x$, $x^n \in X$, implies that $T(x^n) \to T(x)$. The convergence of the sequence $\{x^n\}$ is of course to be taken in the sense of the norm in X and that of $\{T(x^n)\}$ in the sense of the norm in Y.

The reader is perhaps familiar with the so-called ϵ-δ definition of continuity, which says that a transformation T as above is continuous at $x \in X$ if for every $\epsilon > 0$ there is a $\delta > 0$ such that $\| x - \bar{x} \| < \delta$ implies that $\| T(x) - T(\bar{x}) \| < \epsilon$. We show below the equivalence of the two definitions.

Let T be continuous at $x \in X$ in the ϵ-δ sense, and let $x^n \to x$. Given $\epsilon > 0$, there is $\delta = \delta(\epsilon) > 0$ such that $\| x - \bar{x} \| < \delta(\epsilon)$ implies $\| T(x) - T(\bar{x}) \| < \epsilon$. Associated with this value of $\delta(\epsilon)$ there is a number $N(\delta(\epsilon))$ such that, for $n > N(\delta(\epsilon))$, $\| x^n - x \| < \delta(\epsilon)$. Therefore, $\| T(x^n) - T(x) \| < \epsilon$ for $n > N(\delta(\epsilon)) = N_1(\epsilon)$; thus, $T(x^n) \to T(x)$, and ϵ-δ continuity implies continuity as defined above.

Suppose now that T is continuous at $x \in X$, but it is not continuous there in an ϵ-δ sense. Then there is a number $\epsilon > 0$ such that for all $\delta > 0$, there is an $\tilde{x}$ such that $\| x - \tilde{x} \| < \delta$ and $\| T(x) - T(\tilde{x}) \| \geqslant \epsilon$. Consider now any sequence of real numbers $\{\delta_n\}$ which converges to zero. For each n select the corresponding vector $\tilde{x}$ as above, and call this $\tilde{x}_n$. Then $\tilde{x}_n \to x$, since $\| x - \tilde{x}_n \| < \delta_n$, and $\delta_n \to 0$, but $T(\tilde{x}_n)$ does not tend to $T(x)$ since $\| T(x) - T(\tilde{x}_n) \|$ is never less than ϵ. This result obviously contradicts our assumption that T was continuous at x; the two definitions of continuity are seen then to be equivalent.

We leave the proof of the following theorem to the reader.

THEOREM 9.1. If a *linear* transformation mapping X into Y is continuous at a point $x^0 \in X$, then it is continuous at each $x \in X$.

We see, then, that a linear transformation can be defined to be simply *continuous* if it is continuous at any one point of its domain of definition.

It so happens that continuous linear transformations can be characterized in a very simple way, shown in the following definition and Theorem 9.2.

Definition. A transformation $T: X \to Y$ is said to be *bounded* if there exists a constant K such that

$$\| T(x) \| \leqslant K \| x \| \tag{9.2}$$

for all $x \in X$.

Our results of Chapter 2 indicate that a linear transformation mapping a finite-dimensional space into another such space is always bounded; we went there as far as defining a norm for such a transformation, based on an inequality just like (9.2). We could, but will not, define in a similar manner the norm of any bounded, linear transformation.

THEOREM 9.2.   A linear transformation is continuous if and only if it is bounded.

Proof.   Let $T: X \to Y$ be a linear continuous transformation. Assume that it is not bounded so that no constant K can be found to satisfy (9.2). Then for any positive integer n, there is some $x^n \in X$ such that

$$\| T(x^n) \| > n \| x^n \|$$

Let $\hat{x}^n = x^n/n \| x^n \|$; clearly $\| \hat{x}^n \| \to 0$, that is, $x^n \to 0$; since $\| T(\hat{x}^n) \| > 1$, we see that T cannot be continuous. Hence, a continuous linear transformation is bounded.

Let $T: X \to Y$ be a bounded linear transformation. If $x^n \to 0$, since T is bounded, then $Tx^n \to 0$ as well, from which follows the continuity of T at $x = 0$ and hence the continuity of T.

We give some examples of transformations, continuous and discontinuous, in Banach spaces.

i. Let $X = C[0, 1]$. If $x \in C[0, 1]$, define $y = T(x)$ by

$$y(s) = \int_0^1 K(s, t)\, x(t)\, dt, \qquad s \in [0, 1]$$

$K(s, t)$ is a function continuous on $0 \leqslant t, s \leqslant 1$. Obviously $y \in C[0, 1]$ as well, so that the operator T maps this space into itself. This is a linear, continuous operator: Let $\{x^n\} \to x$ in the sense of convergence in $C[0, 1]$, that is, uniformly on $[0, 1]$. Then we can take the limit under the integration sign:

$$\lim_{n \to \infty} \int_0^1 K(s, t)\, x^n(t)\, dt = \int_0^1 K(s, t)\, x(t)\, dt$$

The continuity of T follows.

ii. Let X be the space of real-valued differentiable function on $[0, 1]$ with norm

$$\|f\| = \max_{t \in [0,1]} |f(t)|$$

Let $T(f)(t) = (d/dt) f(t)$. The range of T is the space of bounded functions defined on $[0, 1]$ with sup norm. Consider the sequence $\{f^n\}$,

$f^n(t) = t^n$. Clearly, $\| f^n \| = 1$. However, $T(f^n)(t) = nt^{n-1}$, $\| T(f^n) \| = n = n \| f^n \|$. The linear transformation T is not bounded, since no constant K exists such that $\| T(x) \| \leqslant K \| x \|$ for all $x \in C[0, 1]$; the transformation T is not continuous.

We have already enough background to be able to undertake the task of defining and studying some properties of dynamical systems; this we proceed to do in the following section.

Exercises

*1. Prove that if $x^n \to x$ and $x^n \to \bar{x}$, then $x = \bar{x}$.

*2. Is it true that if $x^n \to x$, $\| x^n \| \to \| x \|$, x, $x^n \in X$? Establish a connection between your answer to this question and the concept of continuity.

*3. Let X be the space R^n. Show that convergence with respect to any norm in R^n is equivalent to convergence of each component of the vectors in R^n.

4. Give an example to show that convergence in $L_2[0, 1]$ does not imply pointwise convergence; that is, if x^n, $x \in L_2[0, 1]$ and $x^n \to x$, then it does not necessarily follow that $| x^n(t) - x(t) |$ tends to zero as n tend to infinity for all $t \in [0, 1]$.

5. Show that every finite-dimensional linear normed space is complete. Show where your reasoning cannot be extended to normed spaces which have a countable number of basis vectors.

*6. Let X be a Banach space, and let $\{ S_n \}$ be a sequence of closed spheres such that $S_{n+1} \subset S_n$, $n = 1, 2, ...$, and that the radius of the spheres tend to zero as n tends to infinity. Show that the intersection of all these spheres, $\bigcap_{n=1}^{\infty} S_n$, is nonempty and consists exactly of one point.

7. Give an example of the situation described in the previous exercise; let $X = C[0, 1]$.

8. Show that if a linear transformation mapping the normed space X into the normed space Y is continuous at $x^0 \in X$, then it is continuous at each $x \in X$.

9.3. Dynamical Systems

We proceed now to abstract the material in Sections 3.7 and 7.4, so as to define the mathematical structure known as a dynamical system. In some cases we shall not impose on these entities the burden of having as many nice properties as the differential systems of Section 3.7 because this would happen to be too restrictive; some interesting infinite-dimensional systems would not qualify then as dynamical systems.

The following are the constituents of a dynamical system:

i. A Banach space X, to be called the *state space*.

ii. A subset T of the space of the real numbers R. The elements of T will be referred as the values of the *time*.

iii. A space U of functions with domains consisting of the intersection of T and half-open intervals on R, and range on a Banach space V. The space U is the space of *inputs* to the dynamical system.

iv. A Banach space Y of outputs of the dynamical system.

v. A *transition function* with domain in the product space

$$T \times T \times X \times U$$

and with range in X, which defines the state $x(t; t_0, x^0; u)$ for any $t \in T$, any initial time $t_0 \in T$, any initial state $x^0 \in X$ and any input defined on $T \cap [t_0, t)$.

vi. An *output function* with domain in the product space $T \times X \times U$ which assigns to each $t \in T$, each $x \in X$ and each $u \in U$ an output $y(t, x, u)$; this function has range in the Banach space Y.

The transition function $x(\cdot, \cdot, \cdot, \cdot)$ of (v) above is to satisfy the following three axioms:

A_1. $x(t_0; t_0, x^0; u) = x^0$, for all $t_0 \in T$, $x^0 \in X$, $u \in U$.   (9.3)

A_2. Consider now three instants of time t_0, t_1, and t_2 such that $t_2 > t_1 > t_0$. If two inputs, u^1 and u^2, are defined on $T \cap [t_0, t_1)$ and $T \cap [t_1, t_2)$ respectively, define an input u as

$$u(t) = \begin{matrix} u^1(t), & t \in T \cap [t_0, t_1) \\ u^2(t), & t \in T \cap [t_1, t_2) \end{matrix}$$

Then the transition function x satisfies

$$x(t_2; t_1, x(t_1; t_0, x^0; u^1); u^2) = x(t_2; t_0, x^0; u)$$ (9.4)

A_3. The transition function must be a continuous function of each of its (four) variables for all fixed values of the other three; or, in our parlance, it is to be a continuous transformation from $T \to X$ for all $t_0 \in T$, all $x^0 \in X$, all $u \in U$; it is to be a continuous transformation from $T \to X$ for all $t \in T$, all $x^0 \in T$, all $u \in U$; it is to be a continuous transformation from $X \to X$ for all $t \in T$, all $t_0 \in T$, all $u \in U$, and, at last, it is to be a continuous transformation from $U \to X$ for all $t \in T$, $t_0 \in T$, $x^0 \in X$. Besides, x is to be a continuous transformation from $T \times T \times X \times U \to X$; the reader should remember that continuity in each variable as described above does not imply continuity in all the variables simultaneously.

As an illustration of these concepts, consider the transition function (3.61). Its continuity in t, t_0, and x^0 is readily determined; let us define a norm in the corresponding space U of inputs by means of the expression

$$\| u \| = \max_j \sup_{\tau \in [t_0, t)} | u_j(\tau)| \tag{9.5}$$

If $\{u^k\}$ tends to u in the norm (9.5), then

$$\lim_{k \to \infty} \int_{t_0}^{t} \phi(t, \lambda) B(\lambda) u^k(\lambda) \, d\lambda = \int_{t_0}^{t} \phi(t, \lambda) B(\lambda) u(\lambda) \, d\lambda$$

since convergence in the norm (9.5) implies uniform convergence of the components of u.

The transition function $x(\cdot, \cdot, \cdot, \cdot)$ considered as a transformation from $T \times T \times R^n \times U \to R^n$, is continuous. Define as a norm in the product space of elements (t, t_0, x^0, u) the following:

$$\|(t, t_0, x^0, u)\| = \max\{| t |, | t_0 |, \| x^0 \|, \| u \|\} \tag{9.6}$$

here the norm of x^0 is any norm in R^n and the norm of u is that in (9.5). It can be proved that (9.6) satisfies the norm axioms in $T \times T \times R^n \times U$. Suppose a sequence $\{(t, t_0, x^0, u)^k\}$ tends to an element (t, t_0, x^0, u) of the product space. Then, if $(t, t_0, x^0, u)^k = (t^k, t_0^k, x^{0k}, u^k)$, $x(t^k; t_0^k, x^{0k}; u^k)$ converges to $x(t; t_0, x^0; u)$; we show this for the zero-input part of the state expression, leaving the (long) complete proof to the reader. Clearly,

$$\| \Phi(t^k, t_0^k) x^{0k} - \Phi(t, t_0) x^0 \| \leqslant \| \Phi(t^k, t_0^k) x^{0k} - \Phi(t^k, t_0^k) x^0 \|$$

$$+ \| \Phi(t^k, t_0^k) x^0 - \Phi(t^k, t_0) x^0 \|$$

$$+ \| \Phi(t^k, t_0) x^0 - \Phi(t, t_0) x^0 \|$$

$$\leqslant \| \Phi(t^k, t_0^k) \| \| x^{0k} - x^0 \|$$

$$+ \| \Phi(t^k, t_0) \| \| \Phi(t_0, t_0^k) - I_n \| \| x^0 \|$$

$$+ \| \Phi(t^k, t_0) - \Phi(t, t_0) \| \| x^0 \|$$

This expression tends to zero, since $\| x^{0k} - x^0 \|$, $\| \Phi(t_0, t_0^k) - I_n \|$ and $\| \Phi(t^k, t_0) - \Phi(t, t_0) \|$ tend to zero as k tend to infinity; note that $\| \Phi(t^k, t_0^k) \|$ and $\| \Phi(t^k, t_0) \|$ are bounded uniformly for all sufficiently large values of k. The transformation $x(\cdot, \cdot, \cdot, \cdot)$ is continuous.

Similar continuity requirements are put in the general case on the function $y(\cdot, \cdot, \cdot)$ which defines the output as a function of the time,

the state and the input. For instance, this is to be a continuous transformation from T into Y for all $x \in X$ and $u \in U$. It is also required
to be a continuous transformation from $T \times X \times U \to Y$.

We specialize now our definition of dynamical systems to a class
of systems which represent better the subject matter of this book,
the *linear* dynamical systems. We say that a dynamical system is linear
if the transition function is linear in the tuple (x^0, u) for all t, t_0 in T
and the output function is linear in (x, u) for all $t \in T$. In other words,

$$x(t; t_0, \alpha x^{01} + \beta x^{02}; \alpha u^1 + \beta u^2) = \alpha x(t; t_0, x^{01}; u^1) + \beta x(t; t_0, x^{02}; u^2)$$
(9.7)

for all t, $t_0 \in T$, all x^{01}, $x^{02} \in X$, all u^1, $u^2 \in U$, all complex α and β;
and also,

$$y(t, \alpha x^1 + \beta x^2, \alpha u^1 + \beta u^2) = \alpha y(t, x^1, u^1) + \beta y(t, x^2, u^2) \qquad (9.8)$$

for all $t \in T$, all x^1, x^2 in X, all u^1, u^2 in U, all complex α and β.

There is an alternative formulation of linearity, which can be derived
easily from the previous definition. Make $\alpha = \beta = 1$, $u^1 = 0$, $u^2 = u$,
$x^{01} = x^0$, $x^{02} = 0$ in (9.7). Then,

$$x(t; t_0, x^0; u) = x(t; t_0, x^0; 0) + x(t; t_0, 0; u) \qquad (9.9)$$

Here $x(t; t_0, \cdot; 0)$ maps X into itself linearly for t, $t_0 \in T$, as required
by (9.7). The function $x(\cdot; t_0, x^0; 0)$ is known as the *state zero-input
response*. Likewise, $x(t; t_0, 0; \cdot)$ maps linearly the subset of U consisting
of those functions in U with domain $T \cap [t_0, t)$ into X for all t, $t_0 \in T$;
the function $x(\cdot; t_0; 0; u)$ is known as the *state zero-state response*.

In a similar way, we can show that

$$y(t, x, u) = y(t, x, 0) + y(t, 0, u) \qquad (9.10)$$

where the right-hand side members are the values of linear transformations mapping, for all $t \in T$, either X or a subset of U (which?)
into the space Y.

When the transition function (9.9) has been substituted for x in (9.10),
we see that, since $y(t, \cdot, 0)$ is a linear map, the resulting expression
for the output consists of a term which is linear in x^0 (the *zero-input*
response) and a term which is linear in u (the *zero-state* response).

We will discuss now some important implications of the requirement
(9.4) when applied to the transition function (9.9). Consider the input
to be zero throughout; our intention is to obtain a necessary condition
for the zero-input portion of (9.9). Clearly,

$$x(t_2; t_1, x(t_1; t_0, x^0; 0); 0) = x(t_2; t_0, x^0; 0)$$

for all $t_2 \geqslant t_1 \geqslant t_0$ in T. Since $x(t; t_0, \cdot; 0)$ is a linear operator mapping X into itself, we can write

$$x(t; t_0, x^0; 0) = G(t, t_0) x^0 \tag{9.11}$$

where $G(t, t_0)$ is, for all $t, t_0, t \geqslant t_0$, a linear operator mapping X into itself. Note that, contrary to our previous practice, we write $G(t, t_0) x^0$ rather than $G(t, t_0)(x^0)$. It follows that the family of operators $\{G(t, \lambda), t, \lambda \in T\}$ satisfies

$$G(t_2, t_1) \, G(t_1, t_0) = G(t_2, t_0) \tag{9.12}$$

for $t_2 \geqslant t_1 \geqslant t_0$.

The role of the operators $G(t, t_0)$ is analogous to that of the transition matrices $\phi(t, t_0)$ which were defined in connection with differential systems with finite-dimensional state spaces; the expression (9.12) bears a relationship with (3.40), the definition of the "group property" of the transition matrices. However, there is a crucial difference between the two requirements: (3.40) *does hold for all* t_0, t_1, and t_2 (which implies, among other things, that $\phi(t, \lambda)$ has an inverse, $\phi(\lambda, t)$), while (9.12) is required to hold only for $t_2 \geqslant t_1 \geqslant t_0$, so that in this more general case one cannot infer the existence of an inverse for the operator $G(t, \lambda)$. In other words while we want our linear dynamical systems to satisfy (9.12), we do not require that the previous states be determined when the present state is known; that is, with regard to the zero-input behavior of the state of our general linear dynamical systems, we do not require uniqueness of trajectory with backward direction of the time, since many previous states and paths may lead to a given state. It is not possible in general to go back in time along a zero-input trajectory in a unique way, one may find points at which there are intersections of many trajectories.

Of course, this situation was encountered already in connection with difference systems in Section 7.4; the trajectories of a difference system with singular matrix A have precisely this property. If we were to require from our general dynamical systems that (9.12) be satisfied for all t_2, t_1, t_0, then we would not be able to admit in this class difference systems with singular matrices A, as well as a wide variety of systems described by partial differential equations. Since we want to be able to consider these systems as being part of our class of dynamical systems, we do not require this, but simply that (9.12) be satisfied for $t_2 \geqslant t_1 \geqslant t_0$. On the other hand, this requirement is sufficient to give dynamical systems enough of a structure so that they have interesting and useful properties.

In what follows, we will study exclusively linear dynamical systems which are *time-invariant* or *constant*; in this case an elegant, intellectually satisfying and useful theory can be constructed, basically by considering the implications of the equality (9.12) and of some of our continuity requirements. We say that a linear dynamical system is *time-invariant* or *constant* if a shift of the time origin changes neither the values of the state or the output, provided that the input is shifted accordingly. Clearly the state zero-input response should be a function of $t - t_0$ only

$$x(t; t_0, x^0; 0) = x(t + \tau; t_0 + \tau; x^0; 0)$$

for all t, t_0, $\tau > 0$ such that t, t_0, $t_0 + \tau$, $t + \tau$ are in the basic set of definition T. We shall consider in the rest of this chapter that the set T is the half-line $[0, \infty)$.

Of course, this requirement implies that $G(t, t_0)$ in (9.11) is a function of $t - t_0$,

$$G(t, t_0) = T(t - t_0)$$

so that (9.12) becomes

$$T(t_2 - t_1)\, T(t_1 - t_0) = T(t_2 - t_0) \tag{9.13}$$

for $t_2 \geqslant t_1 \geqslant t_0$. If we put $t \equiv t_2 - t_1 \geqslant 0$, $s \equiv t_1 - t_0 \geqslant 0$, then $t_2 - t_0 = s + t \geqslant 0$, so that condition (9.13) can be written as

$$T(t)\, T(s) = T(t + s) \tag{9.14}$$

for all $t, s \geqslant 0$.

Of course, (9.3) implies that

$$T(0) = I \tag{9.15}$$

with I the identity operator in X. We have, then, two necessary conditions, (9.14) and (9.15), which are to be satisfied by the family of operators $\{T(t), t \geqslant 0\}$ which describe the state zero-input response of constant linear dynamical systems for $t \geqslant 0$. These operators map the state space X into itself, and have to satisfy several continuity conditions as well, to be discussed below. It is convenient beforehand to remark on the algebraic structure of this set of operators, $\{T(t), t \geqslant 0\}$. A binary, associative, single-valued operation can be defined on this set, so that to each pair of elements $T(t)$, $T(s)$ in it another unique element, $T(s) \circ T(t)$, is associated by means of the equality (9.14)

$$T(s) \circ T(t) = T(t + s) \tag{9.16}$$

This operation is clearly associative:

$$T(s) \circ (T(t) \circ T(\tau)) = (T(s) \circ T(t)) \circ T(\tau) \qquad (9.17)$$

We say that the set $\{T(t),\ t \geqslant 0\}$ is a *semigroup*. Note that, if every element in the set were to have an inverse, then this set would be a *group* [2], since a unit element certainly belongs to the set, namely $T(0) = I$, the identity operator. It should be clear that by not requiring that (9.4) apply for all t_0, t_1, and t_2, and thus not requiring that the operators $T(t)$, $t \geqslant 0$, be invertible, we have given to the set of all these operators a looser structure, that of a semigroup, than it would have otherwise. On the other hand, the corresponding set for a constant differential system with a finite-dimensional state space is a group, since in this case the transition matrices $T(t)$, $t \geqslant 0$, are invertible; thus the name of "group property" given to property (3.51) and, by an extension based on the theory of sets of operators defined by two parameters, to property (3.40) of nonconstant systems. We remark here that there are constant linear dynamical systems with infinite-dimensional state spaces with state zero-input responses which are defined by families of operators $\{T(t),\ t \geqslant 0\}$ which are *groups*, that is, for which all operators $T(t)$, $t \geqslant 0$, do have an inverse.

We must now put continuity conditions on the operators $\{T(t),\ t \geqslant 0\}$, conditions which are of course simple consequences of the continuity requirements put on general dynamical systems above.

Since the state zero-input response $x(t; t_0, x^0; 0) = T(t - t_0)\, x^0$, $t \geqslant t_0$, is to be continuous in x^0, then all operators $T(t)$, $t \geqslant 0$, must be continuous operators mapping X into itself, and thus bounded. We can phrase this conditions by defining $L(X, X)$, the space of all linear bounded operators mapping X into itself, and requiring that all operators $T(t)$, $t \geqslant 0$, be in $L(X, X)$.

Axiom A_3 also requires that, for all $x \in X$, the state zero-input response $T(t)x$ be continuous in t for all $t \in [0, \infty)$; that is,

$$\lim_{t \to t_0} \| T(t)x - T(t_0)x \| = 0 \qquad (9.18)$$

An operator-valued function of the time which satisfies (9.18) is said to be *strongly continuous at* t_0. We are led to define what is known as a *strongly continuous semigroup* of operators.

Definition. A family $\{T(t),\ t \geqslant 0\}$ of bounded linear operators in $L(X, X)$, X being in general a complex Banach space, is said to be a *strongly continuous semigroup* if:

    i. $T(t + s) = T(t)\, T(s)$, $s, t \geqslant 0$.

ii. $T(0) = I$, the identity operator in X.

iii. The function T is strongly continuous for all $t \geqslant 0$.

Some examples of strongly continuous semigroups of operators are shown below.

EXAMPLE

i. Let $X = C$, the set of all complex numbers with the usual norm, and let

$$T(t)z = e^{at}z$$

$z \in C$. The first two axioms in the definition above are obviously satisfied. The continuity of this operator-valued function can easily be checked:

$$T(t)z - T(t_0)z = (e^{at} - e^{at_0})z$$

the absolute value of the complex number in the right-hand side of this equality does tend to zero as $t \to t_0$.

ii. Let $X = C[0, \infty)$ be the space of all bounded continuous functions of a real variable z defined for $z \geqslant 0$. Define, for $x \in C[0, \infty)$,

$$\| x \| = \max_z | x(z)|$$

and define for $t \geqslant 0$ the linear operator $T(t)$ by

$$[T(t)x](z) = x(t + z)$$

Let us show first that this is a bounded operator:

$$\| T(t)x \| = \max_z |[T(t)x](z)| = \max_z | x(t + z)|$$

$$\leqslant \max_z | x(z)| = \| x \|$$

which shows that $T(t)$ is bounded, and thus continuous, for $t \geqslant 0$. Clearly,

$$[T(t + s)x](z) = x(t + s + z) = \{T(t)[T(s)x]\}(z)$$

therefore, $T(t + s) = T(t)\,T(s)$; the first condition in our definition of a strongly continuous semigroup is satisfied. The second condition is obviously satisfied; we prove now the third condition. Let $X \in C[0, \infty)$. Then,

$$[T(t)x - T(t_0)x](z) = x(t + z) - x(t_0 + z)$$

so that

$$\| T(t)x - T(t_0)x \| = \max_z | x(t + z) - x(t_0 + z)| \qquad (*)$$

$x(t + z) - x(t_0 + z)$ is a uniformly continuous function of t for $t \geqslant 0$ for all $z \geqslant 0$; hence, the norm in the equation $(*)$ tends to zero as t tends to t_0. We conclude that $\{T(t)\}, t \geqslant 0\}$ is a strongly continuous semigroup. This is an example of a *translation* semigroup.

iii. It can be proved that the solution of a *diffusion equation*

$$\frac{\partial}{\partial t} x(t, z) = \frac{1}{4} \frac{\partial^2}{\partial z^2} x(t, z)$$

$$x(0, z) = f(z)$$

is given by

$$x(t, z) = \frac{1}{\sqrt{(\pi t)}} \int_{-\infty}^{\infty} e^{-(z-\lambda)^2/t} f(\lambda) \, d\lambda = [T(t)f](z)$$

provided that we define $[T(0)f](z) = f(z)$. Here we choose as our space X the space $C(-\infty, \infty)$ of bounded continuous functions defined on the whole real line with norm

$$\|f\| = \max_z |f(z)|$$

Clearly each $T(t)$ is continuous, since

$$\| T(t)f \| \leqslant \|f\| \cdot \frac{1}{\sqrt{(\pi t)}} \int_{-\infty}^{\infty} e^{-(z-\lambda)^2/t} \, d\lambda = \|f\| \cdot \sqrt{2}$$

Properties i and iii of the semigroup definition are rather cumbersome to prove; since the proofs are readily available in the literature [3], we will omit them.

To summarize our findings, we can say that the state zero-input response of a constant linear dynamical system has to be determined by a strongly continuous semigroup of linear, bounded operators. In the next section we shall go deeply into some implications of this fact, and show how the theory of semigroups becomes highly relevant to the treatment of the solutions of homogeneous systems of the type $\dot{x} = Ax$; here $x \in X$, a Banach space, $\dot{x}$ is a derivative function, to be defined, and A is an operator mapping X into itself. Among other things, we shall obtain conditions on the operator A so that the solution of $\dot{x} = Ax$ is the state zero-input response of a linear constant dynamical system. This and other results will provide a framework in which we shall base our study of systems described by some simple partial differential equations.

Exercises

*1. Complete the proof that the transition function x of a finite-dimensional differential system in a continuous mapping from $T \times T \times R^n \times U$ into R^n.

2. Give an example of a semigroup of operators which is not strongly continuous, but otherwise satisfies all the conditions in the definition given in the text.

*3. We have asked that $\| T(t)x - T(t_0) \|$ tend to zero as t approaches t_0. Is this condition equivalent to asking that

$$\| T(t) - T(t_0) \| \to 0 \tag{*}$$

as $t \to t_0$? Note that the norm in (*) is the norm in $L(X, X)$.

9.4. Semigroups and Infinitesimal Generators

Before following the program set up at the end of the previous section, we must introduce still some more fresh mathematical concepts, this time related to the operations of differentiation and integration of Banach-space-valued functions of the time.

We define the *derivative* of a Banach-space-valued function x of the time which maps an interval in R into a complex Banach space X as another function, to be denoted by $\dot{x}$, which maps the same interval in R into X, such that

$$\dot{x}(t) = \lim_{h \to 0} \frac{1}{h} [x(t + h) - x(t)] \qquad (9.19)$$

provided this limit exists and belongs to the space X. This limit is to be taken in the sense of the norm of the Banach space X; that is, (9.19) implies that

$$\lim_{h \to 0} \left\| \dot{x}(t) - \frac{1}{h} [x(t + h) - x(t)] \right\| = 0 \qquad (9.20)$$

This limit is said to be a *strong limit* (there is another kind, the *weak limit*, which we shall not use; see [1]).

For example, let $X = C(-\infty, \infty)$, the space of functions of a real variable z which are continuous and bounded over the whole of the real line; let the interval of time be, as usual in this chapter, $[0, \infty)$. Then, if $x(t) \in X$, we can write its values as $x(t, z)$, and

$$\{\dot{x}(t)\}(z) = \lim_{h \to 0} \frac{1}{h} [x(t + h, z) - x(t, z)] = \frac{\partial x(t, z)}{\partial t}$$

provided the limit exists and the function $\partial x(t, z)/\partial t$ is, for all $t \in [0, \infty)$, in $C(-\infty, \infty)$, that is, is continuous and bounded. Note that not every function from $[0, \infty)$ into $C(-\infty, \infty)$ has a derivative in the sense described above; in order to have one, it must have a continuous bounded partial derivative with respect to z for all values of the time in $[0, \infty)$.

The introduction of this concept of derivative will allow us to write partial differential equations in an abstract, and thus general, manner. The diffussion equation

$$\frac{\partial x(t, z)}{\partial t} = \alpha \frac{\partial^2 x(t, z)}{\partial z^2}, \qquad x(0, z) = f(z) \qquad (9.21)$$

for instance, can be written as

$$\dot{x}(t) = Ax(t), \qquad t \in [0, \infty), \qquad x(0) = f \qquad (9.22)$$

or even as

$$\dot{x} = Ax, \qquad x(0) = f \tag{9.23}$$

here $X = C(-\infty, \infty)$, $\{\dot{x}(t)\}(z) = \partial x(t, z)/\partial t$ and A is an operator which maps a subset $D(A)$ of the Banach space X into X itself, and which is defined by

$$(Ax)(t, z) = \alpha \frac{\partial^2}{\partial z^2} x(t, z) \tag{9.24}$$

The domain of A, $D(A)$, is the subset of $C(-\infty, \infty)$ composed of all functions which have bounded continuous second derivatives with respect to z for all $t \in [0, \infty)$; note that (9.21) says that the second derivative of $x(t, z)$, whenever it exists, has to be a member of the space $C(-\infty, \infty)$. Of course, the objective of writing an equation such as (9.21) in the abstract manner (9.23) is to apply the theory of the solutions of such an abstract problem, to be developed below, to the study of the solutions of the diffussion equation (9.21).

The theory of the *integral* of a Banach-space-valued function x with domain an interval I of the real line is, in its more general forms, far beyond the scope of our study. We shall, therefore, present a simple theory which is sufficient to handle many problems of interest.

Let X be a Banach space, and $X: I \to X$, I an interval in R. We say that a continuous mapping y of $I \to X$ is a *primitive of x in I* if there exists a *denumerable* set $E \subset I$ such that for any $t \in I - E$, y is differentiable at t and $\dot{y}(t) = x(t)$. If x is such that the one-sided limits $x(t+)$ and $x(t-)$ exist and are finite for all $t \in I$ (such a function is said to be *regulated*; piecewise continuous functions mapping R into itself, such as step functions, square waves, etc., are regulated), we have the following theorem, whose proof we omit [4].

THEOREM 9.3. Any regulated function from an interval I of the reals to a Banach space X has a primitive in I.

It follows immediately from the definition that continuous functions of the time are regulated.

Let now t_1 and t_2 be any two points of an interval I of the reals, and x a regulated function from I to a Banach space X. It is simple to prove that if y_1, y_2 are two primitives of a function x (not necessarily regulated) from I into X, then $y_1 - y_2$ is constant in I. Then, if y_1 and y_2 are two primitives of our regulated function x,

$$y_1(t_2) - y_2(t_2) = y_1(t_1) - y_2(t_1), \qquad \text{for all } t_1, t_2 \in I$$

so that $y_1(t_2) - y_1(t_1) = y_2(t_2) - y_2(t_1)$; this difference is clearly independent of the particular primitive considered, and depends only on x; it will be called the *integral* of x between t_1 and t_2, and written as

$$\int_{t_1}^{t_2} x(t)\, dt \tag{9.25}$$

This integral of a Banach-space-valued function of a real variable can be shown to have many of the formal properties of the Riemann integral in what regards changes of variable, integration by parts, etc. There is an alternative approach to this integral [1], which follows closely the Riemann formulation for real-valued functions, and is based for its definition on the convergence of some associated series; the two approaches can be shown to be equivalent.

Our interest lies mainly in Banach spaces such as $C[0, \infty)$; there the integral (9.25) becomes a simple Riemann integral; indeed if $x \in C[0, \infty)$,

$$\left\{ \int_{t_1}^{t_2} x(t)\, dt \right\}(z) = \int_{t_1}^{t_2} x(t, z)\, dt \tag{9.26}$$

Note that the integral in (9.25) is a function with values in X, so that, in the special case of (9.26), it is to be expected that the integral turns out to be a function of z, the running variable used to define the values of functions in $C[0, \infty]$.

Several properties of the integral (9.25), necessary for some developments to follow, will be introduced as needed.

We are ready to develop the theory of semigroups of linear bounded operators. We define first a fundamental concept. Let $\{T(t),\ t \geqslant 0\}$ be a strongly continuous semigroup, and define the linear, bounded operator A_h by

$$A_h x = \frac{1}{h}\,[T(h)x - x], \qquad h > 0 \tag{9.27}$$

with $x \in X$, a complex Banach space; $T(t)$, $t \geqslant 0$, takes values on $L(X, X)$. Let $D(A)$ be the set of all $x \in X$ for which the (strong) limit

$$\lim_{h \to 0} A_h x$$

exists. Let us define an operator A with domain $D(A)$ by

$$Ax = \lim_{h \to 0} A_h x \tag{9.28}$$

This operator A is to be called the *infinitesimal generator* of the semigroup $\{T(t),\ t \geqslant 0\}$.

Before presenting some properties of A and $D(A)$, we give some examples.

EXAMPLE

i. Let $X = C$, the set of all complex numbers with the usual norm, and let

$$T(t)z = e^{at}z$$

$\{T(t), t \geqslant 0\}$ is a strongly continuous semigroup. Then,

$$A_h z = \frac{1}{h}[e^{ah}z - z]$$

so $Az = az$. Obviously, $D(A) = X = C$.

ii. Let $\{T(t), t \geqslant 0\}$ be the translation semigroup described above. Then,

$$(A_h x)(z) = \frac{1}{h}[x(z + h) - x(z)]$$

so that

$$(Ax)(z) = \frac{dx}{dz}(z)$$

The infinitesimal generator of the translation operator is a differentiation operator. The set $D(A)$ is that of all differentiable functions in $C[0, \infty)$.

Before proving some theorems concerning the operator A and the set $D(A)$, we must make a definition:

Definition. A *subspace* of a normed linear space is a *closed* linear subset of it.

The reader surely remembers that a subspace of a *finite-dimensional* linear space was defined in Chapter 2 without reference to its limit points; indeed, without reference to any norm. It happens that in this case every linear subset is closed under any norm.

We can prove now:

THEOREM 9.4. The set $D(A)$ is a subspace of X and A is linear on $D(A)$.

Proof. Clearly $D(A)$ is a linear subset of X, since if $A_h x^1$ and $A_h x^2$ converge strongly to Ax^1 and Ax^2, then $A_h(\alpha x^1 + \beta x^2)$, with α and β in the field associated with the Banach space X, converges strongly to $A(\alpha x^1 + \beta x^2)$, so that if x^1, x^2 are in $D(A)$, $\alpha x^1 + \beta x^2$ is also a member of this set. The proof that $D(A)$ is closed will be given when proving Theorem 9.7 below. Of course, A is linear on $D(A)$.

The following theorem is basic and represents the foundation of all results to be obtained in this chapter.

THEOREM 9.5. If x^0 is in $D(A)$, then $T(t) x^0$ is in $D(A)$ for $t \geqslant 0$, and

$$\frac{d}{dt} [T(t) x^0] = A[T(t) x^0]$$

Proof. Let $t \geqslant 0$, $h > 0$, and $x^0 \in D(A)$. We prove first that $T(t)$ and A_h commute. Indeed,

$$T(t) A_h x^0 = T(t) \frac{1}{h} [T(h) x^0 - x^0] = \frac{1}{h} [T(t) T(h) x^0 - T(t) x^0]$$

$$= \frac{1}{h} [T(t + h) x^0 - T(t) x^0] = \frac{1}{h} [T(h) T(t) x^0 - T(t) x^0]$$

$$= A_h T(t) x^0$$

It follows that $\lim_{h \to 0} A_h T(t) x^0 = \lim_{h \to 0} T(t) A_h x^0$; therefore $T(t) x^0 \in D(A)$, and $A(T(t) x^0) = \lim_{h \to 0} A_h T(t) x^0 = T(t) A x^0$. Besides,

$$\frac{1}{h} [T(t + h) x^0 - T(t) x^0] = T(t) A_h x^0$$

Letting h tend to zero, we obtain

$$\frac{d}{dt} [T(t) x^0] = T(t) A x^0 = A[T(t) x^0] \qquad (9.29)$$

The theorem follows.

It is clear from (9.29) that $T(t) x^0$ is a solution of $\dot{x} = Ax$ with initial condition $x(0) = x^0$, provided $x^0 \in D(A)$; remember that $T(0) = I$. This single result is basic; dynamical systems with state zero-input behavior described by a semigroup of operators can be seen to have this behavior described also by a homogeneous differential equation $\dot{x} = Ax$, at least for those initial conditions $x^0 \in D(A)$. This result is of extreme interest to us, since descriptions of systems are generally available in a differential form, such as the diffusion equation. Of course, if a description of a system such as $\dot{x} = Ax$ is available, we require A to be the infinitesimal generator of a semigroup; otherwise the system would not be dynamical in the sense of our definition. The following theorems provide necessary conditions which A is to satisfy if it is to be the infinitesimal generator of a semigroup; these conditions will then provide us with means of restricting the class of operators A to be considered.

We prove first a property of the domain $D(A)$; it turns out that this set consists not only of the zero element, but is somewhat larger.

THEOREM 9.6. Let A be the infinitesimal generator of a strongly continuous semigroup $\{T(t), t \geqslant 0\}$, with $T(t)$, $t \geqslant 0$, mapping a complex Banach space X into itself. Then the subspace $D(A)$ is dense in X.

Proof. A set $M \subset X$ is said to be *dense in* X if for every $\epsilon > 0$ and for every $x \in X$ there is an element $\tilde{x} \in M$ such that $\| x - \tilde{x} \| < \epsilon$. In other words, every element of X can be approximated as closely as desired by an element of M. For instance, the set of rational numbers is dense in the real line.

Let now $x \in X$. We prove that the limit

$$\lim_{h \to 0} A_h \int_0^t T(s)x \, ds$$

exists, so that $\int_0^t T(s)x \, ds$ is in $D(A)$. Let $t, h > 0$. Then

$$A_h \int_0^t T(s)x \, ds = \frac{1}{h} \int_0^t (T(h+s)x - T(s)x) \, ds$$

$$= \frac{1}{h} \int_h^{t+h} T(s)x \, ds - \frac{1}{h} \int_0^t T(s)x \, ds$$

$$= \frac{1}{h} \int_t^{t+h} T(s)x \, ds - \frac{1}{h} \int_0^h T(s)x \, ds$$

Consider the integral $(1/h) \int_0^{t+h} T(s)x \, ds$. By definition, this is equal to $(1/h)[y(t+h) - y(t)]$, where y is a primitive of $T(t)x$. (This function is continuous, and thus regulated, so that it has a primitive.) As h tends to zero, $(1/h)[y(t+h) - y(t)]$ tends to $\dot{y}(t) = T(t)x$, so that $(1/h) \int_t^{t+h} T(s)x \, ds$ tends to $T(t)x$; by the same argument, $(1/h) \int_0^h T(s)x \, ds$ tends to $T(0)x = x$, so that $A_h \int_0^t T(s)x \, ds$ tends to $T(t)x - x$ as h tends to zero. It follows that $\int_0^t T(s)x \, ds$ is in $D(A)$.

Since $D(A)$ is a subspace, $(1/t) \int_0^t T(s)x \, ds$ is also in it, for any $t > 0$ and any $x \in X$. However, $(1/t) \int_0^t T(s)x \, ds$ tends to x as t tends to zero; that is, $(1/t) \int_0^t T(s)x \, ds$ can be made to be as close to x as desired by making t small enough. It follows that any neighborhood of any $x \in X$ contains a member of $D(A)$, that is, $(1/t) \int_0^t T(s)x \, ds$ for sufficiently small t, and thus $D(A)$ is dense in X.

This theorem indicates that when considering systems such as $\dot{x} = Ax$, only operators A whose domain is dense in X should be allowed in the right-hand side. For instance, the operator $A = \alpha \, \partial^2/\partial z^2$ in (9.21) has as domain those functions in $C(-\infty, \infty)$ which have bounded continuous second derivatives; this set is dense in $C(-\infty, \infty)$; we conclude that this system may have a solution such as $T(t)x^0$, with $\{T(t), t \geqslant 0\}$, a strongly continuous semigroup. (Indeed, for $\alpha = \frac{1}{4}$, this semigroup is generated by the function $T(\cdot)$ defined by the integral operator with Gaussian kernel given in the previous section.)

Another interesting property, this time of the operator A itself, follows now. We say that a linear operator B whose domain $D(B)$ is a subspace of a Banach space X is *closed* if $x^n \in D(B)$, $x^n \to x$, $Bx^n \to y$, imply that $x \in D(B)$ and $Bx = y$.

It can be proved easily that a bounded linear operator whose domain is a subspace of a Banach space X (and thus is closed) is also closed. Indeed, if $x^n \to x$, $x^n \in D$, D the domain of a bounded transformation B, then x is in D since D is closed; if $Bx_n \to y$, $y = Bx$ since B is bounded and thus continuous.

Not every linear closed operator is bounded, though, as shown by the following example. Let $X = C[0, 1]$, and take B as a differentiation operator, $Bx = \dot{x}$; the domain D of B consists of all functions $x \in C[0, 1]$ for which $\dot{x} \in C[0, 1]$, that is, all functions in $C[0, 1]$ which have a continuous derivative. We proved above that this operator is not bounded. We prove now that it is closed. Let $\{x^n\}$ be a sequence of points in D such that $x^n \to x$ and $Bx^n \to y$; this last requirement means that $\dot{x}^n \to y$. Clearly $x \in D$, since the limit in $C[0, 1]$ implies, in function-theoretic language, that the sequence of functions x^n converges uniformly to the function x; if x^n, $n = 1, 2, \ldots$, have continuous derivatives, the limit also has continuous derivatives and is therefore in D. The sequence $\{Bx^n\}$ also converges uniformly to y, which is therefore the derivative of the original limit function. The operator B is closed.

We prove now:

THEOREM 9.7.   The infinitesimal generator A of a strongly continuous semigroup is closed.

Proof.   Let $x^n \in D(A)$, $n = 1, 2, \ldots$, $\lim_{n \to \infty} x^n = x^0$, $\lim_{n \to \infty} Ax^n = y^0$. We want to prove that $x^0 \in D(A)$ and $Ax^0 = y^0$. By integrating (9.29), we obtain that, if $x \in D(A)$,

$$[T(t) - T(s)]x = \int_s^t T(\lambda)\, Ax\, d\lambda \qquad (9.30)$$

so that, by putting $s = 0$ in (9.30) we get

$$T(t)\, x^0 - x^0 = \lim_{n \to \infty} T(t)\, x^n - x^n$$

$$= \lim_{n \to \infty} \int_0^t T(\lambda)\, Ax^n\, d\lambda$$

Since $\lim_{n\to\infty} T(\lambda)\, Ax^n = T(\lambda)y^0$ uniformly in $[0, t]$, we can write

$$\lim_{n\to\infty} \int_0^t T(\lambda)\, Ax^n\, d\lambda = \int_0^t T(\lambda)\, y^0\, d\lambda$$

it follows that

$$\lim_{t\to 0} A_t x^0 = \lim_{t\to 0} \frac{1}{t}\,[T(t)\,x^n - x^n]$$

$$= \lim_{t\to 0} \frac{1}{t} \int_0^t T(\lambda)\, y^0\, d\lambda = y^0$$

The limit $x^0 \in D(A)$, and $Ax^0 = y^0$. The theorem follows.

Only closed operators A should be considered in the equation $\dot{x} = Ax$ if we want the solution to have a chance of being generated by a semigroup. The operator $\alpha\, \partial^2/\partial z^2$ which appears in (9.21), for instance, qualifies on these grounds:

Let x^1, x^2, ... belong to $D(A)$. The norm to be used is, as before,

$$\| x \| = \max_z |\, x(z)| \tag{*}$$

Let $x^n \to x$ in the sense of the norm $(*)$; this implies uniform convergence of functions. It follows that $x \in D(A)$, that is, that x has a bounded continuous second derivative, and that $Ax^n \to Ax$, that is, that the sequence of functions consisting of the second derivatives of the functions of the original sequence converges to the second derivative of the function limit of the original sequence; the proof is similar to the one used above to show that the differentiation operator in $C[0, 1]$ is closed.

With this result we end for the moment our discussion of semigroups of operators and their infinitesimal generators. Our main conclusion is, simply, that if the homogeneous equation in Banach space $\dot{x} = Ax$ is to have solutions which describe the zero-input behavior of a constant, linear dynamical system, then the operator A is to be closed and its domain $D(A)$ is to be dense in the Banach space X. These are, of course, necessary conditions only. In the next section we introduce a useful concept, that of the resolvent operator.

9.5. Infinitesimal Generators and Resolvents

Let $X = C$, the space of all complex numbers, and let, for $z \in C$,

$$T(t)z = e^{at}z$$

We saw in the previous section that the infinitesimal generator A of the semigroup $\{T(t),\, t \geqslant 0\}$ is given by $Az = az$.

Define an operator $R(\lambda; A)$, to be called the *resolvent* of A, by

$$R(\lambda; A)z = \frac{1}{\lambda - a}\, z$$

where λ, a complex number, is not equal to a. The function $1/(\lambda - a)$ happens to be the Laplace transform of $e^{at} = T(t)$ and e^{at} is the inverse transform of $1/(\lambda - a)$; this suggests that it may be possible to obtain an explicit representation of $T(\cdot)$ in terms of its infinitesimal generator A by means of an inverse Laplace transform suitable defined. Of course, the actual computation of $T(t)$ can also be effected in this way; this approach will furnish us at the same time with an explicit representation for, as well as with means to compute, the values of the operators in $\{T(t),\ t \geq 0\}$.

The discussion at the beginning of this section suggests that we form the following integral:

$$R(\lambda; A)x = \int_0^\infty e^{-\lambda t} T(t)x\, dt \tag{9.31}$$

where λ is a complex variable, $x \in X$, a complex Banach space, and $\{T(t),\ t \geq 0\}$ is a strongly continuous semigroup $T(t) \in L(X, X)$, $t \geq 0$. Note that the integral operation in (9.31) assigns to each $x \in X$ another member of X, which we will call $R(\lambda; A)x$. It can be proved that [5]:

THEOREM 9.8.   The integral (9.31) converges for $\operatorname{Re} \lambda > \alpha$, regardless of $x \in X$, where α is

$$\alpha = \lim_{\omega \to \infty} \frac{1}{\omega} \log \| T(\omega) \| \tag{9.32}$$

We prove now

THEOREM 9.9.   The integral $R(\lambda; A)x$, for $\operatorname{Re} \lambda > \alpha$, satisfies

  i. $(\lambda I - A)\, R(\lambda; A)\, x = x$ for each $x \in X$;

 ii. $R(\lambda; A)(\lambda I - A)\, x = x$ for each $x \in D(A)$.

Proof.   (i) Let $x \in X$, $\operatorname{Re} \lambda > \alpha$. Then, by definition

$$
\begin{aligned}
A_h R(\lambda; A)x &= \frac{1}{h} \int_0^\infty e^{-\lambda t}[T(t + h)x - T(t)x]\, dt \\
&= \frac{1}{h} e^{\lambda h} \int_h^\infty e^{-\lambda t} T(t)\, dt - \frac{1}{h} \int_0^h e^{-\lambda t} T(t)x\, dt - \frac{1}{h} \int_h^\infty e^{-\lambda t} T(t)x\, dt \\
&= \frac{1}{h}(e^{\lambda h} - 1) \int_h^\infty e^{-\lambda t} T(t)x\, dt - \frac{1}{h} \int_0^h e^{-\lambda t} T(t)x\, dt
\end{aligned}
$$

Clearly, this expression converges to $\lambda R(\lambda; A)x - x$; see the proof of Theorem 9.6. Then $R(\lambda; A)\, x \in D(A)$, and, by definition

$$AR(\lambda; A)x = \lambda R(\lambda; A)x - x, \qquad \text{for all } x \in X$$

which is the equality in i.

(ii) If $x \in D(A)$,

$$R(\lambda; A)\, Ax = \int_0^\infty e^{-\lambda t} T(t)[Ax]\, dt$$

$$= \int_0^\infty e^{-\lambda t} \frac{d}{dt}\, [T(t)x]\, dt$$

$$= \lim_{\omega \to \infty} [e^{-\lambda t} T(t)x]_0^\omega + \int_0^\infty e^{-\lambda t} T(t)x\, dt$$

$$= -x + \lambda R(\lambda; A)x$$

which is the equality in ii. The theorem follows.

Theorem (9.9) indicates that $R(\lambda; A)$ is the inverse operator of $\lambda I - A$. It can be proved [5] that $R(\lambda; A)$ is a bounded operator whose domain is dense in X; it is said to be the *resolvent* of the operator A.

The resolvent provides sometimes a simple way of computing $T(t)$ given the infinitesimal generator A, by means of the following procedure:

i. Given A, compute $R(\lambda; A)x$, $x \in X$, by using, for instance, part i of Theorem 9.9.

ii. Since $R(\lambda; A)x$ is the Laplace transform of $T(t)x$, obtain $T(t)x$ by inverting $R(\lambda; A)x$. This step of course would require a rather deep study of direct and inverse Laplace transforms of Banach-space-valued functions. Instead of performing this study, we offer some examples which show the workings of the method, and demonstrate that this last step, of inverting $R(\lambda; A)x$, can easily be performed in some cases of interest to us.

EXAMPLE

Let X be $C[0, \infty)$, and let A be the derivative operator $\partial/\partial z$ with domain $D(A)$ of all those functions in $C[0, \infty)$ which have continuous, bounded derivatives on $[0, \infty)$. Let $x \in X$. Then, by Theorem 9.9i,

$$AR(\lambda; A)x = \lambda R(\lambda; A)x - x$$

Let us call $R(\lambda; A)x = y$, so that $Ay - \lambda y = -x$, that is,

$$\frac{\partial y}{\partial z} - \lambda y = -x \tag{9.33}$$

We solve now this simple equation by means of Laplace transforms. The initial condition $y(0)$ is to be taken as zero; the solution of the equation for $x = 0$ is $y(0) e^{\lambda z}$, which may not be bounded for $z > 0$ unless $y(0) = 0$; remember that we have to take $\mathrm{Re}\,\lambda > \alpha$, see (9.32), and then $\mathrm{Re}\,\lambda$ may be positive. The solution of (9.33) is

$$y(z) = [R(\lambda; A)x](z) = \frac{1}{2\pi i} \int_{C_1} \frac{\hat{x}(s)}{\lambda - s} e^{sz}\, ds \qquad (9.34)$$

where $\hat{x}$ is the Laplace transform of x and C_1 is an appropriate Bromwich path in the complex plane [6], that is, a straight line of equation $\mathrm{Re}\,s = \beta$, with the number β such that all the singularities of $\hat{x}(s)/(\lambda - s)$ are to the left of this line; of course, $\beta > \mathrm{Re}\,\lambda > \alpha$. We perform another inversion, this time with respect to λ, and obtain

$$[T(t)x](z) = \frac{1}{2\pi i} \int_{C_2} [R(\lambda; A)x](z)\, e^{\lambda t}\, d\lambda$$

$$= \frac{1}{2\pi i} \int_{C_2} \left[\frac{1}{2\pi i} \int_{C_1} \frac{\hat{x}(s)}{\lambda - s} e^{sz}\, ds \right] e^{\lambda t}\, d\lambda$$

$$= \frac{1}{2\pi i} \int_{C_1} \hat{x}(s) \left[\frac{1}{2\pi i} \int_{C_2} \frac{e^{sz + \lambda t}}{\lambda - s}\, d\lambda \right] ds$$

$$= \frac{1}{2\pi i} \int_{C_1} \hat{x}(s)\, e^{s(z + t)}\, ds = x(z + t)$$

since $x(z) = (1/2\pi i) \int_{C_1} \hat{x}(s)\, e^{sz}\, ds$. We have obtained a solution to $\partial x/\partial t = \partial x/\partial z$, $x(0, z) = x(z)$, as

$$[T(t)x](z) = x(s + z) \qquad (9.35)$$

The curve C_2 is a straight line of equation $\mathrm{Re}\,\lambda = \beta_1$, $\beta_1 > \mathrm{Re}\,s > \beta$; note that the only finite pole of the integrand

$$e^{sz + \lambda t}/\lambda - s$$

occurs at $\lambda = s$; s is constrained to be on the path C_1.

EXAMPLE

Consider again the diffusion equation

$$\frac{\partial}{\partial t} x(t, z) = \frac{1}{4} \frac{\partial^2}{\partial z^2} x(t, z)$$

Here the basic space is $C(-\infty, \infty)$. The equation for $y = R(\lambda; A)x$ is

$$\frac{1}{4} \frac{\partial^2 y}{\partial z^2} - \lambda y = -x \qquad (9.36)$$

If either $y(0)$ or $(dy/dz)(0)$ are not zero, the corresponding solutions of (9.35) for $x = 0$ are not bounded for all z, so that we take $y(0) = (dy/dz)(0) = 0$, and then

$$y(z) = \frac{1}{2\pi i} \int_{C_1} \frac{\hat{x}(s)}{\lambda - \frac{1}{4}s^2} e^{sz}\, ds$$

and

$$[T(t)x](z) = \frac{1}{2\pi i} \int_{C_2} y(z) \, e^{\lambda t} \, d\lambda$$

$$= \frac{1}{2\pi i} \int_{C_2} \left[\frac{1}{2\pi i} \int_{C_1} \frac{\hat{x}(s)}{\lambda - \frac{1}{4}s^2} e^{sz} \, ds \right] e^{\lambda t} \, d\lambda$$

$$= \frac{1}{2\pi i} \int_{C_1} \hat{x}(s) \left[\frac{1}{2\pi i} \int_{C_2} \frac{e^{sz+\lambda t}}{\lambda - \frac{1}{4}s^2} \, d\lambda \right] ds$$

$$= \frac{1}{2\pi i} \int_{C_1} \hat{x}(s) \, e^{sz+s^2 t/4} \, ds$$

$$= \frac{1}{2\pi i} \int_{C_1} \left[\int_0^\infty x(\tau) \, e^{-s\tau} \, d\tau \right] e^{sz+s^2 t/4} \, ds$$

$$= \int_0^\infty x(\tau) \left[\frac{1}{2\pi i} \int_{C_1} e^{-s\tau+sz+s^2 t/4} \, ds \right] d\tau \qquad (9.37)$$

Let us compute now

$$\frac{1}{2\pi i} \int_{C_1} e^{-s\tau+sz+s^2 t/4} \, ds = \frac{1}{2\pi i} \int_{C_1} e^{s(z-\tau)+ts^2/4} \, ds$$

$$= \frac{1}{\sqrt{(\pi t)}} e^{-(z-\tau)^2/t}$$

which is the same expression given without proof in the previous section. Here, again, C_1 and C_2 are appropriate contours in the corresponding complex planes.

This technique for solving homogeneous equations is sometimes successful, while in many cases the equations derived from Theorem 9.9, i, are as complicated as the original ones. The main value of this approach is theoretical; properties of the solutions of a homogeneous equation $\dot{x} = Ax$ can be obtained from the explicit expressions for these solutions obtained from the results of Theorem 9.9. We shall not pursue this path, which is rather beyond our reach; the reader is referred to [3], [5], and [7] for many interesting results.

Some new results concerning the homogeneous system $\dot{x} = Ax$ are given in the next section, followed by a treatment of nonhomogeneous systems.

Exercises

1. A system is described by the wave equation:

$$\frac{\partial^2 x(z, t)}{\partial t^2} = \frac{\partial^2 x}{\partial z^2} (z, t) \qquad (*)$$

$z \in (-\infty, \infty)$, $t \in [0, \infty)$

$$f(x, 0) = f_1(x), \qquad \frac{\partial f}{\partial x}(x, 0) = f_2(x)$$

a. Put this equation in the usual form $\dot{x} = Ax$. Be careful about the definition of your state space, the norm to be used in it, etc.

b. Show, by means of the method described in the text, that the solution of (*) is

$$f(x, t) = \tfrac{1}{2}[f_1(x + t) + f_1(x - t)] + \tfrac{1}{2} \int_{x-t}^{x+t} f_2(\xi)\, d\xi$$

2. A system is described by the equation:

$$\frac{\partial^2 x(z, t)}{\partial t^2} = \frac{\partial^2 x}{\partial z^2}(z, t) + \tfrac{1}{2}x(z, t)$$

with similar domains of definition and initial conditions as above. Solve for $x(z, t)$

9.6. Homogeneous and Nonhomogeneous Systems

The following theorem summarizes our findings with respect to the homogeneous system $\dot{x} = Ax$, and establishes the important fact that the solution $x(t) = T(t)\, x(0)$ is unique.

THEOREM 9.10. Let A be a closed operator with dense domain mapping a Banach space X into itself. Suppose that this operator A is the infinitesimal generator of a strongly continuous semigroup of operators, $\{T(t), t \geqslant 0\}$. Then the equation

$$\dot{x} = Ax, \qquad x(0) = x^0 \tag{9.38}$$

has as unique solution

$$x(t) = T(t)\, x^0, \qquad t \geqslant 0 \tag{9.39}$$

provided that $x^0 \in D(A)$. This solution has the property that

$$\| x(t) - x^0 \| \to 0$$

as t tends to zero.

Proof. Note that we have to require explicitly that A be the infinitesimal generator of a strongly continuous semigroup; closedness and density of its domain are just necessary conditions. A necessary and sufficient condition related to the concept of resolvent operator is given in [5].

We only have to show that the solution (9.39) is unique. Suppose that there are two solutions, x^1 and x^2, which satisfy (9.38) such that $x^1(0) = x^2(0) = x^0$. Then $\xi = x^1 - x^2$ satisfies

$$\dot{\xi} = A\xi, \qquad \xi(0) = 0 \tag{9.40}$$

We show that this implies that $\xi(t)$ is zero for all $t > 0$. Indeed,

$$\frac{d}{d\sigma}\left[T(t-\sigma)\,\xi(\sigma)\right] = -\frac{d}{d\sigma}\left[T(t-\sigma)\right]\xi(\sigma) + T(t-\sigma)\frac{d}{d\sigma}\xi(\sigma)$$

$$= -AT(t-\sigma)\,\xi(\sigma) + T(t-\sigma)\,A\xi(\sigma)$$

$$= 0, \qquad 0 < \sigma < t$$

hence,

$$T(t-t_1)\,\xi(t_1) = T(t-t_2)\,\xi(t_2) \tag{9.41}$$

for $0 < t_1 < t_2 < t$. Consider now the left-hand side of this equality, $T(t-t_1)\,\xi(t_1)$. As t_1 tends to zero, this function tends to zero as well; since $T(\cdot)$ is a continuous function and ξ is differentiable, thus continuous, (9.41) is valid for $t_1 = 0$, and then

$$T(t-t_2)\,\xi(t_2) = 0 \tag{9.42}$$

for $0 < t_2 < t$. Let us now make t approach t_2, so that

$$\xi(t_2) = 0, \qquad t_2 \geqslant 0 \tag{9.43}$$

and the theorem follows.

A more subtle result is obtained in [8], where it is shown that the conditions of Theorem (9.10) are necessary as well for uniqueness, that is, if there is a solution of (9.38) which is not generated by a semigroup, then there is at least one more solution which satisfies (9.38). This fact shows that the requirement that the solution of (9.38) be generated by a semigroup is crucial, not only in what respects the condition (9.10) on dynamical systems, but also at a more basic level; dynamical systems should have their zero-input behavior defined by a single-valued function $x(\cdot\,; t_0\,, x^0;\, 0)$.

We consider now nonhomogeneous systems, that is, systems with inputs such as the following *driven* diffusion equation

$$\frac{\partial}{\partial t}\,x(t, z) = \alpha\,\frac{\partial^2}{\partial z^2}\,x(t, z) + u(t, z) \tag{9.44}$$

This is an example of the class of systems which in Chapter 1 were referred as being of the *distributed control* type; here the state space will again be taken to be $C(-\infty, \infty)$; for each t, x and u are in this space. Note that we are not imposing boundary conditions, either constant or consisting of a boundary control, on the state; such systems, whose treatment is more complicated, will be considered in the next section.

It is possible to write (9.44) in the abstract, and thus more general, form

$$\dot{x}(t) = Ax(t) + u(t), \qquad t \in [0, \infty) \tag{9.45}$$

where $x(t)$ and $u(t)$ are in $C(-\infty, \infty)$ for $t \in [0, \infty)$, and where A is the operator defined by

$$(Ax)(t, z) = \frac{\partial^2}{\partial z^2} x(t, z)$$

with domain $D(A)$ equal to the subset of $C(-\infty, \infty)$ composed of all functions in this space which have bounded continuous second derivatives. Of course, (9.45) can be written as

$$\dot{x} = Ax + u \tag{9.46}$$

indicating equality of functions of time rather than equality of values.

The objective of this section is to study, with the help of the theory of semigroups developed previously, equations which are a slight generalization of the form of (9.46), that is,

$$\dot{x} = Ax + Bu \tag{9.47}$$

Here x takes values for $t \in [0, \infty)$ on a Banach space X, A is a linear constant operator (not necessarily bounded) with range and domain in X, u takes values in another Banach space V and B is a transformation mapping V into X. Our problem consists in obtaining a solution of (9.47) at $t \geqslant t_0$ for $x(t_0) = x^0$ and for u defined on $[t_0, t)$; the following theorem shows that we can obtain this solution under quite general conditions. Of course, we require that the state zero-input response of (9.47) satisfies the conditions of the previous sections, so that A is a constant, closed operator with dense domain which is the infinitesimal generator of a strongly continuous semigroup $\{T(t), t \geqslant 0\}$.

THEOREM 9.11. In equation (9.47), let: Bu by a regulated function of the time for $t \geqslant t_0$ for all inputs u; moreover, let $Bu \in D(A)$ for all $t \geqslant t_0$ and let $\| ABu \|$ have a finite integral for each finite interval in (t_0, ∞). Then, for every $t \geqslant t_0$ and for every $x^0 \in D(A)$, the unique solution of (9.47) is

$$x(t) = T(t - t_0) x^0 + \int_{t_0}^{t} T(t - \lambda) Bu(\lambda) \, d\lambda \tag{9.48}$$

Proof. Let us make first $t_0 = 0$, and compute

$$\frac{1}{h} [x(t + h) - x(t)] = \frac{1}{h} [T(t + h) x^0 - T(t) x^0]$$

$$+ \frac{1}{h} \int_0^{t+h} T(t + h - \lambda) Bu(\lambda) \, d\lambda - \frac{1}{h} \int_0^t T(t - \lambda) Bu(\lambda) \, d\lambda$$

$$= \frac{1}{h} [T(t + h) x^0 - T(t) x^0] + \frac{1}{h} \int_t^{t+h} T(t + h - \lambda) Bu(\lambda) \, d\lambda$$

$$+ \int_0^t T(t - \lambda) \frac{1}{h} [T(h) - I] Bu(\lambda) \, d\lambda \qquad (9.49)$$

The left-hand side in (9.49) tends to $\dot{x}(t)$ as h tends to zero. The first term in the right-hand side of (9.49) tends, as shown above, to $AT(t) x^0$; the second term tends to $Bu(t)$. The proof that the third term converges to

$$\int_0^t T(t - \lambda) ABu(\lambda) \, d\lambda \qquad (9.50)$$

is not easy, and we omit it [9]; it is here where the assumption of finiteness of the integral of $\| ABu \|$ over every finite interval in (t_0, ∞) is used. Since A is closed and commutes with $T(t - \lambda)$, (9.50) equals

$$A \int_0^t T(t - \lambda) Bu(\lambda) \, d\lambda$$

so that (9.49) actually implies

$$\dot{x}(t) = AT(t) x^0 + Bu(t) + A \int_0^t T(t - \lambda) Bu(\lambda) \, d\lambda$$
$$= Ax(t) + Bu(t)$$

The expression (9.48) is seen to be a solution of (9.47) when $t_0 = 0$. If $t_0 \neq 0$, the proof is identical. The uniqueness of this solution is a trivial consequence of the fact that the solution of $\dot{z} = Az$, $z(0) = 0$, is zero for all t.

An example of the application of this theorem is provided below.

EXAMPLE

Consider the equation (9.44). The solution of this equation is given by (9.48) with $t_0 = 0$ if the function of two variables $u(t, z)$ satisfies the following conditions:

i. For all $t \geqslant 0$ it has bounded continuous second derivatives with respect to z for all z.

ii. u is a regulated function of the time $t \geqslant 0$, that is, for each t, z the limits $u(z, t^+)$ and $u(z, t^-)$ exist.

iii. The integral of

$$\| Au(t) \| = \max_z \left| \frac{\partial^2}{\partial z^2} u(t, z) \right|$$

exists and is finite over every subinterval of $(0, \infty)$.

For instance, an admissible input u is defined by $u(t, z) = \phi(z) v(t)$, with ϕ a function of z with bounded continuous second derivatives, and v a piecewise continuous function of the time which has finite jumps at the discontinuities.

We leave for the reader the task of proving that (9.48), together with a suitably defined output equation, describe indeed a constant, linear dynamical system. It is interesting to point out the differences between our treatments of the homogeneous and nonhomogeneous systems. We derived essentially all the properties of (and restrictions on) the system $\dot{x} = Ax$ from our definition of a constant, linear dynamical system, through the intermediate step of defining a strongly continuous semigroup of operators. Once this was achieved, we modified the right-hand side of $\dot{x} = Ax$ by adding the term Bu, and thus transforming the homogeneous system into a nonhomogeneous one whose properties we studied; it is, under the conditions of Theorem 9.11, a linear constant dynamical system. However, we have not derived properties of and restrictions on this system, other than those associated with the homogeneous system $\dot{x} = Ax$, from the fundamental definitions and axioms; this can indeed be done, and the reader wishing to pursue these matters should consult the papers by A. V. Balakrishnan referred to in the section Supplementary Notes and References at the end of the present chapter. See also [9, 10].

Exercises

1. Show in detail that the system with transition function (9.48), together with a suitable defined output equation, satisfies *all* the axioms of a constant, linear dynamical system.

2. Under what conditions is the transition function (9.48) uniformly continuous in t when the other variables are held constant?

9.7. Systems with Boundary Controls

We present now a technique to study some dynamical system described by simple partial differential equations together with initial and boundary conditions. The basic method will be explained through an example. Consider the diffusion equation

$$\frac{\partial}{\partial t} x(t, z) = \frac{\partial^2}{\partial z^2} x(t, z) + u(t, z) \tag{9.51}$$

together with the initial condition

$$x(z, 0) = f(z), \qquad 0 \leqslant z \leqslant L \tag{9.52}$$

and the *boundary conditions*

$$x(0, t) = h(t), \qquad x(L, t) = g(t) \tag{9.53}$$

Here h and g are differentiable functions of the time and z takes values in the closed interval $[0, L]$. Of course,

$$x(0, 0) = f(0) = h(0), \qquad x(L, 0) = f(L) = g(0) \tag{9.54}$$

In the form in which this problem has been posed, we cannot apply the concepts developed in the previous sections; the zero-input solutions of (9.51) which satisfy (9.52) and (9.53) for different initial functions f are not a linear subset of a linear space. However, it is possible to transform the problem so than we can apply those concepts and techniques. Indeed, let $x(z, t)$ be a solution of (9.51), (9.52), and (9.53), and define

$$\bar{x}(z, t) = x(z, t) - \frac{z}{L} g(t) + \frac{1}{L} (L - z) h(t) \tag{9.55}$$

then

$$\bar{x}(0, t) = 0, \qquad \bar{x}(L, t) = 0 \tag{9.56}$$

while

$$\bar{x}(z, 0) = x(z, 0) - \frac{z}{L} g(0) + \frac{1}{L} (L - z) h(0) \tag{9.57}$$

and

$$\frac{\partial}{\partial t} x(z, t) = \frac{\partial}{\partial t} \bar{x}(z, t) + \frac{z}{L} \dot{g}(t) - \frac{1}{L} (L - z) \dot{h}(t)$$

$$\frac{\partial^2}{\partial z^2} x(z, t) = \frac{\partial^2}{\partial z^2} \bar{x}(z, t) \tag{9.58}$$

so that, at last,

$$\frac{\partial}{\partial t} \bar{x}(z, t) = \frac{\partial^2}{\partial z^2} \bar{x}(z, t) - \frac{z}{L} \dot{g}(t) + \frac{1}{L} (L - z) h(t) + u(t, z) \tag{9.59}$$

The original problem is equivalent to a new problem with zero boundary conditions and some extra input functions. What has been gained? Simply that the set of real-valued, continuous functions defined on $[0, L]$ which vanish at $z = 0$ and $z = L$ is a linear space, over the real numbers for instance, which becomes of course the state space X of the system (9.59). Here $D(A)$ is the set of all functions defined on $[0, L]$ which vanish at $z = 0$ and $z = L$ and which have continuous second derivatives on this interval. The solution of (9.59) subject to (9.56) and (9.57) can now proceed as in the previous sections, provided that we can obtain an explicit expression for the semigroup $\{T(t),\, t \geqslant 0\}$. This is a quite simple task; we shall use x rather than $\bar{x}$, so that our problem consists in solving

$$\frac{\partial}{\partial t} x(t, z) = \frac{\partial^2}{\partial z^2} x(t, z)$$

$$x(0, z) = f(z), \qquad x(t, 0) = 0, \qquad x(t, L) = 0 \tag{9.60}$$

We shall look for a solution of (9.60) in the form of a series,

$$x(t, z) = \sum_{k=1}^{\infty} F_k(t)\, G_k(z) \tag{9.61}$$

each of whose terms satisfies the differential equation in (9.60) and vanishes at $z = 0$ and $z = L$. Therefore, for some constants λ_k^2,

$$\frac{\dot{F}_k(t)}{F_k(t)} = \frac{G_k''(z)}{G_k(z)} = -\lambda_k^2 \tag{9.62}$$

The overdot indicates, as usual, differentiation with respect to the time, while the prime $(')$ denotes differentiation with respect to the variable z. In order to satisfy the boundary conditions on $G_k(z)$, $k = 1, 2, ..., \lambda_k$ must equal $k\pi/L$:

$$F_k(t) = A_k \exp\left\{ \frac{-k^2\pi^2}{L^2}\, t \right\} \tag{9.63}$$

$$G_k(z) = B_k \sin \frac{k\pi}{L}\, z \tag{9.64}$$

and

$$x(t, z) = \sum_{k=1}^{\infty} C_k \exp\left\{ \frac{-k^2\pi^2}{L^2}\, t \right\} \sin \frac{k\pi}{L}\, z \tag{9.65}$$

We determine the constants C_k by means of the initial condition

$$x(0, z) = f(z) = \sum_{k=1}^{\infty} C_k \sin \frac{k\pi}{L} z \tag{9.66}$$

so that

$$C_k = \int_0^{L} f(z) \sin \frac{k\pi}{L} z \, dz \tag{9.67}$$

or, at last,

$$x(t, z) = [T(t)f](z)$$

$$= \sum_{k=1}^{\infty} \left[\int_0^{L} f(z) \sin \frac{k\pi}{L} z \, dz \right] \exp\left[\frac{-k^2\pi^2}{L^2} t \right] \sin \frac{k\pi}{L} z \tag{9.68}$$

Note that the semigroup property $T(t + s) = T(t) T(s)$, $t, s \geqslant 0$, of the family of operators defined in (9.68) follows trivially from the exponential nature of the functions F_k and G_k, $k = 1, 2, \ldots$.

Many problems with simple boundary conditions can be treated by the technique developed in this section. The reader will find a different approach to this class of problems in [10].

9.8. Hilbert Spaces

In order to study the controllability of systems such as (9.47) we will extend some concepts first encountered in Section 2.9, those of inner product spaces and adjoint transformations defined on them.

Let X be a linear space over the complex numbers. We say that a function from $X \to C$, with values $(x, y) \in C$, $x, y \in X$, is an *inner product* if

   i. $(x, y) = \overline{(y, x)}$, $x, y \in X$;

   ii. $(x + y, z) = (x, z) + (y, z)$, $x, y, z \in X$;

   iii. $(\lambda x, y) = \lambda(x, y)$, $\lambda \in C$, $x, y \in X$;   (9.69)

   iv. $(x, x) \geqslant 0$,   with $(x, x) = 0$ if and only if $x = 0$.

If X is a linear space over R, the definition of an inner product proceeds similarly, with only some trivial changes. Of course, the inner products defined in Section 2.9 for C^n and R^n satisfy the corresponding axioms. We show below some examples of inner products in other spaces. A space in which an inner product is defined becomes an *inner product space*.

EXAMPLE

i.   The space $L_2[0, 1]$, defined in Section 2.8 as the space of real-valued functions defined on $[0, 1]$ such that they have an integrable square, becomes a (real) inner product space by defining for all $f, g \in L_2[0, 1]$,

$$(f, g) = \int_0^1 f(t)\, g(t)\, dt$$

Note that the expression $(f, f)^{1/2}$ is the norm (2.39) in $L_2[0, 1]$.

ii.   Let X be the space of *complex valued*, continuous functions defined on $[a, b]$. We define the inner product here as

$$(f, g) = \int_a^b f(t)\, \bar{g}(t)\, dt, \qquad f, g \in X \tag{9.70}$$

iii.   Let X be the space of sequences of complex numbers $(\alpha_1, \alpha_2, ..., \alpha_n, ...)$, such that

$$\sum_{n=1}^{\infty} |\alpha_n|^2 < \infty \tag{*}$$

Then, we define as inner product of $x = (\alpha_1, ..., \alpha_n, ...)$ and $y = (\beta_1, ..., \beta_n, ...)$,

$$(x, y) = \sum_{n=1}^{\infty} \alpha_n \bar{\beta}_n$$

This expression can be shown to converge because of the inequality $(*)$ above.

As in Section 2.9, we say that $(x, x)^{1/2}$ is, for any $x \in X$, the norm *generated* by the inner product (x, y), $x, y \in X$. This norm can be shown quite readily to satisfy the definition (2.31) of a norm by using the definition (9.69) of an inner product.

Every inner product space can be, therefore, normed by means of the norm generated by the inner product. This space, considered now as a normed space, will or will not be complete. If it is complete, we give it a special name.

Definition.   An inner product space which is complete with respect to the norm generated by the inner product is said to be a *Hilbert space*.

EXAMPLE

The space $L_2[0, 1]$ is complete, as mentioned in Section 9.2, and thus a Hilbert space. The space of complex valued, continuous functions defined on $[a, b]$ is not complete when considered as a normed space with the norm generated by the inner product (9.70). A fundamental sequence in this space will not, in general, converge to a continuous function.

In the following section, the state space will be considered to be a Hilbert space; in the matter of controllability we can get rather sharper results by the use of the inner product than if we had limited ourselves to state spaces which are merely Banach spaces. We go deeper now into the subject of *adjoint transformations* in Hilbert spaces.

As in Section 9.3 let $L(X, Y)$ be the set of bounded (and thus continuous), linear transformations mapping X, a Hilbert space, into Y, also a Hilbert space. Let H be in $L(X, Y)$; note that we are assuming that the domain of H is all of X. (As a matter of fact, the theory below can be generalized to the case when D_H is only dense in X). We shall write Hx instead of $H(x)$ as in most of the book. Define

$$D' = \{y \in Y : \text{there is } z \in X \text{ such that } (Hx, y) = (x, z) \text{ for all } x \in X\} \quad (9.71)$$

The set D' consists of those elements in Y for which an element $z \in X$ exists such that $(Hx, y) = (x, z)$. We will show that this set is nonempty and that, indeed, it equals the whole space Y. With this in mind, we will first define a special class of functions, the *bounded linear functionals*, then we will prove an important property of these functionals, the Riesz representation theorem.

Definition. A linear function $f: X \to C$ mapping a complex Hilbert space into the complex numbers is known as a *linear functional*. If this function is bounded (and thus continuous) in the sense of Definition 9.2, then we say that it is a *bounded* linear functional.

Our main interest in these functions lies in the following important theorem.

THEOREM 9.12 (Riesz representation theorem). Let $f: X \to C$ be a bounded linear functional on a Hilbert space H. Then there is a unique element $z \in X$ such that $f(x) = (x, z)$ for all $x \in X$.

Proof. Consider the null space of f, $N = \{x \in X, f(x) = 0\}$. It can be proved, and is proposed as an exercise at the end of this section, that, if N is not all of X, then there is a nonzero element w in X such that $(w, x) = 0$ for all $x \in N$. (Of course, if $N = X$, $f(x) = 0 = (x, 0)$ for all $x \in X$, so that the vector z is simply the zero vector in X.) We show now that there is an scalar $\alpha \in C$ such that $f(x) = (x, \alpha w)$ for all $x \in X$, that is, such that $z = \alpha w$.

Assume first that x is a multiple of w, that is, $x = \beta w$, $\beta \in C$. If we put

$$\alpha = \overline{f(w)}/\| w \|^2 \quad (9.72)$$

then

$$(x, z) = (\beta w, \alpha w) = \beta \bar{\alpha} \| w \|^2 = \beta f(w) = f(x)$$

so that at least for those elements of X which are multiples of w the vector αw with α defined in (9.72) satisfies the contention of the theorem. We show that this is true *for all* $x \in X$. Let $x = x - \gamma w + \gamma w$, with $\gamma = f(x)/f(w)$. Since $f(x - \gamma w) = f(x) - \gamma f(w) = 0$, $x - \gamma w \in N$, and since $(y, w) = 0$ for all $y \in N$, $(x - \gamma w, \alpha w) = 0 = f(x - \gamma w)$. Besides, γw is a multiple of w so that $f(\gamma w) = (\gamma w, \alpha w)$; at last, since f is linear, $f(x) = f(x - \gamma w) + f(\gamma w) = (x, \alpha w)$, $x \in X$. The uniqueness of $z = \alpha w$ follows from the fact that if $f(x) = (x, \alpha w) = (x, y)$ for all $x \in X$, $(x, \alpha w - y) = 0$ for all $x \in X$, which implies $\alpha w - y = 0$. The theorem follows.

It is a simple matter now to show that the set D' in (9.71) is all of Y.

THEOREM 9.13.   Let $H \in L(X, Y)$, X and Y being complex Hilbert spaces. Then for every $y \in Y$ there is a unique $z \in X$ such that $(Hx, y) = (x, z)$.

Proof.   Start by defining a function $f_y : X \to C$ by the equality

$$f_y(x) = (Hx, y) \tag{9.73}$$

for all $x \in X$, $y \in Y$. For fixed $y \in Y$, this is a linear functional. We show it is a bounded linear functional. Since H is bounded, there is a constant K such that $\| Hx \| \leqslant K \| x \|$ for all $x \in X$. Thus,

$$| f_y(x) | = |(Hx, y)| \leqslant \| Hx \| \| y \| \leqslant K \| y \| \| x \|$$

This shows that, according to the definition (9.2) of boundedness, f_y is bounded, and thus continuous. Note that the same symbol is being used for norms of elements of X and Y. These norms are, of course, the norms generated by the corresponding inner products. Note as well that in the derivation above we have used the Cauchy–Schwarz inequality, which can be proved for general inner product spaces just as we proved it for R^n and C^n in Theorem 2.13.

The theorem now follows trivially, since, by the Riesz representation theorem, a unique vector z, depending on y, exists such that

$$f_y(x) = (Hx, y) = (x, z) \tag{9.74}$$

for all $x \in X$.

We are at last in a position to define the adjoint transformation of H. We mentioned in the proof of the above theorem that the (unique) vector z in (9.74) does depend in general on the vector y. We define the transformation H', the *adjoint* of H, mapping Y into X, by means of the relation

$$H'(y) = z \qquad (9.75)$$

That is, $H': Y \to X$ is defined by $(Hx, y) = (x, H'y)$ for all $x \in X$, $y \in Y$. This transformation is obviously well defined; suppose that elements z^1, z^2 exist such that $(Hx, y) = (x, z^1) = (x, z^2)$ for all $x \in X$. Then $(x, z^1 - z^2) = 0$ for all $x \in X$, which implies $z^1 - z^2 = 0$.

It is quite simple to show that H' is a linear transformation; we leave this proof to the reader. It is not, however, so simple to show that H' is a bounded and thus continuous, transformation [1]. We shall not present the proof here, since we shall not need this property in our treatment of controllability.

EXAMPLE

Consider the operator H mapping $L_2[0, 1]$ into itself, defined by

$$f(t) = (Hx)(t) = \int_0^1 K(t, s) \, x(s) \, ds$$

that is, $f = Hx$, $f, x \in L_2[0, 1]$. The function $K(\cdot, \cdot)$ is continuous. For any $y \in L_2[0, 1]$,

$$(f, y) = (Hx, y) = \int_0^1 f(t) \, y(t) \, dt$$

$$= \int_0^1 y(t) \left(\int_0^1 K(t, s) \, x(s) \, ds \right) dt$$

$$= \int_0^1 x(s) \left(\int_0^1 K(t, s) \, y(t) \, dt \right) ds$$

$$= \int_0^1 x(s) \, g(s) \, ds = (x, g) = (x, H'y)$$

with

$$g(s) = \int_0^1 K(t, s) \, y(t) \, dt = (H'y)(s), \qquad y \in L_2[0, 1]$$

We have succeeded in characterizing the adjoint operator of H.

Exercises

*1. Let X be a complex inner product space. Show from the definition of inner product that

    i. (x, x) is real for all $x \in X$.

    ii. $(x, \alpha y) = \bar{\alpha}(x, y)$, $x, y \in X$, $\alpha \in C$.

*2. Show that the norm generated by the inner product in an inner product space X does satisfy the definition (2.31).

3. Show that, if X is a *real* inner product space,

$$(x, y) = \tfrac{1}{4} \| x + y \|^2 - \tfrac{1}{4} \| x - y \|^2, \qquad x, y \in X$$

Derive a similar expression when X is a complex inner product space.

*4. Show in detail that, if $y \in X$, an inner product space, and $(y, x) = 0$ for all $x \in X$, then $y = 0$.

*5. Let $f: X \to C$ be a linear functional on X, a Hilbert space. If $N = \{x \in X, f(x) = 0\}$ is not all of X, show that there is a nonzero element $w \in X$ such that $(w, x) = 0$ for all $x \in N$.

*6. Show that the adjoint operator H' is linear.

9.9. Controllability

We examine now the controllability of systems with infinite dimensional state spaces. In the finite dimensional case, we defined in Chapter 5 a constant system to be controllable if, essentially, an admissible control can be associated with each element x of the state space so that the state can be brought, by the action of this control, from the value 0 at $t = 0$ to the value x in finite time. This definition, while perfectly adequate for finite dimensional systems, is too restrictive when used in the present, wider setting; most infinite dimensional systems are not, according to this definition, controllable. However, many of these systems do have a weaker, but still significant, property; given an arbitrary element x in the state space, admissible controls can be found which bring the state as close as desired to the state x in finite time. This new concept, to be called *weak controllability*, will be investigated in this section.

Consider again the system

$$\dot{x} = Ax + Bu \qquad (9.76)$$

where X, the state space, is now a Hilbert space, A is a linear constant operator with range and domain in X, u takes values in another Hilbert space V, and B is a transformation mapping V into X. We assume that all the conditions of Theorem 9.11 are satisfied, so that, for instance, A is the infinitesimal generator of a strongly continuous semigroup $\{T(t),\ t \geqslant 0\}$, and the zero-state response of (9.76) is defined by

$$x(t; 0, 0; u) = \int_0^t T(t - \lambda)\, Bu(\lambda)\, d\lambda, \qquad t \geqslant 0 \qquad (9.77)$$

We assume that the controls u belong to a class U of admissible controls. For instance, if we were trying to control the temperature along a bar by varying the input heat rate at one end, we may require that the admissible functions representing these controls are square-integrable over any finite interval of time.

We are in a position to define the property of weak controllability.

Definition. The system (9.76) is said to be *weakly controllable at time* $t \geqslant 0$ if, given $\epsilon > 0$ and $x \in X$, an admissible control $u \in U$ can be found such that

$$\| x(t; 0, 0; u) - x \| < \epsilon \qquad (9.78)$$

The norm here is of course the one generated by the inner product in X.

Definition. The system (9.76) is said to be *weakly controllable* if it is weakly controllable at some finite time $t \geqslant 0$.

Note that a controllable finite dimensional system is also weakly controllable. We state and partially prove the main theorem on weak controllability. With this purpose, we write (9.77) as

$$x(t) = H_t u \qquad (9.79)$$

with $u \in U$ and $x(t) \in X$.

THEOREM 9.14. The system (9.76) is weakly controllable at $t \geqslant 0$ if and only if $H_t'x = 0$ implies that $x = 0$.

Proof. Note that the theorem says, in fact, that the system (9.76) is weakly controllable at $t \geqslant 0$ if and only if the null space of H_t' equals the zero vector in X.

We shall prove only the necessity of the contention of the theorem: the proof of sufficiency is rather beyond our reach (See [5, p. 479], where this proof is presented in a different setting). Assume then that (9.76) is weakly controllable at t. Given $x \in X$ and $\epsilon > 0$, there is therefore $u \in U$ such that $\| H_t u - x \| < \epsilon$. Suppose now that $H_t'x = 0$. We show that $x = 0$. Indeed, for any $u \in U$,

$$0 = (u, H_t'x) = (H_t u, x)$$

Thus

$$\epsilon^2 > \| H_t u - x \|^2 = (H_t u - x, H_t u - x) = \| H_t u \|^2 + \| x \|^2$$

Therefore, $\| x \| < \epsilon$ for any arbitrary $\epsilon > 0$, and then $\| x \| = 0$, $x = 0$. Necessity follows.

EXAMPLE [11]

Consider again the difussion equation

$$\partial x(t, z)/\partial t = \partial^2 x(t, z)/\partial z^2 \tag{9.80}$$

together with the initial condition

$$x(0, z) = 0, \qquad 0 \leqslant z \leqslant 1$$

and boundary conditions

$$\partial x(t, z)/\partial z = 0 \qquad \text{for} \quad z = 0$$
$$\partial x(t, z)/\partial z = u(t) \qquad \text{for} \quad z = 1$$

This system can be interpreted as the dynamical description of a bar of unit length insulated at one end ($z = 0$) and heated by a heat source of strength $u(t)$, $t \geqslant 0$, at the other end ($z = 1$). Initially, at $t = 0$, the bar has temperature zero. Is it possible to reach any temperature pattern $f(z)$, $0 \leqslant z \leqslant 1$, at a finite time, by using a control u in some appropriate manner?

Of course, the problem is not yet well posed, since we have specified neither the state space nor the class U of admissible controls. We shall take as state space $L_2[0, 1]$; we shall fix t at an arbitrary positive value, take as U the space $L_2[0, t]$, and investigate the weak controllability of (9.80) at time t. If this system is weakly controllable at t, it should be possible to find a control $u \in L_2[0, t]$ so that any function in $L_2[0, 1]$ is approximated arbitrarily closely by the state function of (9.80) at time t resulting from the application of the control u.

We obtain first the explicit solution (9.79) for the system (9.80). This can be done just as in Section 9.7; the result is

$$x(t, z) = \int_0^t h(z, t - \lambda) u(\lambda) \, d\lambda \tag{9.81}$$

with

$$h(z, t) = 1 + 2 \sum_{k=1}^{\infty} \cos k\pi z \exp(-k^2\pi^2 t) \tag{9.82}$$

The adjoint of the transformation H_t defined by (9.81) and (9.82) is given by

$$(H_t'g)(\lambda) = \int_0^1 h(z, t - \lambda) g(z) \, dz \tag{9.83}$$

for $g \in L_2[0, 1]$; this expression can be obtained in a similar way as that in the last example of Section 9.8. We test now whether $H_t'g = 0$ implies $g = 0$. Suppose $H_t'g = 0$. Since this function is in $L_2[0, t]$, its values are zero for almost all $\lambda \in [0, t]$; that is,

$$\int_0^1 g(z) \left[1 + 2 \sum_{k=1}^{\infty} \cos k\pi z \exp(-k^2\pi^2(t - \lambda)) \right] dz = 0 \tag{9.84}$$

for almost all $\lambda \in [0, t]$. Since the set of functions defined by

$$\{\exp(-k^2\pi^2(t - \lambda)), \quad k = 0, 1, 2, \ldots\}$$

is linearly independent over the interval $[0, t]$, the equality (9.84) is satisfied if and only if the scalars multiplying these functions are zero, that is,

$$\int_0^1 g(z) \cos k\pi z \, dz = 0, \qquad k = 0, 1, 2, \dots \tag{9.85}$$

This, at last, implies that $g(z) = 0$ almost everywhere on $[0, 1]$, or as a member of $L_2[0, 1]$, $g = 0$. The system (9.80) is weakly controllable at any $t > 0$, and it is therefore weakly controllable.

Before leaving this subject, the reader should be cautioned that the method to test weak controllability implied by Theorem 9.14 is not the final answer to this matter. More desirable is a method for which the explicit expression (9.77) is not needed; that is, one would want to be able to tell whether a system is or is not controllable just by studying some properties of the operator A and the transformation B. This was done, of course, in Chapter 5 in a much simpler setting. The reader wishing to pursue these matters should read the papers by Fattorini mentioned in the Supplementary Notes and References section at the end of this chapter.

Exercise

*1. Show that if a constant finite dimensional system is not controllable, then it is not weakly controllable. Deduce that the concept of weak controllability is quite irrelevant to the study of finite dimensional systems.

9.10. Conclusion

We have come to the end of our text, and it is well that some thought be given to how the material presented in this book should be used in the professional life of the reader.

The author would consider his work to be a failure if the study of this book did not prompt the imaginative and insightful student to pursue further some of the ideas presented. This compulsion to research should, however, be oriented properly. The benefits to be obtained from investigation into the properties of linear finite dimensional systems are limited; there are of course many problems to be solved, but a large proportion of them are quite artificial and of little engineering or mathematical interest. An exception is, perhaps, the area of stability of time-varying linear finite-dimensional systems.

The young researcher should concentrate his efforts on infinite dimensional systems. In doing so, he would follow a pattern of development in the best scientific tradition; the good scientist, after all, does not spend his life trying to extend marginally the work of others, but boldly explores new avenues of thought. The researcher in the theory of systems should not try to extend marginally concepts such as controllability, structure, or synthesis of finite-dimensional linear systems; he should investigate new systems, performing different tasks, perhaps then finding other properties worth defining and studying.

Note that such a program of investigation into infinite-dimensional systems cannot be carried out with the background presented in this text. The student should, before attempting any contribution to this body of knowledge, acquire a solid mathematical foundation equivalent to a full year of graduate study.

SUPPLEMENTARY NOTES AND REFERENCES

§9.3. Our definition of dynamical system is neither the most general nor the most useful, even if it is perfectly adequate for our purposes. A more modern and general approach to dynamical systems is based upon the concept of the *attainability map*, which specifies the set of states attainable at time t starting in state x at time zero. For some results in this area as well as references on it, the reader is referred to

E. Roxin, On generalized dynamical systems defined by contingent equations, *J. Differential Equations* **1** (1965), 188–205.

D. Eggert and P. Varaiya, Affine dynamical systems, *J. Comput. System Sci.* **1** (1967) 330–348.

§9.4. The theory of semigroups, resolvents, etc., is covered very well in the texts [3], [5], and [7].

§9.7. A nice summary of the theory of partial differential equations appears in

W. L. Brogan, Optimal control theory applied to systems described by partial differential equations. In "Advances in Control Systems," Vol. 6, C. T. Leondes, ed., pp. 221–316, Academic Press, New York, 1968.

The following text goes quite deeply into many matters related to the subject of this chapter:

A. Friedman, "Partial differential equations." Holt, New York, 1969.

See also

D. L. Russell, Optimal regulation of linear symmetric hyperbolic systems with finite-dimensional controls, *SIAM J. Control* **4** (1966), 276–294.

§9.8. Other references on controllability are

H. O. Fattorini, Some remarks on complete controllability, *SIAM J. Control* **4** (1966), 686–694.

H. O. Fattorini, On complete controllability of linear systems, *J. Differential Equations* **3** (1967), 391–402.

A. Friedman, Optimal control in Banach spaces, *J. Math. Anal. Appl.* **19** (1967), 35–55.

With reference to the subject of stability of such systems, the reader should consult the book by Hahn and the book by Massera and Schäffer referred to at the end of Chapter 8.
The problem of determining state equations for these systems from input–output knowledge has also been studied, mainly by Balakrishnan. See [10] and

A. V. Balakrishnan, A theory of linear systems with infinite dimensional state space. In "Proc. Symp. System Theory," pp. 69–88, Polytechnic Press of the Polytechnic Institute of Brooklyn, New York, 1965.

A. V. Balakrishnan, On the state space theory of linear systems, *J. Math. Anal. Appl.* **14** (1966), 371–391.

A. V. Balakrishnan, Foundations of the state-space theory of continuous systems, I, *J. Comput. System. Sci.* **1** (1967), 91–116.

REFERENCES

1. L. A. Liusternik and V. J. Sobolev, "Elements of Functional Analysis." Ungar, New York, 1961.
2. A. A. Albert, "Fundamental Concepts of Higher Algebra." Univ. of Chicago Press, Chicago, Illinois, 1956.
3. K. Yosida, "Functional Analysis." Springer, Berlin, 1968.
4. J. Dieudonné, "Foundations of Modern Analysis." Academic Press, New York, 1960.
5. N. Dunford and J. Schwartz, "Linear Operators, Part I." Wiley (Interscience), New York, 1957.
6. A. Papoulis, "The Fourier Integral and its Applications." McGraw-Hill, New York, 1962.
7. E. Hille and R. S. Phillips, "Functional Analysis and Semigroups." Am. Math. Soc., Providence, Rhode Island, 1957.
8. R. S. Phillips, A note on the abstract Cauchy problem, *Proc. Nat. Acad. Sci. U.S.A.* **40** (1954), 244–248.
9. A. V. Balakrishnan, Optimal control problems in Banach spaces, *J. SIAM Control* **3** (1965), 152–180.
10. A. V. Balakrishnan, Semigroup theory and control theory. In *Proc. IFIP Congress* 1965, 157–164.
11. M. Vidyasagar and T. J. Higgins, A controllability criterion for stable linear systems, *IEEE Trans. Automatic Control* **AC-15** (1970), 391–392.

APPENDIX

An Introduction to Computational Methods

A.1. Elementary Row and Column Operations

Many computational procedures consist of a succession of elementary operations on whole rows and columns of matrices. We study row operations first. Let X be the space of $m \times n$ matrices with entries in the field C. An elementary row operation is a function mapping X into itself; we will be interested in three instances of these operations:

i. The multiplication of the elements of a row of a matrix $A \in X$, for instance the jth row, by a scalar $c \in C$, $c \neq 0$. The resulting matrix can be thought as being obtained from A by multiplying this matrix by the $m \times m$ matrix

$$
j \begin{bmatrix}
1 & & & & & & & \\
 & \cdot & & & & & & \\
 & & \cdot & & & & & \\
 & & & \cdot & & & & \\
 & & & & 1 & & & \\
\cdot & \cdot & \cdot & \cdot & \cdot & \cdot & \cdot & c & & & & \\
 & & & & & 1 & & \\
 & & & & & & \cdot & \\
 & & & & & & & \cdot \\
 & & & & & & & & \cdot \\
 & & & & & & & & & 1
\end{bmatrix}, \qquad c \neq 0 \qquad \text{(A.1)}
$$

ii. The addition to the elements of the jth row, the elements of the ith row multiplied by $c \in C$. If $i \leqslant j$, the $m \times m$ matrix which characterizes this operation is

$$
j \begin{bmatrix} 1 & & & & \\ & \cdot & & & \\ \cdots c \cdot & & & & \\ & & \vdots & \cdot & \\ & & \vdots & & 1 \end{bmatrix} \tag{A.2}
$$
$$ i$$

while, if $j \leqslant i$, this matrix becomes

$$
\begin{matrix} & & i & \\ \end{matrix}
$$
$$
\begin{bmatrix} 1 & & \vdots & \\ & \cdot & \vdots & \\ & \cdot c \cdots & & j \\ & & \cdot & \\ & & & 1 \end{bmatrix} \tag{A.3}
$$

iii. The interchange of rows i and j. Suppose that $i > j$; then the $m \times m$ matrix which characterizes this operation is

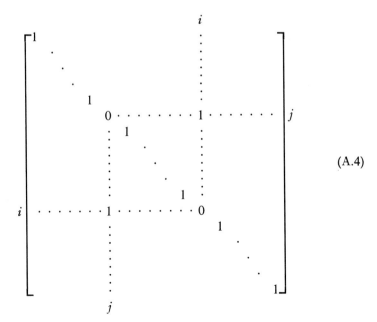

$$\tag{A.4}$$

Each of these operations has an inverse which is also an elementary row operation; for instance, the inverse of the operation i is the multiplication of the jth row by $1/c$. Therefore, the matrices (A.1), (A.2), (A.3), and (A.4) are nonsingular.

Definition. It is said that an $m \times n$ matrix B is *row equivalent* to an $m \times n$ matrix A if B can be obtained from A by a finite sequence of elementary row operations. Therefore, we have:

THEOREM A.1.   Row-equivalent matrices have the same rank.

We will see below how, to test the rank of a matrix, it is convenient to simplify it by means of a sequence of elementary row operations, so that the rank of the new matrix can be determined by inspection.

A similar set of *elementary column operations* can be defined on matrices. It is left to the reader to determine matrices which characterize operations on columns such as i, ii, and iii above. Note that this time the matrices have to postmultiply the original matrix. These operations have inverses, so that the matrices characterizing them are nonsingular. We have, therefore, if we define column-equivalent matrices in a similar way as above, the following theorem:

THEOREM A.2.   Column-equivalent matrices have the same rank.

We present now an application of these ideas to the problem, frequently referred to in the text, of determining the rank of a matrix, say an $m \times n$ matrix with complex coefficients. Of course, the rank of a matrix equals the dimension of the row space of A; if this matrix is constructed so that its row vectors equal a given set of m n-vectors, then by determining its rank, we determine in fact the maximum number of linearly independent row vectors of A, and thus of linearly independent vectors in our set.

The procedure for solving this problem consists in transforming the matrix A by means of elementary row operations (we could also use elementary column operations) into an *echelon matrix*, which is characterized by:

i. The leading entry in each nonzero row is 1.

ii. In each row after the first, the number of zeros preceding this leading entry is larger than the corresponding number of zeros in the preceding row.

We show the transformation procedure by means of some examples.

EXAMPLE

$$\begin{bmatrix} 1 & -1 & 1 & 3 \\ 2 & -5 & 3 & 10 \\ 3 & 3 & 1 & 1 \end{bmatrix} \begin{matrix} ② \rightarrow ② - 2① \\ ③ \rightarrow ③ - 3① \end{matrix} \begin{bmatrix} 1 & -1 & 1 & 3 \\ 0 & -3 & 1 & 4 \\ 0 & 6 & -2 & -8 \end{bmatrix}$$

$$\begin{matrix} ② \rightarrow -\frac{1}{3}② \\ ③ \rightarrow ③ + 2② \end{matrix} \begin{bmatrix} 1 & -1 & 1 & 3 \\ 0 & 1 & -\frac{1}{3} & -\frac{4}{3} \\ 0 & 0 & 0 & 0 \end{bmatrix}$$

The second matrix was obtained from the first by replacing the second row by the sum of the second row and $(-2) \times$ (the first row) and the third row by the sum of the third row and $(-3) \times$ (the first row), as indicated in the notation; similarly the third matrix was obtained from the second by the operations indicated.

EXAMPLE

$$\begin{bmatrix} 2 & 4 & 8 & 10 & 14 \\ 1 & 2 & 3 & 4 & 5 \\ -1 & -2 & 0 & 2 & 1 \end{bmatrix} \begin{matrix} ① \rightarrow \frac{1}{2}① \\ ② \rightarrow ② - \frac{1}{2}① \\ ③ \rightarrow ③ + \frac{1}{2}① \end{matrix} \begin{bmatrix} 1 & 2 & 4 & 5 & 7 \\ 0 & 0 & -1 & -1 & -2 \\ 0 & 0 & 4 & 7 & 8 \end{bmatrix}$$

$$② \rightarrow -② \quad \begin{bmatrix} 1 & 2 & 4 & 5 & 7 \\ 0 & 0 & 1 & 1 & 2 \\ 0 & 0 & 4 & 7 & 8 \end{bmatrix}$$

$$③ \rightarrow ③ - 4② \quad \begin{bmatrix} 1 & 2 & 4 & 5 & 7 \\ 0 & 0 & 1 & 1 & 2 \\ 0 & 0 & 0 & 3 & 0 \end{bmatrix}$$

$$③ \rightarrow \frac{1}{3}③ \quad \begin{bmatrix} 1 & 2 & 4 & 5 & 7 \\ 0 & 0 & 1 & 1 & 2 \\ 0 & 0 & 0 & 1 & 0 \end{bmatrix}$$

The reader should be able to see the pattern of operations from these two examples; after the leading nonzero entry of a row has been made unity, this entry is used to transform into zeros all the lower entries in the same column. These operations have no effect on the preceding columns, which have entries of zero in all places affected.

Since the original matrix and the echelon matrix are related by a series of row operations, the rank of both matrices is equal; the rank of the echelon matrix is very simple to compute, being equal to the number of nonzero rows present in it. In the first example above, the original matrix has rank 2, while in the second it has rank 3. If the matrix whose rank is to be determined has many short rows, it may be more convenient to determine the rank of its transpose by the row operations explained above.

Another illustration of the usefulness of elementary row operations is provided by the computation of the inverse of an $n \times n$, nonsingular matrix A with complex coefficients. Consider the $n \times 2n$ matrix $[A\ I_n]$, and transform this by means of elementary row operations until the matrix A has been replaced by the identity matrix I_n. If we call M the matrix corresponding to the whole sequence of operations, then

$$M[A\ I_n] = [MA\ M] = [I_n\ A^{-1}]$$

remember that $MA = I_n$. The place of I_n in $[A\ I_n]$ is taken, at the end, by the matrix A^{-1}.

EXAMPLE

$$A = \begin{bmatrix} 6 & 5 \\ 4 & 3 \end{bmatrix}$$

$$[A\ I_2] = \begin{bmatrix} 6 & 5 & 1 & 0 \\ 4 & 3 & 0 & 1 \end{bmatrix} \quad ① \to \tfrac{1}{6}① \quad \begin{bmatrix} 1 & \tfrac{5}{6} & \tfrac{1}{6} & 0 \\ 4 & 3 & 0 & 1 \end{bmatrix}$$

$$② \to ② + (-4)① \quad \begin{bmatrix} 1 & \tfrac{5}{6} & \tfrac{1}{6} & 0 \\ 0 & -\tfrac{1}{3} & -\tfrac{2}{3} & 1 \end{bmatrix}$$

$$② \to (-3)② \quad \begin{bmatrix} 1 & \tfrac{5}{6} & \tfrac{1}{6} & 0 \\ 0 & 1 & 2 & -3 \end{bmatrix}$$

$$① \to ① + (-\tfrac{5}{6})② \quad \begin{bmatrix} 1 & 0 & -\tfrac{3}{2} & \tfrac{5}{2} \\ 0 & 1 & \underbrace{2 \quad -3}_{A^{-1}} \end{bmatrix}$$

Other numerical methods for the inversion of matrices can be found in [1] and [2].

Yet another, and final, illustration, is provided by the computation of the matrix V in Section 6.4. There an $n_0 \times n$ matrix V is sought such that $VS = I_{n_0}$; S is a known, $n \times n_0$ matrix. Construct a matrix $[S\ I_n]$, and by a sequence of row operations take S into the form

$$\begin{bmatrix} I_{n_0} \\ S_1 \end{bmatrix}$$

Then, if the matrix N represents the sequence of row operations,

$$N[S\ I_n] = \begin{bmatrix} I_{n_0} & N_1 \\ S_1 & N_2 \end{bmatrix} \begin{matrix} \} \ n_0 \\ \} \ n-n_0 \end{matrix}$$

where N has been partitioned into the matrices N_1 and N_2 as shown. Then,

$$NS = \begin{bmatrix} N_1 \\ N_2 \end{bmatrix} S = \begin{bmatrix} N_1 S \\ N_2 S \end{bmatrix} = \begin{bmatrix} I_{n_0} \\ S_1 \end{bmatrix} \qquad \text{(A.5)}$$

so that $N_1 S = I_{n_0}$ and N_1 is the matrix V sought.

EXAMPLE

Let S be as in an example in Section 6.4,

$$S = \begin{bmatrix} 0 & 0 & 1 \\ 1 & 0 & 0 \\ 0 & 0 & 1 \\ 1 & 0 & 0 \\ 0 & 1 & 0 \end{bmatrix}$$

The sequence of transformations is as follows; here $n_0 = 3$, $n = 5$.

$$\begin{bmatrix} 0 & 0 & 1 & 1 & 0 & 0 & 0 & 0 \\ 1 & 0 & 0 & 0 & 1 & 0 & 0 & 0 \\ 0 & 0 & 1 & 0 & 0 & 1 & 0 & 0 \\ 1 & 0 & 0 & 0 & 0 & 0 & 1 & 0 \\ 0 & 1 & 0 & 0 & 0 & 0 & 0 & 1 \end{bmatrix} \begin{matrix} ① \to ② \\ ② \to ⑤ \\ \\ \\ \end{matrix} \begin{bmatrix} 1 & 0 & 0 & 0 & 1 & 0 & 0 & 0 \\ 0 & 1 & 0 & 0 & 0 & 0 & 0 & 1 \\ 0 & 0 & 1 & 0 & 0 & 1 & 0 & 0 \\ 1 & 0 & 0 & 0 & 0 & 0 & 1 & 0 \\ 0 & 1 & 0 & 0 & 0 & 0 & 0 & 1 \end{bmatrix}$$

$$V = \begin{bmatrix} 0 & 1 & 0 & 0 & 0 \\ 0 & 0 & 0 & 0 & 1 \\ 0 & 0 & 1 & 0 & 0 \end{bmatrix}$$

The operations will generally be more complicated than in this example.

A.2. The Computation of Eigenvalues, Eigenvectors, and the Matrix $(\lambda I_n - T)^{-1}$

This section consists of a single theorem, which will enable us to determine numerically the eigenvalues and eigenvectors of a matrix as well as the matrix $(\lambda I_n - T)^{-1}$, of importance in the reduction of T to diagonal or Jordan canonical form. Of course, the determination of eigenvectors corresponding to a known eigenvalue is a simple matter, consisting only of the solution of systems of linear algebraic equations; alternatively, once the matrix $(\lambda I_n - T)^{-1}$ has been computed, eigenvectors of T can be found by means of a partial fraction expansion, as indicated in Section 4.5.

It is, therefore, sufficient for the purposes of this section to develop an algorithm for computing the coefficients of the characteristic equation of a matrix T (from which the eigenvalues can be obtained by a variety of numerical methods) and the matrix coefficients of the powers of the parameter λ in the numerators of the matrix $(\lambda I_n - T)^{-1}$; this matrix is known to be the ratio of a polynomial of order $(n - 1)$ in λ with matrix coefficients and the characteristic polynomial of T, as shown in Section 4.5. This algorithm is provided by the following theorem [$1, 3$].

THEOREM A.3.   Let T be an $n \times n$ constant matrix with complex entries. If the characteristic polynomial of T is

$$\lambda^n + a_{n-1}\lambda^{n-1} + \cdots + a_1\lambda + a_0 \tag{A.6}$$

and $(\lambda I_n - T)^{-1}$ is written as

$$(\lambda I_n - T)^{-1} = \frac{\lambda^{n-1}T_{n-1} + \lambda^{n-2}T_{n-2} + \cdots + \lambda T_1 + T_0}{\lambda^n + a_{n-1}\lambda^{n-1} + \cdots + a_1\lambda + a_0} \tag{A.7}$$

then:

i. The coefficients of (A.6) and of the numerator of (A.7) are given by

$$
\begin{aligned}
a_{n-1} &= -\text{tr}(T) & T_{n-1} &= I_n \\
a_{n-2} &= -\tfrac{1}{2}\text{tr}(T_{n-2}T) & T_{n-2} &= T_{n-1}T + a_{n-1}I_n \\
&\;\;\vdots & &\;\;\vdots \\
a_{n-k} &= -\frac{1}{k}\text{tr}(T_{n-k}T) & T_{n-k} &= T_{n-k+1}T + a_{n-k+1}I_n \\
&\;\;\vdots & &\;\;\vdots \\
a_1 &= -\frac{1}{n-1}\text{tr}(T_1T) & T_0 &= T_1T + a_1I_n \\
a_0 &= -\frac{1}{n}\text{tr}(T_0T) &
\end{aligned}
\tag{A.8}
$$

ii. The matrix $T_0T + a_0I_n$ is the null matrix 0_n .

Proof.   It is clear from (A.7) that,

$$(\lambda^{n-1}T_{n-1} + \lambda^{n-2}T_{n-2} + \cdots + \lambda T_1 + T_0)(\lambda I_n - T)$$

$$= \lambda^n(T_{n-1}) + \lambda^{n-1}(T_{n-2} - T_{n-1}T) + \lambda^{n-2}(T_{n-3} - T_{n-2}T)$$

$$+ \cdots + \lambda^{n-k}(T_{n-k-1} - T_{n-k}T) + \cdots + \lambda(T_0 - T_1T) + (-T_0T)$$

$$= (\lambda^n + a_{n-1}\lambda^{n-1} + \cdots + a_1\lambda + a_0)I_n \tag{A.9}$$

for all λ which are not eigenvalues of T. Thus, the coefficients of the powers of λ in (A.9) must be equal; we obtain in this way the second column of equalities in (A.8), as well as the contention in ii.

We proceed to show the validity of the first column of equalities in (A.8) by means of mathematical induction. It is known from Chapter 4 that $a_{n-1} = -\text{tr}(T)$. We show the validity of $a_{n-k} = -(1/k)\,\text{tr}(T_{n-k}T)$ by assuming the validity of the expressions $a_{n-2} = -\frac{1}{2}\,\text{tr}(T_{n-2}T)$ up to and including $a_{n-k+1} = -[1/(1+k)]\,\text{tr}(T_{n-k+1}T)$. Indeed, it follows from (A.8) that

$$T_{n-k}T = T^k + a_{n-1}T^{k-1} + a_{n-2}T^{k-2} + \cdots + a_{n-k+1}T$$

so that

$$\text{tr}(T_{n-k}T) = \text{tr}(T^k) + a_{n-1}\,\text{tr}(T^{k-1}) + \cdots + a_{n-k+1}\,\text{tr}(T)$$

$$= \sum_{i=1}^{n} (\lambda_i^{\,k} + a_{n-1}\lambda_i^{\,k-1} + \cdots + a_{n-k+1}\lambda_i)$$

where $\lambda_1, \ldots, \lambda_n$ are the eigenvalues, not necessarily distinct, of the matrix T. The result sought is a simple consequence of the following identity

$$\sum_{i=1}^{n} (\lambda_i^{\,k} + a_{n-1}\lambda_i^{\,k} + \cdots + a_{n-k+1}\lambda_i) = -ka_{n-k} \qquad \text{(A.10)}$$

$k = 1, 2, \ldots, n$, known as Newton's formula. We can prove this by considering the characteristic polynomial of T,

$$f(\lambda) = \lambda^n + a_{n-1}\lambda^{n-1} + \cdots + a_1\lambda + a_0$$
$$= (\lambda - \lambda_1)(\lambda - \lambda_2) \cdots (\lambda - \lambda_n)$$

so that, for $\lambda \neq \lambda_i$, $i = 1, \ldots, n$,

$$\frac{\partial f}{\partial \lambda} = \frac{f}{\lambda - \lambda_1} + \frac{f}{\lambda - \lambda_2} + \cdots + \frac{f}{\lambda - \lambda_n} \qquad \text{(A.11)}$$

Since $f(\lambda_i) = 0$, $i = 1, \ldots, n$, we can write

$$f(\lambda) = (\lambda^n - \lambda_i^{\,n}) + a_{n-1}(\lambda^{n-1} - \lambda_i^{\,n-1}) + \cdots + a_1(\lambda - \lambda_i)$$

thus

$$\frac{f(\lambda)}{\lambda - \lambda_i} = \sum_{q=0}^{n-1} \lambda^{n-1-q}\lambda_i^{\,q} + a_{n-1}\sum_{q=0}^{n-2} \lambda^{n-2-q}\lambda_i^{\,q} + \cdots + a_1$$

$$= \lambda^{n-1} + (\lambda_i + a_{n-1})\,\lambda^{n-2} + (\lambda_i^{\,2} + a_{n-1}\lambda_i + a_{n-2})\,\lambda^{n-3} + \cdots \qquad \text{(A.12)}$$

so that, from (A.11)

$$\frac{df}{d\lambda} = n\lambda^{n-1} + \left(\sum_{i=1}^{n} \lambda_i + na_{n-1} \right) \lambda^{n-2}$$

$$+ \left(\sum_{i=1}^{n} \lambda_i^2 + a_{n-1} \sum_{i=1}^{n} \lambda_i + na_{n-2} \right) \lambda^{n-3} + \cdots \qquad (A.13)$$

this expression can be compared to

$$\frac{df}{d\lambda} = n\lambda^{n-1} + (n-1) a_{n-1}\lambda^{n-2} + (n-2) a_{n-2}\lambda^{n-3} + \cdots + a_1$$

so that

$$(n-1) a_{n-1} = \sum_{i=1}^{n} \lambda_i + na_{n-1}$$

$$(n-2) a_{n-2} = \sum_{i=1}^{n} \lambda_i^2 + a_{n-1} \sum_{i=1}^{n} \lambda_i + na_{n-2}$$

$$\vdots$$

$$(n-k) a_{n-k} = \sum_{i=1}^{n} \lambda_i^k + a_{n-1} \sum_{i=1}^{n} \lambda_i^{k-1} + \cdots + a_{n-k+1} \sum_{i=1}^{n} \lambda_i + na_{n-k}$$

then,

$$a_{n-1} = - \sum_{i=1}^{n} \lambda_i$$

$$a_{n-2} = -\frac{1}{2} \left(\sum_{i=1}^{n} \lambda_i^2 + a_{n-1} \sum_{i=1}^{n} \lambda_i \right)$$

$$\vdots$$

$$a_{n-k} = -\frac{1}{k} \left(\sum_{i=1}^{n} \lambda_i^k + a_{n-1} \sum_{i=1}^{n} \lambda_i^{k-1} + \cdots + a_{n-k+1} \sum_{i=1}^{n} \lambda_i \right)$$

$$(A.14)$$

Newton's formula (A.10) has been shown to be correct. The theorem follows.

Note that this theorem provides an iterative method for the computation of the coefficients of the characteristic equation of a matrix and of the numerator of the inverse matrix T. A check on the accuracy of the computations is provided by the value of the matrix $T_0 T + a_0 I_n$, which should result in the null matrix 0_n.

It is sometimes necessary to improve on the values of the eigenvalues resulting from solving the characteristic equation of a matrix, because the roots of this equation may be very sensitive to the value of some of the coefficients. In this case, iterative methods exist [1, 2] which permit the computation of the eigenvalues up to any degree of accuracy. The values obtained from the numerical solution of the characteristic equation may be used as initial values in the iterative process chosen.

A.3. The Computation of the Transition Matrices of Constant Systems

Probably the best way of estimating the entries of the transition matrix of a constant system by means of a computer is to apply methods of numerical integration to the corresponding systems of differential equations; indeed, this approach is not limited to constant systems, but can as well be applied to time-varying systems. This method, however, often produces just a string of numbers which give little insight into the properties of the corresponding system; it is preferable sometimes to compute the transition matrix of a constant system by other methods, involving either the reduction of the matrix A in $\dot{x} = Ax$ to a Jordan canonical form, or the use of the Hamilton–Cayley theorem for expressing higher powers of A in terms of lower ones, or Sylvester's theorem. The application of the Hamilton–Cayley theorem to this end was covered essentially in Chapter 4, while we shall not treat Sylvester's theorem [4]; we proceed therefore to discuss some numerical aspects of the reduction of A to Jordan canonical form and of the subsequent computation of the transition matrix. Many of these aspects have been covered in Chapter 4 and in Section A.2; we summarize our procedure as consisting of the following steps:

i. Compute the coefficients of the characteristic polynomial (A.6) and of the numerator of the matrix (A.7) by means of the procedure of Theorem A.3.

ii. By means of some numerical technique, solve for the roots of the characteristic equation of A.

iii. Make a partial fraction expansion of the matrix (A.7) as in (4.23) or (4.26). This step involves the determination of all the matrices M_{ij}

defined in Section 4.5; this is effected simply by making a partial fraction expansion of each entry in the matrix (A.7).

iv. Carry on the procedure of Section 4.5 to construct from these matrices M_{ij} the matrix S which describes the new basis with respect to which the matrix A takes the Jordan canonical form.

v. Compute the transition matrix using equations (4.20) and (4.39). Of course, the matrix A may be diagonalizable, which would simplify somehow the computations.

A.4. Positive Definite Matrices

We present here a result which is helpful in the characterization of positive definite matrices, and thus in the study of the stability of constant systems. The application of this result involves the determination of the sign of some determinants associated with the matrix which is under test for positive definiteness; the determinants themselves can be computed, for instance, by the application of the recursive method described in Section A.2 in connection with the computation of the coefficients of the characteristic equation of a matrix T; it is well known that the constant term a_0 in this equation equals the determinant of the matrix $(-T)$. Our results are contained in the following theorem:

THEOREM A.4. Let F be a real, symmetric matrix. Then F is positive definite if and only if the determinants

$$f_{11}, \quad \begin{vmatrix} f_{11} & f_{12} \\ f_{21} & f_{22} \end{vmatrix}, \quad \cdots \quad \begin{vmatrix} f_{11} & \cdots & f_{1n} \\ \vdots & & \vdots \\ f_{n1} & \cdots & f_{nn} \end{vmatrix} \quad (A.15)$$

are positive.

Proof. We prove first that F is positive definite if and only if all its eigenvalues are positive. Let $x^1, ..., x^n$ be orthogonal eigenvectors of F with a norm of unity; they form an orthonormal basis for R^n, so that

$$x = \sum_{i=1}^{n} a_i x^i, \quad a_i = (x, x^i), \quad i = 1, ..., n, \quad x \in R^n$$

and

$$(Fx, x) = \left(\sum_{i=1}^{n} a_i F x^i, \sum_{i=1}^{n} a_i x^i \right)$$

$$= \sum_{i=1}^{n} \lambda_i |a_i|^2 \quad (A.16)$$

where λ_i, $i = 1, ..., n$, are the eigenvalues of F. If F is positive definite, then all eigenvalues are positive; suppose $\lambda_k < 0$, $1 \leqslant k \leqslant n$. Then $(Fx^k, x^k) = \lambda_k < 0$, which contradicts our assumption that F is positive definite. If all eigenvalues of F are positive, (Fx, x) is positive for all x, as implied by (A.16).

Let F_k be the symmetric matrix formed from the first k rows and k columns of F. If F is positive definite, then the determinant of F is positive, since it equals the product of all eigenvalues of F, which are themselves positive. The determinants of F_k, $k = 1, ..., n - 1$, are also positive, because $(x, F_k x) = (x, Fx)$ for all vectors $x \in X$ for which $x_{k+1} = \cdots = x_n = 0$. The determinants in (A.15) are positive.

Assume now that all the determinants in (A.15) are positive, and suppose that F_k, $k < n$, is positive definite. We will show that F_{k+1} also shares this property, and, since $F_1 = f_{11}$ is positive definite by hypothesis, we conclude by induction that $F_n = F$ is positive definite. Let $\mu_1, ..., \mu_k$ be the eigenvalues of F_k, and $\nu_1, ..., \nu_{k+1}$ those of F_{k+1}. It is known [5] that

$$\nu_1 \geqslant \mu_1 \geqslant \nu_2 \geqslant \mu_2 \geqslant \cdots \geqslant \nu_k \geqslant \mu_k \geqslant \nu_{k+1} \tag{A.17}$$

we assume of course that $\mu_1 \geqslant \cdots \geqslant \mu_k$, $\nu_1 \geqslant \cdots \geqslant \nu_{k+1}$; the expression (A.17) is a result of the "inclusion" principle [5]. Of course, (A.17) implies that all the eigenvalues of F_{k+1}, with the possible exception of ν_{k+1}, are positive; since the determinant of this matrix is positive, then $\nu_1 \nu_2 \cdots \nu_{k+1} > 0$, which implies at last that $\nu_{k+1} > 0$, and F_{k+1} is then positive definite. The theorem follows.

REFERENCES

1. V. N. Faddeeva, "Computational Methods of Linear Algebra." Dover, New York, 1959.
2. L. Fox, "An Introduction to Numerical Linear Algebra." Oxford Univ. Press (Clarendon), London and New York, 1964.
3. L. A. Zadeh and C. A. Desoer, "Linear System Theory." McGraw-Hill, New York, 1963.
4. P. M. DeRusso, R. J. Roy, and C. M. Close, "State Variables for Engineers." Wiley, New York, 1965.
5. J. N. Franklin, "Matrix Theory." Prentice-Hall, Englewood Cliffs, New Jersey, 1968.

Author Index

Numbers in parentheses are reference numbers and indicate that an author's work is referred to although his name is not cited in the text. Numbers in italics show the page on which the complete reference is listed.

A

Albert, A. A., 277(2), *309*
Allen, D. R., *20*
Anderson, B. D. O., 255(6), 259(6), 260(6), *264*
Antonsiewicz, H. A., *174*, 253(5), *264*
Athans, M., *173*

B

Bachman, G., 58(4), *70*
Balakrishnan, A. V., *219*, 295(9), 296(10), 299(10), 309, *309*
Barnett, S., 246(2), 248, *264*
Bellman, R., *21*, 74(1), *101*, *137*, 245(1), *264*
Bertram, J. E., *263*
Birkhoff, G., 33(2), 37(2), *70*, 116(3), *137*
Boltyanskii, V. G., *101*
Bona, B. E., *21*
Bongiorno, J. J., Jr., 213(6), *219*
Bonnel, R. D., *174*
Braun, L., *19*
Bridgland, T. F., 250(4), *264*
Brockett, R. W., *22*, *263*

C

Carlin, H. J., *20*
Cesari, L., *263*
Chang, A., 157, 158(8), *175*
Chen, C. T., 151, 153(6), 164(6), 165(6), 166(6), 166(11), 174(6), *174*, *175*, 236(1), *238*
Chu, Hsin, *173*
Close, C. M., *20*, 319(4), *321*
Coddington, E. A., *101*
Conti, R., *174*
Cooke, K. L., *21*
Cullen, C. G., *173*
Cunningham, W. J., 4(4), *22*

D

Davis, P., 253(5), *264*
DeRusso, P. M., *20*, 319(4), *321*
Desoer, C. A., *21*, 151, 153(6), 164(6), 165(6), 166(6, 11), 174(6), *174*, *175*,

Subject Index

A

Adjoint operator, 65–66, 301–303
  system, 92–93, 162
  transformation, 301–303
Analytic system, 178
  synthesis, 206–209
Attainable set, 170–171, 174

B

Banach space, 268, 271
Basis, 37
  change, 102, 104–107
  ordered, 38

C

Cartesian product, 27
Casorati, 230
Cauchy sequence, 267
Characteristic equation, 107
Complete space, 268
Completion, 268
Composite systems, 164–166
Computational methods, 310–321
Continuity, 269–271
Control systems, 2–3
  optimal, 20
  time-lag, 21

Controllability, 14, 138–159, 173–175
  of composite systems, 164–166
  of constant systems, 142, 148–152
  of difference systems, 234–237
  necessary and sufficient conditions,
    143–147
  output, 167
  stability and, 256–261
  synthesis and, 185
  of time-varying systems, 157–159
  total, 167–169
  transfer functions and, 155–157
  transformations and, 152–155
  uniform, 256
  weak, 304–307
Coordinates, 38

D

Delay, 21, 221
Diagonalization, 120–125
Difference equations, 221–223
Difference systems, 200–238
  constant, 234
  controllability, 234–237
  homogeneous, 224–230
  nonhomogeneous, 230–233
  observability, 234–237
  stability, 262–263
  transition matrix, 226–228

ELECTRICAL SCIENCE

A Series of Monographs and Texts

Editors

Henry G. Booker

UNIVERSITY OF CALIFORNIA AT SAN DIEGO
LA JOLLA, CALIFORNIA

Nicholas DeClaris

UNIVERSITY OF MARYLAND
COLLEGE PARK, MARYLAND

C. L. Sheng. Threshold Logic. 1969

Dale M. Grimes. Electromagnetism and Quantum Theory. 1969

Robert O. Harger. Synthetic Aperture Radar Systems: Theory and Design. 1970

M. A. Lampert and P. Mark. Current Injection in Solids. 1970

W. V. T. Rusch and P. D. Potter. Analysis of Reflector Antennas. 1970

Amar Mukhopadhyay. Recent Developments in Switching Theory. 1971

A. D. Whalen. Detection of Signals in Noise. 1971

J. E. Rubio. The Theory of Linear Systems. 1971

In Preparation

Keinosuke Fukunaga. Introduction To Statistical Pattern Recognition

Jacob Klapper and John T. Frankle. Phase Lock and Frequency Feedback Systems: Principles and Techniques